Macromolecular Symposia

192

WILEY-VCH

pp. 1–292

March 2003

Macromolecular Symposia publishes lectures given at international symposia and is issued irregularly, with normally 14 volumes published per year. For each symposium volume, an Editor is appointed. The articles are peer-reviewed. The journal is produced by photo-offset lithography directly from the authors' typescripts.
Further information for authors can be obtained from:
Editorial office "Macromolecular Symposia"
Wiley-VCH
Boschstrasse 12, 69469 Weinheim,
Germany
Tel. +49 (0) 62 01/6 06-2 38 or -5 81; Fax +49 (0) 62 01/6 06-3 09 or 5 10;
E-mail: macro-symp@wiley-vch.de
http://www.ms-journal.de
Suggestions or proposals for conferences or symposia to be covered in this series should also be sent to the Editorial office at the address above.

Macromolecular Symposia:
Annual subscription rates 2003 (print only or online only)*
Germany, Austria € 1318; Switzerland SFr 2168; other Europe € 1318; outside Europe US $ 1568.
Macromolecular Package, including Macromolecular Chemistry & Physics (18 issues), Macromolecular Bioscience (12 issues), Macromolecular Rapid Communications (18 issues), Macromolecular Theory & Simulations (9 issues) is also available. Details on request.
* For a 5 % premium in addition to **Print Only** or **Online Only**, Institutions can also choose both print and online access.
Packages including Macromolecular Symposia and Macromolecular Materials & Engineering are also available. Details on request.
Single issues and back copies are available. Please, inquire for prices.

Orders may be placed through your bookseller or directly at the publishers:
WILEY-VCH Verlag GmbH & Co. KGaA, P. O. Box 10 11 61, 69451 Weinheim, Germany, Tel. +49 (0) 62 01/6 06-400, Fax +49 (0) 62 01/60 61 84. E-mail: service@wiley-vch.de

Macromolecular Symposia (ISSN 1022-1360) is published with 14 volumes per year by WILEY-VCH Verlag GmbH & Co. KGaA, P. O. Box 10 11 61, 69451 Weinheim, Germany. Air freight and mailing in the USA by Publications Expediting Inc., 200 Meacham Ave., Elmont, NY 11003, USA. Application to mail at Periodicals Postage rate is paid at Jamaica, NY 11431, USA. US POSTMASTER please send address changes to: Macromolecular Symposia, c/o Wiley-VCH, III River Street, Hoboken, NJ 07030, USA.

Printing: Strauss Offsetdruck, Mörlenbach. Binding: J. Schäffer, Grünstadt

The Pacific Polymer Federation

In conjunction with

Sociedad Polimérica de México

Universidad Nacional Autónoma de México

Invited lectures presented at the

7th Pacific Polymer Conference

held in Oaxaca, Oaxaca, México

December 3–7, 2001

Symposium Editor

Professor Takeshi Ogawa

Instituto de Investigaciones en Materiales
Universidad Nacional Autónoma de México
Circuito Exterior, Ciudad Universitaria, México DF 04510, México

International Scientific and Organizing Committee

W. Cook (Australia), F. Rabagliati (Chile), K. Horie (Japan), T. Kajiyama (Japan), S. C. Kim (Korea), J. E. McGrath (USA), W. H. Daly (USA), T. Lodge (USA), F. Wang (China), H. Chan (Singapore), S. Tantayanon (Thailand)

National Organizing Committee

T. Ogawa (Chairman), G. Burillo (Secretariat), E. Bucio (Treasurer), A. Licea, R. Guerrero

---

ISBN 3-527-30696-x

The **tables of contents** of the published issues are displayed on the WWW.

This service as well as further information on our journals can be found at the following WWW address:

http://www.ms-journal.de

The service is free for everybody.

Contents of Macromolecular Symposia 192

7th Pacific Polymer Conference
Oaxaca (México), 2001

* The asterisk indicates the name of the author to whom inquiries should be addressed.

Author Index

Preface

The Pacific Polymer Conference is held once every two years at different places in the Pacific Basin Countries. The 1st conference was held in Hawaii in 1989, followed by the 2nd (Otsu, Japan, 1991), the 3rd (Gold Coast, Australia, 1993), the 4th (Hawaii, 1995), 5th (Kyongju, Korea, 1997), 6th (Guangzhou, China, 1999) and 7th (Oaxaca, México, 2001). The 8th will be held in Bangkok, Thailand.

The Pacific Polymer Federation was founded in October 1987 by the polymer groups of USA, Australia and Japan, and it consists of the members from 15 different Pacific rim countries.

The Pacific Polymer Conference is one of the most important international congresses on polymers, covering every field of polymer science, and many very high quality papers are presented. At the 7th meeting at Oaxaca, over 400 participants were registered at the beginning, but because of the September 11 atrocity, many cancellations were made. However, more than 320 papers were presented at the meeting. The participants enjoyed the meeting, which was of high academic quality, as well as the pleasant Spanish colonial city of Oaxaca and the prehispanic ruins of the Zapoteca civilization.

This volume consists of some of the invited lectures of the 7th meeting. The editor wishes to express his gratitude to the invited lecturers who contributed to this volume. The organizing committee is grateful to the National University of México for support for the conference.

Takeshi Ogawa
Instituto de Investigaciones en Materiales,
Universidad Nacional Autónoma de México

Macromol. Symp. **192**, *1–12 (2003)*

Synthesis of Novel Architectures by Radical Polymerization with Reversible Addition Fragmentation Chain Transfer (RAFT Polymerization)

Graeme Moad,[1,2]* *Roshan T.A. Mayadunne,*[2] *Ezio Rizzardo,*[1,2] *Mellissa Skidmore,*[2] *San H. Thang*[1]

[1]CSIRO Molecular Science
Bag 10, Clayton South, Victoria 3169, Australia
[2]CRC for Polymers, CSIRO Molecular Science
Bag 10, Clayton South, Victoria 3169, Australia

Summary: Radical polymerization with **R**eversible **A**ddition-**F**ragmentation chain **T**ransfer - RAFT polymerization can be adapted to produce a wide variety of block copolymers and other polymers of complex architecture. Such polymerizations require use of a suitable compound [Z-(C=X)X-R], which is selected to have a very high transfer constants in radical polymerization of the desired monomers, and appropriate choice of reaction conditions. Block and multiblock polymers, stars and grafts are available through use of precursors containing multiple thiocarbonylthio groups. Stars and microgels are also available through self assembly and crosslinking of RAFT synthesized block copolymer micelles and by a varient of the 'arm-first' approach wherein of a diene monomer is (co)polymerized in the presence of a preformed RAFT-made polymer. This paper briefly reviews RAFT polymerization then examines the scope for utilizing the technology in the construction of novel architectures.

Introduction

In recent years, considerable effort has been expended to develop free radical processes that display the characteristics of a living polymerization.[1-6] Living polymerizations provide molecular weights that are predetermined by reagent concentrations and conversion, yield narrow molecular weight distributions and, most importantly, the polymer products can be reactivated for chain extension or block copolymer synthesis. Recently we have described a new method for achieving this end.[7-11] The key step in the process is **R**eversible **A**ddition-**F**ragmentation chain **T**ransfer. Accordingly, we designated the method **RAFT** polymerization.[11]

Some of the advantages of RAFT polymerization, over competing technologies [atom transfer radical polymerization (ATRP),[5,6] nitroxide mediated polymerization (NMP)[4]], stem from the fact that it is tolerant of a very wide range of functionality in monomer and solvent (*e.g.* -

 CCC 1022-1360/00/$ 17,50+.50/0

OH, -COOH, $CONR_2$, $-NR_2$, SO_3Na). This means that it is applicable to a vast range of monomer types and that polymerizations can be successfully carried out under a wide range of reaction conditions (bulk, solution, suspension, emulsion). The chain transfer process can be effective over a wide temperature range (>20-150°C). Indeed, with some limitations imposed by the need to limit termination reactions, the reaction conditions employed in RAFT polymerization are typical of those used for conventional free radical polymerization.

This paper will illustrate application of RAFT polymerization to the synthesis of polymers of various architectures (di-, tri-, multi-block, star, etc.). The aforementioned tolerance to functionality in the monomer means that there are tractable routes available to many structures that were previously not directly accessible. It is important to first consider the RAFT process to gain some understanding on how to select the RAFT agent appropriately so as to minimize side reactions and reaction times and maximize block purity

RAFT Polymerization

Scheme 1.

RAFT polymerization provides the characteristics usually associated with living polymerization. The overall process results in monomer units being inserted into RAFT agent structure as shown in Scheme 1. Reaction conditions should usually be chosen such that the fraction of initiator-derived chains (should be greater than or equal to the number of chains formed by radical-radical termination) is negligible. Molecular weights can then be estimated using the relationship (1):[12]

Calculated Degree of Polymerization ~ [monomer consumed]/[RAFT agent] (1)

Positive deviations from equation (1) indicate incomplete usage of RAFT agent while negative deviations indicate that other sources of polymer chains are significant (*e.g.* the initiator). Polydispersities depend on the chain transfer constants associated with both the initial and the polymeric RAFT agent. The initial RAFT agent should be chosen such that is rapidly consumed during the initial stages of the polymerization. The rate of consumption of the RAFT agent can usually be predicted using the relationship (2).

$$-\frac{d[\mathbf{2}]}{d[\mathrm{M}]} \approx C_{tr} \frac{[\mathbf{2}]}{[\mathrm{M}]+C_{tr}[\mathbf{2}]+C_{-tr}[\mathbf{4}]} \qquad (2)$$

Where $C_{tr} = k_{tr}/k_p$, $C_{-tr} = k_{-tr}/k_i$, $k_{tr} = k_{add} \times \frac{k_\beta}{k_{-add}+k_\beta}$ and $k_{-tr} = k_{-\beta} \times \frac{k_{-add}}{k_{-add}+k_\beta}$. Other parameters are defined in Scheme 2.

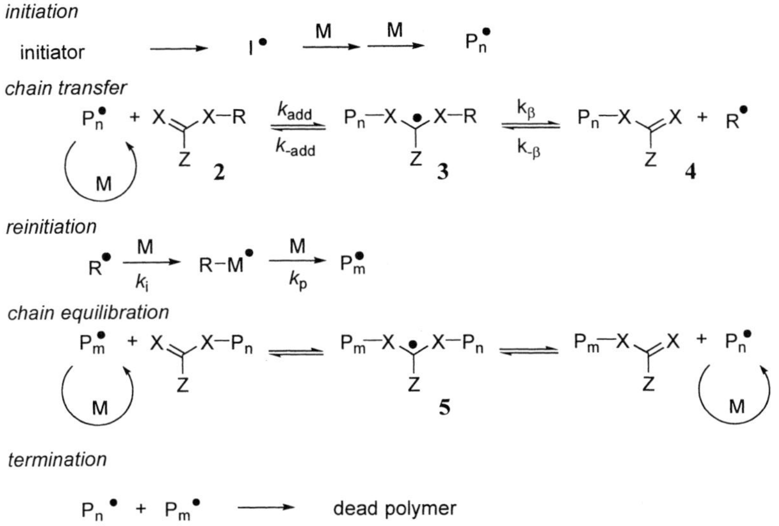

Scheme 2.

Or, with knowledge of the rate of consumption of RAFT agent this, equation (2) can be solved to provide transfer constants.[12] In this kinetic treatment it is assumed that the adduct species **3** and **5** give no chemistry other than fragmentation to give a radical and RAFT agent. Slow fragmentation or consumption of these species in side reactions will give retardation.[8,12-15] A fraction of dead chains determined by the number of chains formed from the initiator and the termination mechanism will always be formed. It is obviously important in block synthesis to maintain a very high fraction of living ends throughout the process. The common industrial practice of adding initiator or increasing initiation rate towards the end of a polymerization to drive the process towards 100% conversion should be avoided. This is particularly important in the intermediate stages of block syntheses involving sequential monomer addition.

RAFT Agents

A wide variety of reagents with the capacity to react by RAFT have now been reported. Examples of RAFT agents include macromonomers (*e.g.* **6** and **7**)[16-18] and thiocarbonylthio compounds (*e.g.* **1** and **8-14**).[8,9,11]

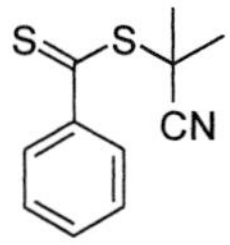

8

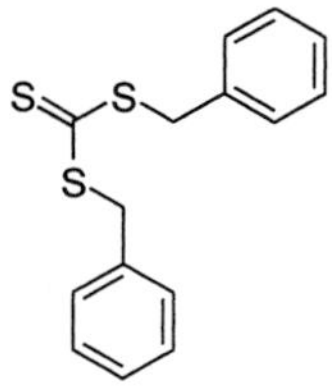

9

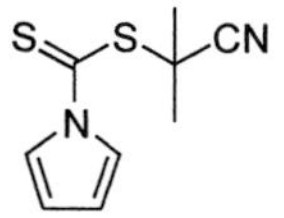

10

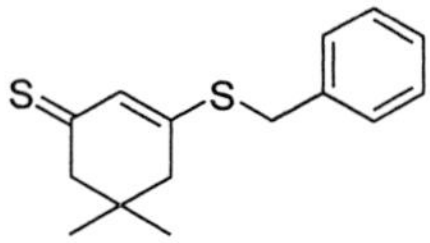

11

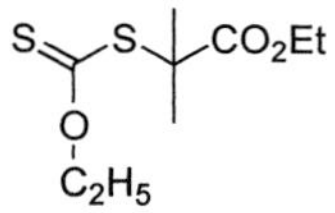

12

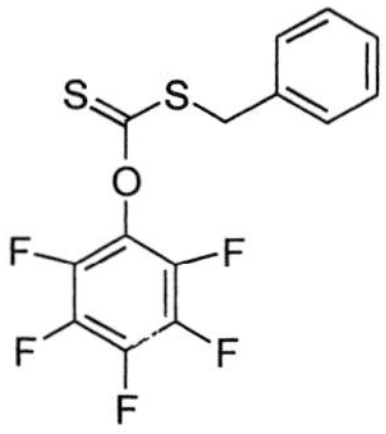

13

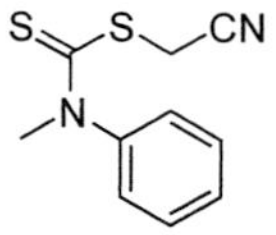

14

The generic features common to all RAFT agents can be summarized as shown in Figure 1.

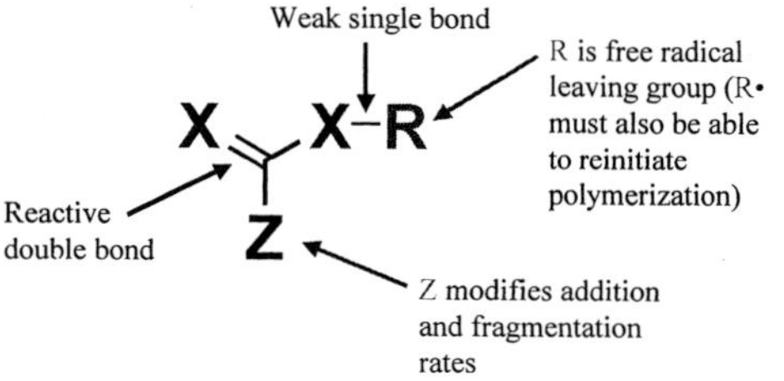

Figure 1. Generic RAFT Agent Structure.

The RAFT agent should to be chosen with attention to the particular polymerization process (the monomers and the reaction conditions) and according to required end group functionality.[9]

Design feature to take into account are:

(a) The RAFT agent (**2**, **4**) should have a high transfer constant in the monomers to be polymerized.

(b) The intermediates (**3** and **5**) should fragment rapidly and give no side reactions.

(c) The expelled radicals (R·) should efficiently reinitiate polymerization.

The magnitude and origin of the substituent effects on the rates of addition and fragmentation are discussed in detail elsewhere. Macromonomers ($X=CH_2$ *e.g.* **6**, **7**) are suited to the synthesis of methacrylic and similar polymers and copolymers. However, transfer constants are generally <1.0 and narrow polydispersities can only be achieved using feed addition protocol.[18-20]

As a general guideline, dithioesters (*e.g.* **8**)[11] and trithiocarbonates (*e.g.* **1**, **9**)[9,21] and certain dithiocarbamates[22,23] (*e.g.* **10**) are the preferred RAFT agents for (meth)acrylic and styrenic monomers in that their use affords narrow polydispersity polymers in a batch polymerization process. When the monomer is vinyl acetate, N-vinyl pyrrolidone[9,23] or similar Z is chosen to give a less stable intermediate (less side reactions, less retardation). Xanthates (*e.g.* **12**) and dithiocarbamates such as **14** are preferred. Within each class of thiocarbonylthio compound it is possible to tune the activity through the choice of substituents on Z.[22-25]

Blocks by Sequential Monomer Addition

In its simplest form, block synthesis simply requires sequential addition of monomers in a polymerization process.[22,26-28] The reaction scheme is then identical to that shown in Scheme 2 where R is a preformed polymer chain. Di-, tri- and multiblock polymers can be formed in this way. Block copolymers based on non radically polymerized chains can be formed by appropriate functionalization of a chain end.[26,29]

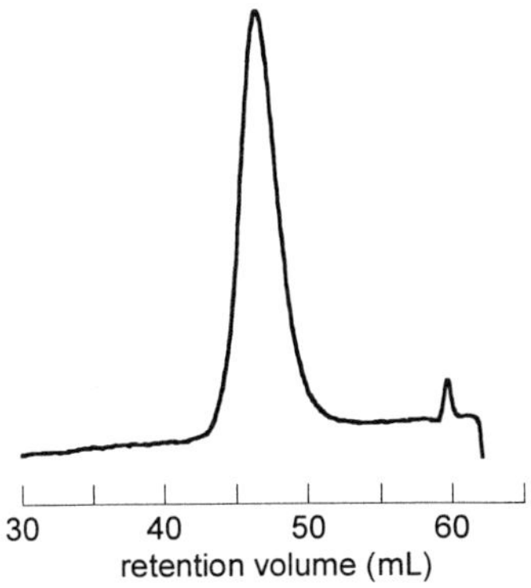

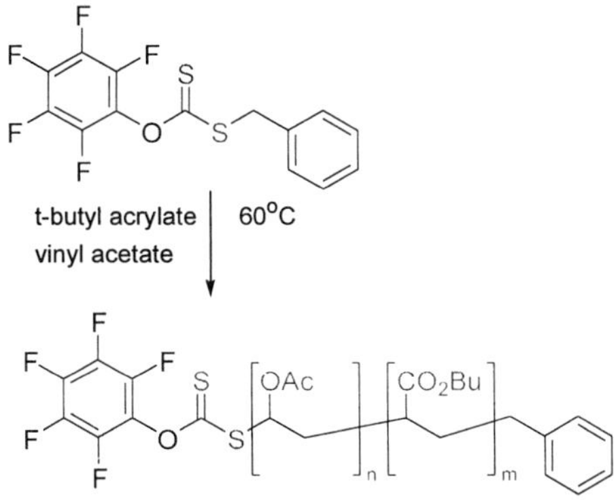

Figure 2. GPC trace showing poly(*t*-butyl acrylate-*co*-vinyl acetate-*block*-poly(vinyl acetate) (M_n16500 M_w/M_n 1.31) prepared by batch gradient copolymerization at 60°C with AIBN initiator.

In copolymerization of monomers with disparate reactivity ratios compositional drift with conversion will necessarily provide a gradient copolymer. Use of monomers with appropriately disparate reactivity ratios enables synthesis of block copolymers in a one step batch polymerization. In copolymerization of vinyl acetate and *t*-butyl acrylate all propagating radicals prefer to add *t*-butyl acrylate (r_{VAc}~0.026, r_{nBA}~5.93). Thus, the first formed polymer is predominantly composed of *t*-butyl acrylate units and the product following exhaustion of this monomer is poly(*t*-butyl acrylate-*co*-vinyl acetate-*block*-poly(vinyl acetate).[24] It is important in this experiment to chose a RAFT agent which is compatible with both the acrylate ester and vinyl acetate. The xanthate **10** was used in this case to provide a block copolymer with polydispersity ~1.3 (see Figure 2). The blocky structure of the product was immediately apparent from the ^{1}H NMR spectrum. Styrene and maleic anhydride show a strong tendency to alternate in the chain. Copolymerizations with excess styrene yield poly(styrene-*co*-maleic anhydride)-*block*-poly(styrene).[29,30]

Multiarm Polymer Synthesis

Triblocks from Difunctional RAFT Agents

Triblocks can also be made using difunctional RAFT agent (for example compounds **15**, **16**).[21,26,31] Possibly the simplest compounds of this class are the trithiocarbonates.[21] The use of a difunctional RAFT agent as a triblock precursor helps ensure that both arms are of the same length and composition.

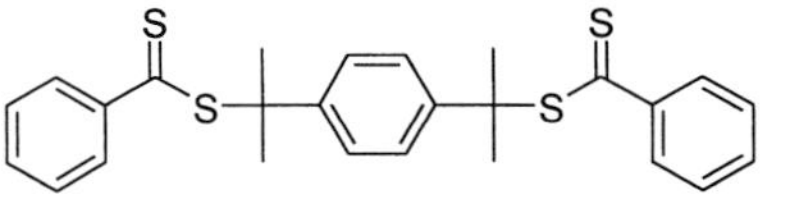

15

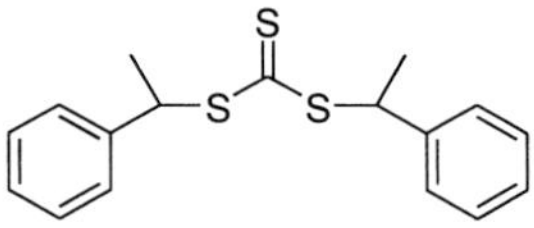

16

Some recent work has been orientated towards using this strategy for the synthesis of triblocks intended as thermoplastic elastomers based on a „soft" acrylate mid section and „hard" styrene or styrene-co acrylonitrile outer blocks (Scheme 3). The RAFT agent was chosen such that the thiocarbonylthio functions would remain at the chain ends (note that methyl is a poor free radical leaving group). In the case shown (Scheme 3), an α,ω-bis(methylsulfanylthio-carbonylsulfanyl)-poly(butyl acrylate) M_n=77000, M_w/M_n=1.05 was successfully converted to a triblock copolymer M_n=171000, M_w/M_n=1.12. A series of such block copolymers has been made to enable a full structure property correlation. It was found that narrow polydispersities for both the mid- and end blocks were critical to achieving good mechanical properties.

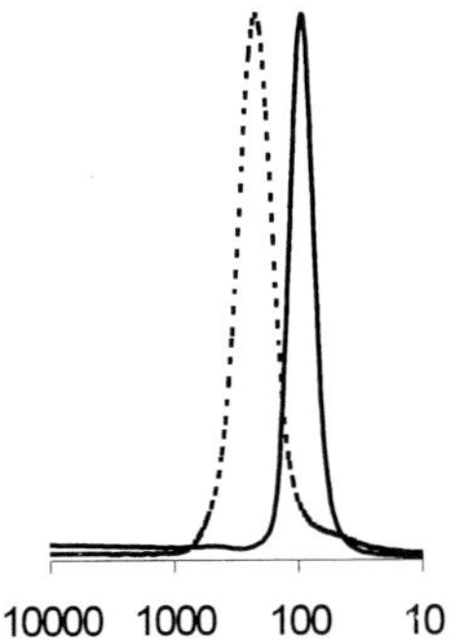

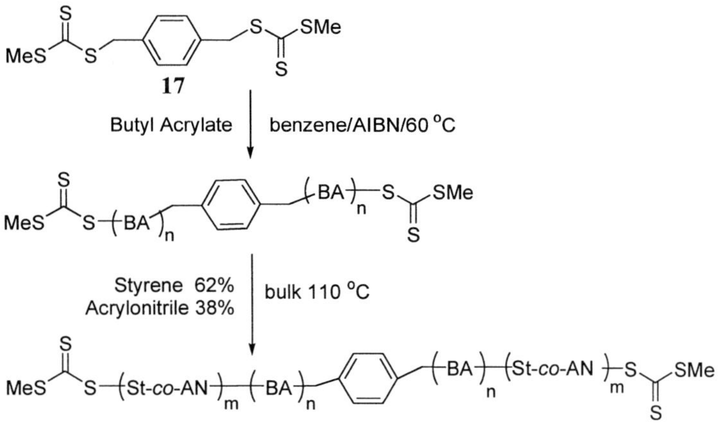

Figure 3. Molecular weight weight distributions for poly(butyl acrylate (—) and derived triblock copolymer (- - -).

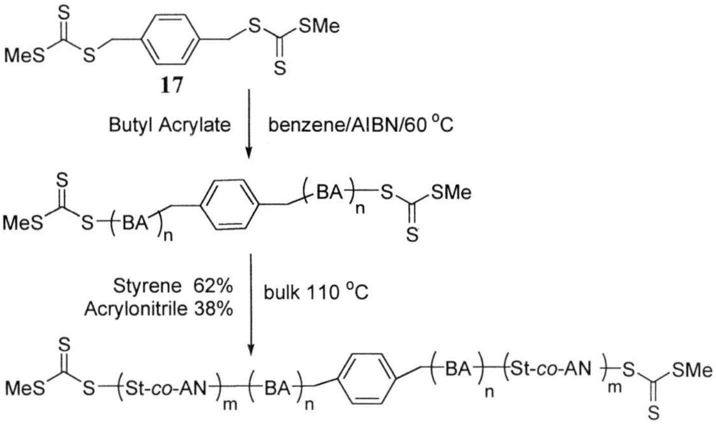

Scheme 3.

Star Synthesis

Stars from Multifunctional RAFT Agents

This form of star synthesis can be seen as an extension of triblock syntheses described above where the number of thiocarbonylthio groups exceeds two.[8,9,32-34] The multifunctional compound may be a small organic compound, a macromolecular species, a dendrimer, a particle, or indeed, any object possessing multiple thiocarbonylthio groups (though here the distinction between star and graft and other copolymers may become blurred). We can envisage two limiting forms of star growth depending on the orientation of the thiocarbonylthio group with respect to the core.[8,9]

Away from Process

In this case the thiocarbonylthio groups remain attached to the core (Z') (see Scheme 4). All propagating radicals are linear chains detached from the core. No radical centers are at any stage attached to the core.

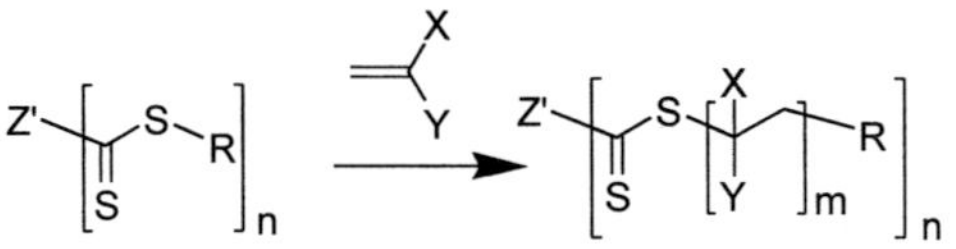

example

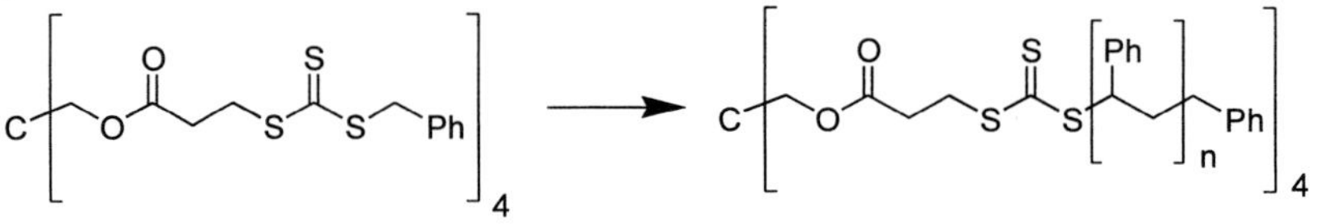

Scheme 4.

Advantages: no star-star coupling.

Disadvantages: cleavage of thiocarbonylthio group results in destruction of the star. This property may be turned to an advantage in some circumstances. For example, polymer supported polymer synthesis.

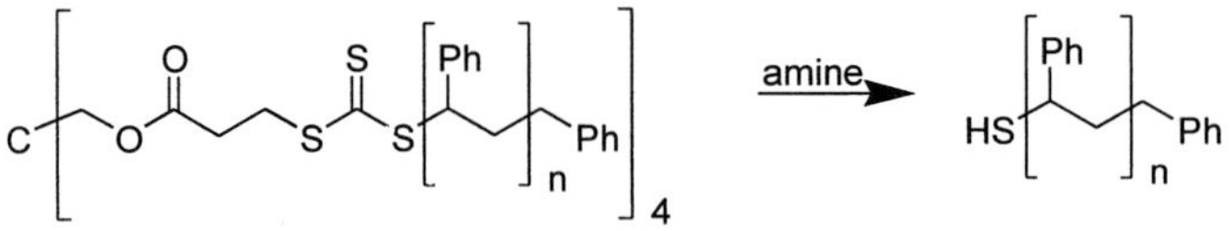

Scheme 5.

Thus, the four-arm star shown in Scheme 4 is cleaved by treatment with a primary or secondary amine to provide a narrow polydispersity linear polymer (Scheme 5).

Attached to Process

In this case, most thiocarbonylthio groups remain attached to the periphery of the star. (Scheme 6). Most propagating radicals (other than initiator-derived chains) remain attached to the core (R').

Although no multi-RAFT agents are commercially available, precursor species are readily obtained. The RAFT agents shown in Schemes 5 and 6 were prepared from commercially available thiols and halo-compounds respectively. We have recently reported[35] methods to quantitatively (>95% overall yield) convert (more readily available) multi-hydroxy compounds to both types of multi-RAFT agents (Scheme 7). These methods have been successfully applied to make (n=) 3,4,6 and 8-armed stars of narrow polydispersity. Star blocks are also possible.

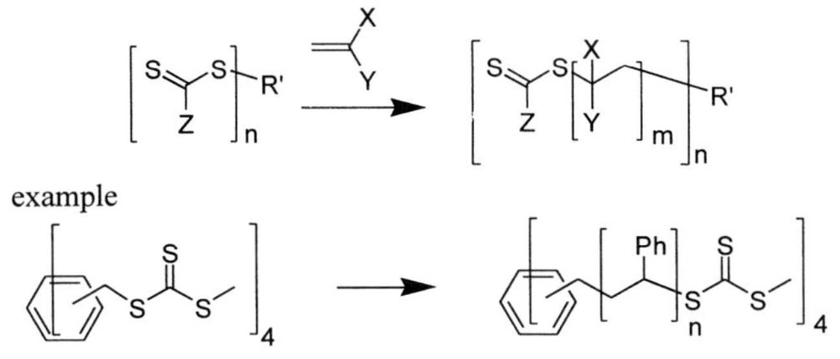

Scheme 6.

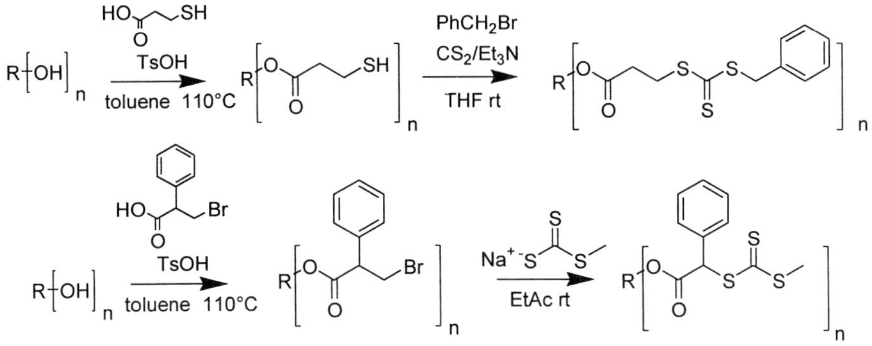

Scheme 7.

In the above-mentioned approaches to stars, the core has been synthesized first. There are a number of alternative strategies available whereby the arms of the star are synthesized first. The arm length and polydispersity are determined by the properties of that precursor polymer.

Stars by Self-Assembly of Block Copolymers

Suitably constructed hydrophilic-hydrophobic (or solvophilic-solvophobic) block copolymers possess the ability to self-assemble to form micelles (or other supramolecular structures). Such supramolecular assemblies may only be stable in a particular medium and/or over a specific concentration range. Various crosslinking mechanisms can be used to stitch these structures together to form a stable star polymer, microgel (or other structure). The core, the shell or mid block section of the micelle may be crosslinked to stabilize the structure.[36-38] Star synthesis would usually require crosslinking of the micelle's core (cf. Scheme 8). Care is needed in selecting reaction conditions so as to avoid intermicelle crosslinking and gel formation.

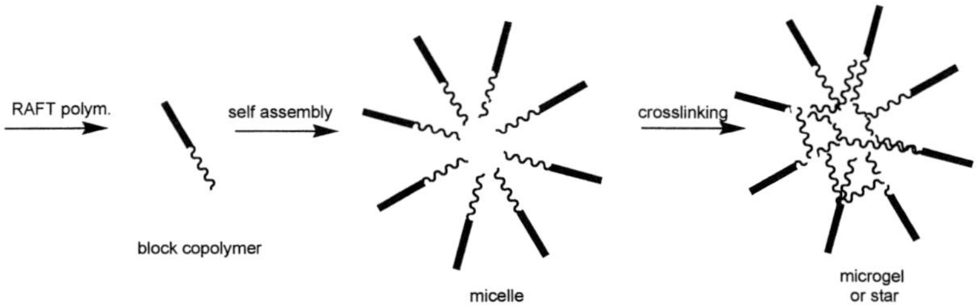

Scheme 8.

In this application, RAFT polymerization opens new possibilities, over more conventional strategies, mainly because of the much wider range of monomers that can be polymerized.[39] For example, monomers with unprotected acid functionality (methacrylic acid, acrylic acid) can be successfully converted to narrow polydispersity blocks. We have applied this approach to create novel encapsulants for dyes and rheology control agents. Many other applications are possible.[39] Since it is possible to maintain the thiocarbonyl thio groups during the crosslinking step further elaboration of the supramolecular structures by RAFT polymerization is also possible.

Stars by Copolymerization of Multifunctional Monomers

The so called 'arm-first' process involves making a living polymer then using this to initiate (co)polymerization of a di (or higher) functional monomer which forms a crosslinked core (Scheme 9). The methodology has been used to make star shaped molecules using NMP[40-44] or ATRP.[45-47]

Again RAFT polymerization offers new scope to this process.[48] The RAFT process involves (co)polymerization of an appropriate diene monomer (*e.g.* divinylbenzene) in the presence of a polymer formed by RAFT polymerization.

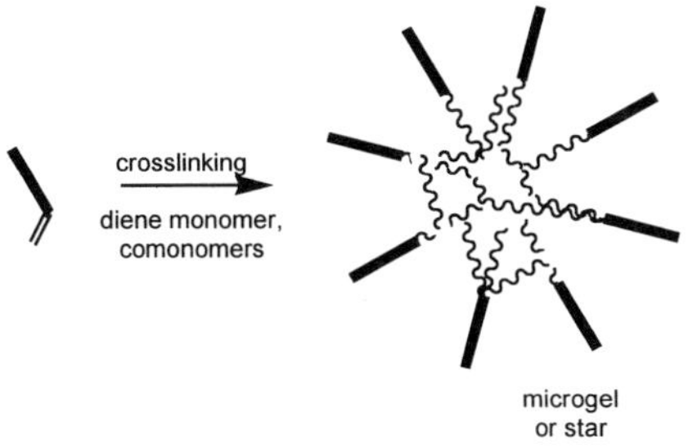

Scheme 9.

Conclusions

Living radical polymerization provides ready access a wide variety of polymer architectures. RAFT polymerization significantly extends the scope of existing methods allowing the synthesis of novel structures and compositions.

Acknowledgements

We acknowledge DuPont Performance Coatings for their support of this work. Many people were involved in devising and conducting the experiments described. Their names are given in the references

[1] K. Matyjaszewski *ACS Symp. Ser.* **1998**, *685*, 2-30.
[2] K. Matyjaszewski *ACS Symp. Ser.* **2000**, *768*, 2-26.
[3] G. Moad, D.H. Solomon *The Chemistry of Free Radical Polymerization* Pergamon: Oxford, 1995.
[4] C.J. Hawker, A.W. Bosman, E. Harth *Chem. Rev.* **2001**, *101*, 3661-88.
[5] M. Kamigaito, T. Ando, M. Sawamoto *Chem. Rev.* **2001**, *101*, 3689-745.
[6] K. Matyjaszewski, J. Xia *Chem. Rev.* **2001**, *101*, 2921-90.
[7] E. Rizzardo, J. Chiefari, Y.K. Chong, F. Ercole, J. Krstina, J. Jeffery, T.P.T. Le, R.T.A. Mayadunne, G.F. Meijs, C.L. Moad, G. Moad, S.H. Thang *Macromol. Symp.* **1999**, *143*, 291-307.
[8] G. Moad, J. Chiefari, J. Krstina, A. Postma, R.T.A. Mayadunne, E. Rizzardo, S.H. Thang *Polym. Int.* **2000**, *49*, 933-1001.
[9] E. Rizzardo, J. Chiefari, R.T.A. Mayadunne, G. Moad, S.H. Thang *ACS Symp. Ser.* **2000**, *768*, 278-96.
[10] T.P. Le, G. Moad, E. Rizzardo, S.H. Thang Int. Patent Appl. WO 9801478 (*Chem. Abs.* (1997) 128: 115390)
[11] J. Chiefari, Y.K. Chong, F. Ercole, J. Krstina, J. Jeffery, T.P.T. Le, R.T.A. Mayadunne, G.F. Meijs, C.L. Moad, G. Moad, E. Rizzardo, S.H. Thang *Macromolecules* **1998**, *31*, 5559-62.
[12] G. Moad, J. Chiefari, C.L. Moad, A. Postma, R.T.A. Mayadunne, E. Rizzardo, S.H. Thang *Macrmol. Symp.* **2002**, in press.
[13] M.J. Monteiro, H. de Brouwer *Macromolecules* **2001**, *34*, 349-52.
[14] C. Barner-Kowollik, J.F. Quinn, T.L.U. Nguyen, J.P.A. Heuts, T.P. Davis *Macromolecules* **2001**, *34*, 7849-57.
[15] C. Barner-Kowollik, J.F. Quinn, D.R. Morsley, T.P. Davis *J. Polym. Sci., Part A: Polym. Chem.* **2001**, *39*, 1353-65.
[16] J. Krstina, C.L. Moad, G. Moad, E. Rizzardo, C.T. Berge, M. Fryd *Macromol. Symp.* **1996**, *111*, 13-23.

[17] J. Krstina, G. Moad, E. Rizzardo, C.L. Winzor, C.T. Berge, M. Fryd *Macromolecules* **1995**, *28*, 5381-5.
[18] G. Moad, F. Ercole, C.H. Johnson, J. Krstina, C.L. Moad, E. Rizzardo, T.H. Spurling, S.H. Thang, A.G. Anderson *ACS Symp. Ser* **1998**, *685*, 332-60.
[19] G. Moad, C.T. Berge, J. Krstina, E. Rizzardo, C.L. Winzor,*20th Australian Polymer Symposium,* **1995**, 14-5.
[20] G. Moad, J. Krstina, C.L. Moad, E. Rizzardo, C.T. Berge, M. Fryd,*2nd IUPAC-Sponsered International Symposium on Free Radical Polymerization: Kinetics and Mechanisms,* **1996**, 37.
[21] R.T.A. Mayadunne, E. Rizzardo, J. Chiefari, J. Krstina, G. Moad, A. Postma, S.H. Thang *Macromolecules* **2000**, *33*, 243-5.
[22] R.T.A. Mayadunne, E. Rizzardo, J. Chiefari, Y.K. Chong, G. Moad, S.H. Thang *Macromolecules* **1999**, *32*, 6977-80.
[23] M. Destarac, D. Charmot, X. Franck, S.Z. Zard *Macromol. Rapid Commun.* **2000**, *21*, 1035-9.
[24] J. Chiefari, R.T. Mayadunne, G. Moad, E. Rizzardo, S.H. Thang PCT Int. Appl. WO 9931144 (*Chem. Abs.*
[25] J. Chiefari, R.T.A. Mayadunne, C.L. Moad, G. Moad, A. Postma, E. Rizzardo, S.H. Thang *Macromolecules* **2002**, submitted.
[26] B.Y.K. Chong, T.P.T. Le, G. Moad, E. Rizzardo, S.H. Thang *Macromolecules* **1999**, *32*, 2071-4.
[27] Y. Mitsukami, M.S. Donovan, A.B. Lowe, C.L. McCormick *Macromolecules* **2001**, *34*, 2248-56.
[28] A. Kotzev, A. Laschewsky, R.H. Rakotoaly *Macromol. Chem. Phys.* **2001**, *202*, 3257-67.
[29] H. De Brouwer, M.A.J. Schellekens, B. Klumperman, M.J. Monteiro, A.L. German *Journal of Polymer Science Part a-Polymer Chemistry* **2000**, *38*, 3596-603.
[30] M.Q. Zhu, L.H. Wei, M. Li, L. Jiang, F.S. Du, Z.C. Li, F.M. Li *Chem. Commun.* **2001**, 365-6.
[31] D. Taton, A.Z. Wilczewska, M. Destarac *Macromol. Rapid Commun.* **2001**, *22*, 1497-503.
[32] M. Stenzel-Rosenbaum, T.P. Davis, V. Chen, A.G. Fane *J. Polym. Sci., Part A: Polym. Chem.*, **2001**, *39*, 2777-83.
[33] T. Tsujii, M. Ejaz, K. Sato, A. Goto, T. Fukuda *Macromolecules* **2001**, *34*, 8872-8.
[34] M. Baum, W.J. Brittain *Macromolecules* **2002**, *35*, 610-5.
[35] R.A. Mayadunne, G. Moad, E. Rizzardo *Tetrahedron Lett.* **2002**, submitted.
[36] K.L. Wooley *J. Polym. Sci., Part A: Polym. Chem.* **2000**, *38*, 1397-407.
[37] S.Y. Liu, S.P. Armes *J. Amer. Chem. Soc.* **2001**, *123*, 9910-1.
[38] M.L. Becker, E.E. Remsen, K.L. Wooley *J. Polym. Sci., Part A: Polym. Chem.* **2001**, *39*, 4152-66.
[39] C.T. Berge, J. Chiefari, J.W. Johnson, A. Mau, R. Mayadunne, C.L. Moad, G. Moad, E. Rizzardo, G. Swiegers, S. Thang, G. Wilson WO 200177198 (*Chem. Abs.* (2001) 135:304605)
[40] S. Abrol, P.A. Kambouris, M.G. Looney, D.H. Solomon *Macromol. Rapid Commun.* **1997**, *18*, 755-60.
[41] A.W. Bosman, A. Heumann, G. Klaerner, D. Benoit, J.M.J. Frechet, C.J. Hawker *Journal of the American Chemical Society*, **2001**, *123*, 6461-2.
[42] T. Tsoukatos, S. Pispas, N. Hadjichristidis *J. Polym. Sci., Part A: Polym. Chem.* **2001**, *39*, 320-5.
[43] S. Abrol, M.J. Caulfield, G.G. Qiao, D.H. Solomon *Polymer* **2001**, *42*, 5987-91.
[44] A.J. Pasquale, T.E. Long *J. Polym. Sci., Part A: Polym. Chem.* **2001**, *39*, 216-23.
[45] J.H. Xia, X. Zhang, K. Matyjaszewski *Macromolecules* **1999**, *32*, 4482-4.
[46] K.Y. Baek, M. Kamigaito, M. Sawamoto *Macromolecules* **2001**, *34*, 215-21.
[47] K.Y. Baek, M. Kamigaito, M. Sawamoto *J. Polym. Sci., Part A: Polym. Chem.* **2002**, *40*, 633-41.
[48] C.T. Berge, M. Fryd, J.W. Johnson, G. Moad, E. Rizzardo, C. Scopazzi, S.H. Thang WO 200002939 (*Chem. Abs.* (2000) 132:93843)

Styrene Copolymerization Using Diphenylzinc-Additive Initiator Systems: Styrene/*p*-Substituted Styrenes

Franco M. Rabagliati, Mónica A. Pérez, Rodrigo A. Cancino, Marcelo A. Soto, Francisco J. Rodríguez, Carlos J. Caro*

Grupo Polímeros, Departamento Ciencias Químicas, Facultad Química y Biología. Universidad de Santiago de Chile, Casilla 40, Correo 33, Santiago, Chile
E-mail: frabagli@lauca.usach.cl

Summary: Combined systems including diphenylzinc (Ph_2Zn), a metallocene, and methylaluminoxane (MAO), have been employed to initiate the copolymerization of styrene (S) with p-alkylsubstituted styrenes and with α-olefins. The copolymerization processes depend largely on the comonomer, the nature of the metallocene included in the initiator system, the presence of Ph_2Zn, the polymerization temperature and the solvent used. Titanocenes produced true copolymers for S/p-substituted styrene, but not in S/α-olefin copolymerization. On the other hand zirconocenes either did not copolymerize S/p-substituted styrene or produced very low conversions, while they succeeded in copolymerizing S/α-olefin, depending on the particular zirconocene employed. A low p-methylstyrene (p-MeS) content in the S/p-MeS copolymer and a low p-tertbutylstyrene (p-Bu^tS) content in the S/p-Bu^tS copolymer decreased Tm, making them easier to process material than s-PS.

Introduction

Amorphous polystyrene or crystal-polystyrene, is a very important commercial polymer with many and diversified applications due to its very reliable properties. It is a good electrical insulator, has excellent optical clarity, and is easy to process. On the other hand, it presents some deficiencies: it is attacked by organic solvents, has poor weatherability, is brittle with low impact resistance, and has low upper temperature limit. With the discovery of syndiotactic polystyrene, s-PS, by Ishihara,[1,2] and by the same time Pellecchia[3] a new type of polystyrene with outstanding properties was obtained: high melting point (Tm = 270°C), high crystallinity, high solvent and thermal resistance.[4-6] Despite these excellent properties, s-PS is still a brittle material with low impact resistance, and its processing requires a rather high temperature.

With the aim to clarify the mechanism of stereospecific polymerization of styrene, and on the other hand, and in order to prepare a material with the properties s-PS but with decreased deficiencies, there have been many studies on the copolymerization of styrene.[10-14]

 CCC 1022-1360/00/$ 17,50+.50/0

Cardi et al.[15] reported the S/p-MeS copolymerization using a $Ti(EtO)_4$-MAO initiator system obtaining the corresponding copolymer with a composition enriched in p-MeS with respect to the composition in the feed. Grassi et al.,[16] established that the I^+ substituents on the benzene ring of styrene enhanced the reactivity to a greater extent than that observed for isotactic polymerization, and that the stereospecificity is also effected by ring substituents even when they are in the para position. They concluded that the copolymer of S with p-MeS obtained using Bz_4Ti-MAO as initiator system is cosyndiotactic, and considered that this fact seems also to confirm that the coordination of the monomer to the catalytic center is actually a determining factor in the stereoregulation.[16]

Recently, Schwecke and Kaminsky[17] reported on syndiotactic copolymers of styrene with p-MeS, p-Bu^{t}S and α-MeS obtained using a $CpTiF_3$-MAO initiator system. The authors emphasized the fact that small amounts of p-Bu^{t}S in S/p-Bu^{t}S copolymer are needed to reduce the Tm by up to 40°C with respect to that of s-PS, whereas 20% of p-MeS in the S/p-MeS copolymer S/p-MeS decreased the melting point by up 60°C. This was considered an advantage regarding the processing of these materials.

We have been working on the polymerization[18-20] of S and its copolymerization with p-alkylstyrenes (S/p-alkylstyrene), and with α-olefins (S/α-olefin)[21-24] initiated by combined systems including diphenylzinc, a metallocene, and methylaluminoxane. Our findings indicate that these systems are active initiators of styrene homopolymerization as well as of S/p-alkylstyrene and of S/α-olefin copolymerization. We have found that the polymerization processes depend on the temperature, the metallocene, Ph_2Zn/metallocene molar ratio, as well as on the polarity of the solvent employed. For copolymerization, the synthesis also depends on the nature of the comonomer employed and on the S/comonomer composition in the initial feed.

This paper deals with new experimental results on S polymerization and on the copolymerization of S/para-alkylstyrenes. For comparison, some example of S/long-chain-α-olefin copolymerizations are also included.

Experimental

Homo- and copolymerization experiments were carried out under argon atmosphere in a 100 cm^3 Schlenk tube equipped with a magnetic stirrer. Solvent toluene (to complete 25 cm^3), MAO Ph_2Zn, and metallocene toluene-solution, were sequentially charged by syringe under argon pressure. Polymerization was initiated by injecting the styrene or simultaneously the required amount of styrene and the comonomer. The reactions were kept at 60°C under stirring for the required length of time. Polymerization was finished by adding a mixture of hydrochloric acid and methanol. The polymers, coagulated in the acidified methanol, were recovered by filtration after washing several times with methanol, and dried in vacuum at 60°C.

Viscosities were measured either in chloroform or in o-dichlorobenzene depending on the solubility of the polymer, and intrinsic viscosities were determined by the one-point method.[19] For chloroform soluble polymer, viscosity measurements were carried out in chloroform at 25°C and viscosity-average molecular weights (M_v) for a-PS, were calculated according to equation,[20] $|\eta| = 1.12x10^{-4}\, M_v^{0.73}$, which is reported to be valid for the 7-150x10^4 molecular weight range. For s-PS fractions and insoluble in chloroform copolymers, viscosities were measured in o-dichlorobenzene at 135°C.

DSC analyses were performed on a Rheometrics Scientific DSC apparatus with samples placed under a nitrogen atmosphere. The samples (3 to 4 mg) were heated at a rate of 10°C/min, and after cooling to room temperature they were reheated at the same rate. The reported Tg and Tm were those obtained in the second scan.

NMR spectra were recorded on a Bruker AMX-300 spectrometer at 70°C, operating at 300.1 and 75.5 MHz for ^{1}H and ^{13}C respectively. The polymers and copolymers were dissolved in deuterated benzene (C_6D_6, 5% w/v). A total of 64 and 4000 scans with 16K and 32K data points and with a relaxation delay of 1 and 2 seconds were collected for ^{1}H and ^{13}C respectively. Chemical shifts were calibrated to tetramethylsilane (TMS) used as internal reference.

The NMR spectra of samples soluble at room temperature were recorded on a Bruker Avance DRX-300 spectrometer operating at 300.13 MHz. Polymers were dissolved in deuterated chloroform and TMS was used as internal reference.

Results

Table 1, shows the results of S/p-MeS and of S/p-Bu^{t}S copolymerizations initiated by the $CpTiCl_3$-MAO system for various S/comonomer molar ratios as well as for the homopolymerization of S, p-MeS, and p-Bu^{t}S. These results indicate that conversion to polymer increases as the molar ratio of p-MeS or of p-Bu^{t}S increases in the initial feed. The reported activity ratio corresponds to the ratio between the activity for the particular copolymerization with respect to the activity of styrene homopolymerization, which arbitrarily was given a value of 1. The activity ratio increases with the increase of p-alkylstyrene in the initial feed. From these figures it is clear that para-alquilsubstituted styrenes are more reactive than styrene in concordance with an I^{+} effect of the para-substituent, which facilitate the coordination of monomer through its π-vinyl electrons to the active species of the initiator system.

Table 1 Styrene/p-alkylstyrene copolymerization initiated by $CpTiCl_3$-MAO in toluene after 6 hours at 60 °C.[a)]

Initial Feed S/p-alkylS Mol/mol	Conversion[b)] %	Activity ratio[c)]	$\vert\eta\vert$[d)] dL/g	T_g °C	T_m °C
S/p-MeS					
S only	30.4	1.0	0.25	97.5	245.7
95/5	34.3	1.1	0.26	98.1	225.1
75/25	38.7	1.3	0.23	100.6	n.s.
50/50	40,7	1.4	0.23	102.1	n.s.
25/75	49.5	1.8	0.20	104.4	n.s.
5/95	45.8	1.7	0.20	106.0	n.s.
p-MeS only	57.3	2.1	0.21	105.4	n.s.
S/p-Bu^{t}S					
S only	30.5	1.0	0.25	93.5	255.0
95/5	45.3	1.0	0.35	101.1	n.s.
75/25	73.0	1.7	0.33	110.0	n.s.
50/50	82.0	2.1	0.43	123.0	n.s.
25/75	100.0	2.9	0.40	127.4	n.s.
5/95	n.p.	-------	------	------	------
p-Bu^{t}S only	100.0	3.2	0.63	134.0	n.s.

[a)] Polymerization conditions: Total volume = 25 mL, [S] + [p-MeS] = 2.1 mol/L, [Al] = 0.33 mol/L, [$CpTiCl_3$] = 2.0E-04 mol/L.
[b)] Based on comonomers in initial feed.
[c)] Referred to that of styrene arbitrarily defined as equal to 1.
[d)] Measured in o-diclorobenzene at 135°C.
n.p. = not performed; n.s. = no signal

Table 2, shows the results obtained for this copolymerization when the initiator system, Ph_2Zn-$CpTiCl_3$-MAO, included diphenylzinc. The inclusion of Ph_2Zn resulted in an improved conversion to polymer while, activity ratio, intrinsic viscosity and thermal behavior remained

Table 2 Styrene/p-alkylstyrene copolymerization initiated by Ph_2Zn-$CpTiCl_3$-MAO in toluene after 6 hours at 60 °C.[a)]

Initial Feed S/p-alkylS Mol/mol	Conversion[b)] %	Activity ratio[c)]	$\mid \eta \mid$[d)] dL/g	T_g °C	T_m °C
S/p-MeS					
S only	32.8	1.0	0.35	97.1	254.8
95/5	39.2	1.2	0.34	95.1	231.8
75/25	44.2	1.4	0.33	98.5	n.s.
50/50	41.3	1.3	0.30	100.9	n.s.
25/75	58.4	2.0	0.26	103.9	n.s.
5/95	60.6	2.1	0.23	102.6	n.s.
p-MeS only	69.3	2.1	0.23	106.3	n.s.
S/p-Bu^{t}S					
S only	42.0	1.0	0.20	100.7	262.0
95/5	54.1	1.3	0.22	103.0	n.s.
75/25	68.2	1.8	0.23	118.9	n.s.
50/50	70.2	2.1	0.23	129.6	n.s.
25/75	n.p.	------	------	------	n.s.
5/95	n.p.	------	------	------	n.s.
p-Bu^{t}S only	94.0	3.5	0.22	141.6	n.s.

a) Polymerization conditions: Total volume = 25 mL, [S] + [p-MeS] = 2.1 mol/L, [Al] = 0.33 mol/L, [$CpTiCl_3$] = 2.0E-04 mol/L.
b) Based on comonomers in initial feed.
c) Referred to that of styrene arbitrarily defined as equal to 1.
d) Measured in o-diclorobenzene at 135°C.
n.p. = not performed; n.s. = no signal

almost unchanged when compared to the initiator system without Ph_2Zn, accounting for the formation of more active species than that produced in the former initiator system. The same as in the case of the $CpTiCl_3$-MAO system, the initiator system including Ph_2Zn copolymerized S/p-MeS = 95/5 producing crude copolymer which shows a crystalline nature with a Tm value lower than that of s-PS. The lowering of Tm has been pointed out recently by Schwecke and Kaminsky[17] when copolymerizing styrene with p-alkylstyrene initiated by $CpTiF_3$-MAO system.

Figure 1 shows the DSC thermograms of S and p-MeS homopolymers and of the S/p-MeS copolymers at various S/p-MeS molar ratios in the initial feed, obtained using the Ph_2Zn-$CpTiCl_3$-MAO initiator system. Only PS and S/p-MeS (95/5) copolymer showed a crystalline melting temperature.

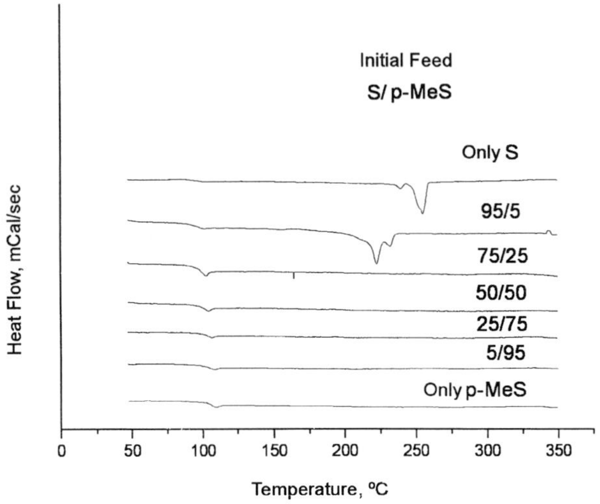

Fig. 1: DSC thermograms of S/p-MeS copolymers obtained using the Ph_2Zn-$CpTiCl_3$-MAO initiator system, in toluene after 6 hours at 60°C.

The use of the $IndTiCl_3$, Ind_2ZrCl_2 and $Et(Ind)_2ZrCl_2$ metallocenes combined with Ph_2Zn and MAO for S, p-Bu^tS and S/p-Bu^tS polymerization are presented in Table 3. From these results it can be seen that conversion to polymer or copolymer is much affected by the nature of metallocene. The metallocene included showed conversion to polymer notoriously lower than the ones produced by Ph2Zn-$CpTiCl_3$-MAO, which is in accordance with previous findings in the sense that the active species producing s-PS corresponds to species including reduced Ti^{+3}. On the other hand, we have already pointed out that zirconocene produced PS mostly amorphous with a low s-PS content.

For styrene/α-olefin copolymerizations using Ph_2Zn-metallocene-MAO systems it was found that titanocene either did not copolymerize S/α-olefin or produce very low conversions. On the other hand, when using zirconocenes such as Ind_2ZrCl_2 or better yet $Et(Ind)_2ZrCl_2$ true S/α-olefin copolymers were obtained. Figure 2 shows the ^{1}H-NMR spectra for S/1-$C_{10}H_{20}$ and S/1-$C_{16}H_{32}$ copolymers obtained using the Ph_2Zn-Ind_2ZrCl_2-MAO initiator system. In each case the corresponding CH_3-group signal at 0.9 ppm can be seen, which together with the other signals indicates the incorporation of both styrene and 1-alkene comonomers to the polymer chains. DSC analysis of such copolymers showed only one Tg signal, indicating the presence of a single product. The NMR and DSC analyses are indicative of a true copolymer

Table 3 Styrene/p-Bu^{t}S copolymerization initiated by Ph2Zn-Metallocene-MAO in toluene after 6 hours at 60 °C.[a)]

Initial Feed S/p-alkylS Mol/mol	Conversion[b)] %	Activity ratio[c)]	$\lvert\eta\rvert$[d)] dL/g	T_g °C	T_m °C
$IndTiCl_3$					
S only	1.8	1.0	0.30	100.3	n.s.
95/5	2.6	1.5	0.28	101.3	n.s.
75/25	2.7	1.7	0.32	112.7	n.s.
50/50	3.5	2.5	0.39	125.0	n.s.
25/75	4.6	3.6	0.46	132.4	n.s.
5/95	6.5	5.5	0.40	138.1	n.s.
p-Bu^{t}S only	7.3	6.2	0.42	n.p.	------
Ind_2ZrCl_2					
S only	2.2	1.0	0.18	107.9	n.s.
95/5	4.7	2.2	0.15	113.4	n.s.
75/25	5.2	2.7	0.18	121.5	n.s.
50/50	5.3	3.1	0.17	131.5	n.s.
25/75	5.3	3.4	0.15	141.8	n.s.
5/95	6.5	4.5	0.16	151.2	n.s.
p-Bu^{t}S only	9.7	6.5	0.18	152.4	n.s.
$Et(Ind)_2ZrCl_2$					
S only	2.1	1.0	0.12	97.9	n.s.
95/5	0.4	0.2	0.12	103.2	n.s.
75/25	2.6	1.4	0.12	110.5	n.s.
50/50	4.8	2.8	0.12	134.9	n.s.
25/75	4.8	3.2	0.12	137.3	n.s.
5/95	5.0	3.6	0.12	154.2	n.s.
p-Bu^{t}S only	5.1	3.7	0.14	157.0	n.s.

[a)]Polymerization conditions: Total volume = 25 mL, [S] + [p-MeS] = 2.1 mol/L, [Al] = 0.33 mol/L, [$CpTiCl_3$] = 2.0E-04 mol/L.
[b)] Based on comonomers in initial feed.
[c)] Referred to that of styrene arbitrarily defined as equal to 1.
[d)] Measured in o-diclorobenzene at 135°C.
n.p. = not performed; n.s. = no signal

instead of a mixture of both PS and P(1-alkene) homopolymers.

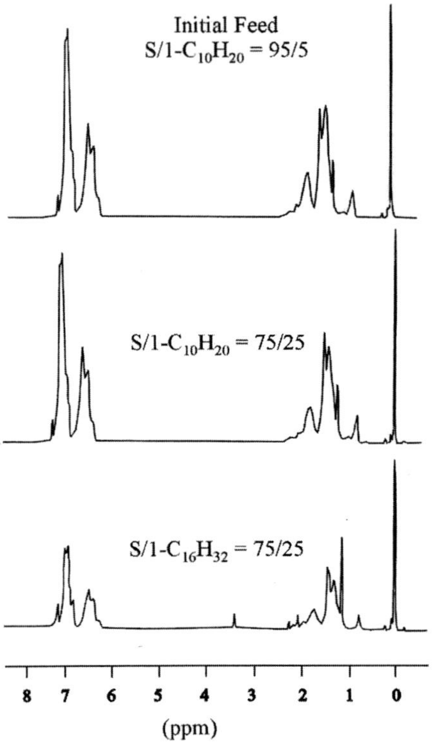

Fig. 2: ^{1}H-NMR in $CDCl_3$ at 25°C of S/1-$C_{10}H_{20}$ and S/1-$C_{16}H_{32}$ copolymers obtained using Ph_2Zn-Ind_2ZrCl_2-MAO system. Taken from reference 22.

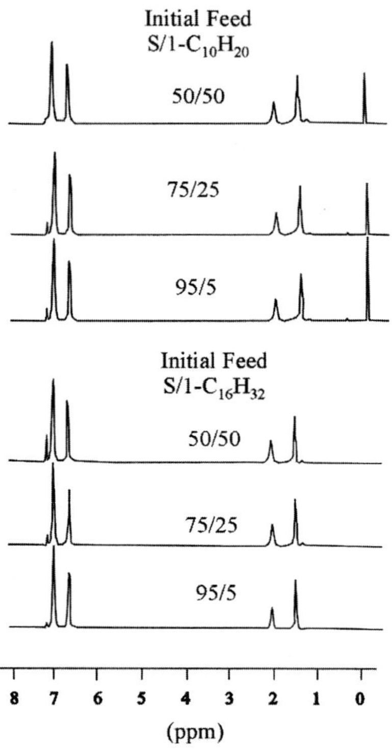

Fig. 3: ^{1}H-NMR in C_6D_6 at 70°C of S/1-$C_{10}H_{20}$ and S/1-$C_{16}H_{32}$ polymerization products obtained using Ph_2Zn-$CpTiCl_3$-MAO system. Taken from reference 22.

Figure 3 shows the ^{1}H-NMR spectra of the crude products obtained after S/1-$C_{10}H_{20}$ and S/1-$C_{16}H_{32}$ were polymerized using the Ph_2Zn-$CpTiCl_3$-MAO initiator system. No CH_3-group signal was detected, only those corresponding to PS and particularly those of s-PS in concordance with $CpTiCl_3$ behavior.

Table 4 shows the results obtained for S/α-olefin copolymerization initiated by Ph_2Zn-$Et(Ind)_2ZrCl_2$-MAO at an initial S/α-olefin molar ratio = 75/25 in the feed. It is seen that this initiator system is more effective for S/α-olefin than for S/p-alkylS copolymerization, in accordance with the activity of zirconocenes towards olefin homopolymerization. In these results it is also important to note the lower Tg values with regard to the Tg of s-PS. This behavior can be attributed to a weaker phenyl-phenyl group interaction due to a spacer effect when α-olefin incorporate between styrene units along the polymer chains.

Table 4 Styrene/α-olefin (75/25) copolymerization initiated by Ph_2Zn-Et(Ind)$_2$ZrCl_2-MAO in toluene after 6 hours at 60°C.[a)]

α-Olefin	Conversion[b)] %	Activity ratio[c)]	$\|\eta\|$[d)] dL/g	Tg °C	Tm °C
1-$C_{10}H_{20}$	1.00	2.1	n.d.	81.8	n.s.
1-$C_{16}H_{32}$	2.51	2.0	0.09	81.2	n.s.
1-$C_{18}H_{36}$	2.35	2.0	n.d.	81.8	n.s.

a) Polymerization conditions: Total volume = 25 mL, [S] + [comonomer] = 2.1 mol/L, [Al] = 0.33 mol/L, [Et(Ind)$_2$ZrCl_2] = 2.0E-04 mol/L.
b) Based on comonomers in initial feed.
c) Referred to that of styrene arbitrarily defined as equal to 1.
d) Measured in o-diclorobenzene at 135°C.
n.p. = not performed; n.s. = no signal

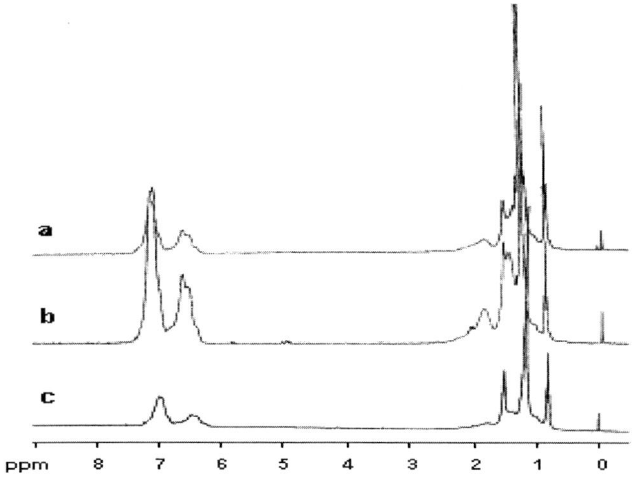

Fig. 4: ^{1}H-NMR of copolymers S/1-alkene (75/25) obtained in toluene after 6 hours at 60°C: a) S/1-$C_{18}H_{36}$, initiator system: Et(Ind)$_2$ZrCl_2-MAO, b) S/1-$C_{16}H_{32}$, initiator system: Et(Ind)$_2$ZrCl_2-MAO, c) S/1-$C_{10}H_{20}$, initiator system: Ph_2Zn-Et(Ind)$_2$ZrCl_2-MAO.

Figure 4 shows the ^{1}H-NMR spectra of copolymers obtained using the Ph_2Zn-Et(Ind)$_2$ZrCl_2-MAO initiator system included in Table 4. As in cases shown in Figure 2, there is the typical signal at ca. 0.9 ppm, which indicates the presence of a CH_3-group together with those corresponding to styrene. Again, these NMR spectra and the DSC thermograms indicate the presence of true styrene/α-olefin copolymers.

Conclusions

We can conclude from the present results that systems $CpTiCl_3$-MAO and Ph_2Zn-$CpTiCl_3$-MAO are effective initiators for styrene and p-methylstyrene homopolymerization as well as for S/p-MeS copolymerization at various S/p-MeS molar ratios in the initial feed. Furthermore, these initiator systems did not lead to S/α-olefin copolymerization nor did they initiate the homopolymerization of the corresponding α-olefin, but only to very low conversion to PS.

From our results we can conclude that systems including a zirconocene copolymerize S/α-olefin, and conversion to copolymer increases as the proportion of α-olefin in the initial feed increases. This difference with titanocenes can be attributed to zirconium atom, which is smaller than titanium and also has a lower tendency to form cationic species.

Furthermore these results also confirm the cationic pathway in the homo- and the copolymerization of styrene with p-alkylsubstituted styrene when a titanocene participates in the initiator system. The presence of electron-donating substituent group at the para position of the styrene phenyl ring enhance the nucleophilicity of para-alkylsubstituted styrene, making it a more reactive monomer as compared to styrene itself.

The present results are in agreement with previous conclusions in the sense that styrene polymerization through Ph_2Zn-Titanocene-MAO proceeds by coordination of monomer to active species followed by insertion of the monomer in the growing polymer chain-metal atom link, in a cationic pathway for chain propagation.

Further experimental work is under progress from which conclusive results are expected.

Acknowledgements

Financial support from the Departamento de Investigaciones Científicas y Tecnológicas, Universidad de Santiago de Chile, DICYT-USACH, Grant 05-9741-RC, and from Fondo Nacional de Desarrollo Científico y Tecnológico, FONDECYT, Grants 198-1135 and 101-0036, are gratefully acknowledged. The authors also thank Miss Ana M. Cavieres and Miss Maricel Cerda for the DSC measurements.

[1] N. Ishihara, T. Seimiya, M. Kuramoto, M. Uoi. Macromolecules, **19,** 2464 (1986).

[2] N. Ishihara, M. Kuramoto, M. Uoi, Macromolecules, **21,** 3356 (1988).
[3] C. Pellecchia, P. Longo, A. Grassi, P. Ammendola, A. Zambelli, Makromol. Chem., Rapid Commun. **8,** 277 (1987).
[4] F. De Candia, G. Romano, R. Russo, V. Vittoria, Colloid Polym. Sci., **268,** 720 (1990).
[5] C.M. Hsiung, M. Cakmak, Int. Polym. Process, **7,** 51 (1992).
[6] G. Guerra, V.M. Vitagliano, C. de Rosa, P. Corradini, V. Petraccone, Macromolecules, **23,** 1539 (1990).
[7] K. Soga, H. Nakatani, T. Monoi, Macromolecules, **23,** 953 (1990).
[8] A. Grassi, P. Longo, A. Proto, A. Zambelli, Macromolecules, **22,** 104 (1989).
[9] P. Longo, A. Grassi, L. Oliva, Makromol. Chem., **191,** 2387 (1990).
[10] C. Pellecchia, A. Proto, A. Zambelli, Macromolecules, **25,** 4450 (1992).
[11] A. Zambelli, A. Proto, P. Longo, P. Oliva, Macromol. Chem. Phys., **195,** 2623 (1994).
[12] L. Oliva, L. Caporaso, C. Pellecchia, A. Zambelli, Macromolecules, **28,** 4665-4667 (1995).
[13] B.K. Hong, K.H. Kim, J.C. Cho, W.H. Jo, Macromolecules, **31,** 9081 (1998).
[14] M. Galimberti, M. Destro, O. Fusco, F. Piemontesi, I. Camurati, Macromolecules, **32,** 258 (1999).
[15] N. Cardi, R. Pó, L. Abbondanza, L. Abis, G. Conti, Macromol. Symp., **102,** 123 (1996).
[16] A.Grassi, P. Longo, A. Proto, A. Zambelli, Macromolecules, **22,** 104 (1989).
[17] C.Schwecke, W. Kaminsky, Macromol. Rapid Commun., **22,** 508 (2001).
[18] F.M. Rabagliati, M.A. Pérez, R. Quijada, Polym. Bull., **41,** 441 (1998).
[19] F.M. Rabagliati, M. Pérez, R. Cancino, R. Quijada, Polym. Int. **48,** 681 (1999).
[20] F.M. Rabagliati, M.A. Pérez, R.A. Cancino, M.A. Soto, F.J. Rodríguez, A.G. León, H.A. Ayal, R. Quijada, Bol. Soc. Chil. Quím. **45,** 219 (2000).
[21] F.M. Rabagliati, M.A. Pérez, M.A. Soto, A. Matínez de Ilarduya, S. Muñoz-Guerra, Eur. Polym. J., **37,** 1001 (2001)
[22] F.M. Rabagliati, R.A. Cancino, F.J. Rodríguez, Polym. Bull., **46,** 427 (2001).
[23] F.M. Rabagliati, M.A. Pérez, R.A. Cancino, M.A. Soto, F.J. Rodríguez, J.C. Caro, A.G.León, H.A. Ayal, R. Quijada, Macromol. Symp., **168,** 31 (2001).
[24] F.M. Rabagliati, C.J. Caro, M. Pérez, Bol. Soc. Chil. Quím., accepted.

Living Ring-Opening Polymerization of Cyclic Thiocarbonate

Takeshi Endo,[1,]* *Nobukatsu Nemoto,*[2] *Fumio Sanda*[3]

[1] Department of Polymer Science and Engineering, Faculty of Engineering, Yamagata University, 4-3-16 Jonan, Yonezawa, Yamagata 992-8510, Japan
[2] Department of Materials Chemistry and Engineering, College of Engineering, Nihon University, Tamura-machi, Koriyama, Fukushima 963-8642, Japan
[3] Department of Polymer Chemistry, Graduate School of Engineering, Kyoto University, Kyoto 606-8501, Japan

Summary: This work deals with the cationic ring-opening polymerization of a cyclic thiocarbonate, 5,5-dimethyl-1,3-dioxane-2-thione (**1**). The polymerization was carried out with 2 mol% of trifluoromethanesulfonic acid, methyl trifluoromethanesulfonate, boron trifluoride etherate, or triethyloxonium tetrafluoroborate as an initiator to afford the polythiocarbonate with the narrow molecular weight distribution (M_n = 11200-31000, M_w/M_n = 1.04-1.15). The molecular weight of the obtained polymer could be controlled by the feed ratio of the monomer to the initiator and increased when the second monomer was added to the polymerization mixture after quantitative consumption of **1** in the first stage, supporting that the cationic ring-opening polymerization of **1** proceeded via a living process.

Introduction

It has been extensively studied that cyclic carbonates undergo ring-opening polymerization both with cationic and anionic initiators.[1] We have also found that cyclic carbonates show volume expansion during polymerization, which can be accounted for by the difference in strength of the intermolecular interaction between monomers and polymers.[2] Recently, sulfur-containing polymers such as polythioethers[3] and polythioesters[4] are gaining importance due to their excellent optical and thermal properties. In the course of the study of the ring-opening polymerization of cyclic carbonates, we have designed a sulfur analogue of cyclic carbonate, i.e., cyclic monothiocarbonate. We have reported the first anionic ring-opening polymerization of a six-membered cyclic monothiocarbonate, 1,3-oxathian-2-one, to afford poly(trimethylene thiocarbonate).[5] Polythiocarbonate with the similar structure can be obtained by the cationic ring-opening polymerization of 1,3-dioxane-2-thione as reported by

 CCC 1022-1360/00/$ 17,50+.50/0

Kricheldorf et al.[6] Unfortunately, the obtained polythiocarbonate is poorly soluble in common organic solvents, therefore, the detailed polymerization behavior is not sufficiently reported. This article deals with the controlled cationic ring-opening polymerization of a six-membered cyclic monothiocarbonate, 5,5-dimethyl-1,3-dioxan-2-thione (**1**) as described in Scheme 1.

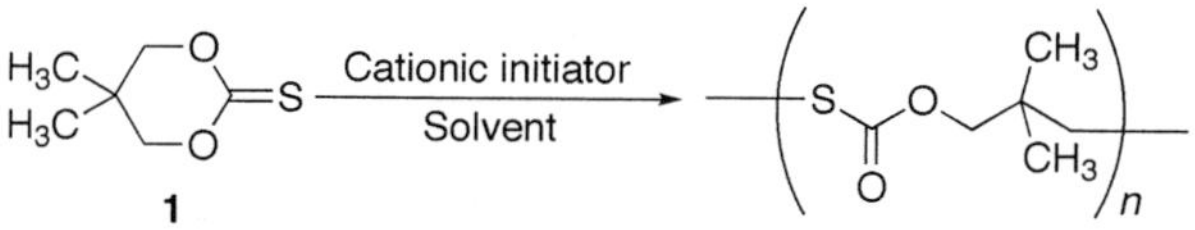

Scheme 1.

Results and Discussion

Monomer **1** was synthesized by the reaction of 2,2-dimethylpropane-1,3-diol with an equivalent amount of thiophosgene in the presence of 2 equiv. of 2,3-dimethyl-1-phenyl-3-pyrazolin-5-one in chloroform at 50°C for 12 h in 45% yield. The cationic polymerization of **1** was carried out with 2 mol% of trifluoromethanesulfonic acid (TfOH), methyl trifluoromethanesulfonate (TfOMe), boron trifluoride etherate (BF_3OEt_2), or triethyloxonium tetrafluoroborate (Et_3OBF_4) as an initiator at 30-80 °C under a dry nitrogen atmosphere as summarized in Table 1. The consumption of the monomer was achieved within 8 h in runs 1 and 2, and within 3 h in runs 3-5. There is no significant difference between the GPC profiles at 8 h and that at 24 h (conditions: monomer, 2.00 mmol; initiator, 0.040 mol; solvent, 2.00 mL of CH_2Cl_2; temperature, 30 °C), supporting the absence of back-biting reaction.

Table 1. Cationic ring-opening polymerization of **1**[a)]

run	Solvent	initiator	time h	temp. °C	conv. [b)] %	M_n [c)]	M_w/M_n [c)]
1	CH_2Cl_2	Et_3OBF_4	24	30	> 99	13200	1.04
2	$CHCl_3$	Et_3OBF_4	20	40	> 99	16800	1.04
3	PhCl	TfOMe	12	80	> 99	11200	1.15
4	PhCl	TfOH	12	80	> 99	13900	1.13
5	PhCl	BF_3OEt_2	12	80	> 99	31000	1.08

a) Monomer, 0.25 mmol; solvent, 0.25 mL; initiator, 0.005 mmol.
b) Determined by ^1H NMR spectroscopy measured in $CDCl_3$.
c) Estimated from GPC eluted with THF based on polystyrene standards.

In all cases in Table 1, polythiocarbonates with the narrow molecular-weight-distribution could be obtained, which exhibited good solubility in common organic solvents such as tetrahydrofuran (THF), chloroform, dichloromethane, and acetone.

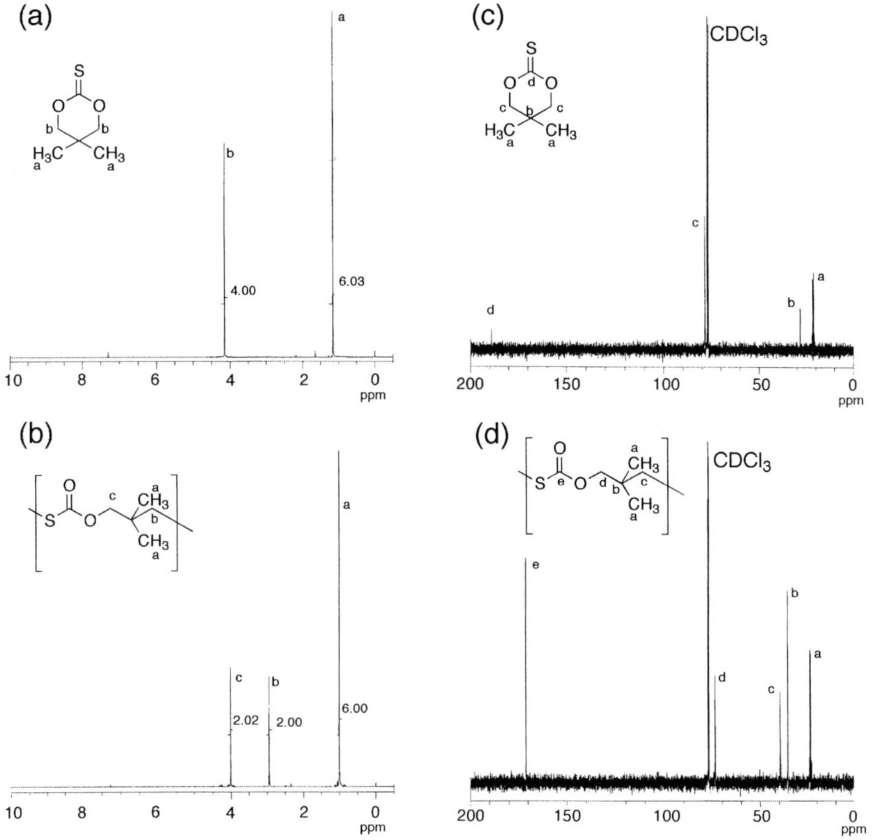

Figure 1. (a) ^{1}H NMR spectra (solvent, $CDCl_3$; 500 MHz) of **1**, (b) polymer obtained in the polymerization in run 3 in Table 1, (c) ^{13}C NMR spectra (solvent, $CDCl_3$; 125 MHz) of **1**, (d) polymer obtained in the polymerization in run 3 in Table 1.

The conversion estimated by ^{1}H NMR spectroscopy was nearly equal to the yield of the polymer. The difference in the observed M_n's among the initiators used would be due to the difference in initiation efficiency. The difference in between the observed and expected M_n's would be owing to the low initiation efficiency and/or the different solvation state in THF between the polythiocarbonate and the polystyrene standards. The obtained polymer could be isolated by precipitation with hexane and the structure was examined by ^{1}H NMR, ^{13}C NMR, and IR spectroscopy. Figure 1 illustrates the ^{1}H and ^{13}C NMR spectra of **1** and the polymer obtained in the polymerization in run 3 in Table 1. The ^{1}H NMR spectrum of the polymer (Figure 1b) showed three singlet signals at 1.01, 2.94, and 4.07 ppm, which were assignable to methyl group, α- and γ-methylenes of the thiocarbonate sulfur, respectively. The ^{13}C NMR spectrum of the polymer (Figure 1d) showed the corresponding three methylene carbon signals at 35.5, 39.5, and 73.6 ppm, the methyl carbon one at 23.4 ppm, and the carbonyl carbon one at 171.0 ppm. The assignments of the observed signals were confirmed by DEPT experiment. The ^{13}C NMR spectrum of **1** (Figure 1c) showed the thiocarbonyl carbon signal at 189.1 ppm, indicating the progress of the cationic ring-opening polymerization accompanying isomerization of thiocarbonyl group into carbonyl group. The isomerization was supported by a C=O absorption peak at 1717 cm^{-1} in the IR spectrum of the polymer. Scheme 2 illustrates a plausible mechanism for the ring-opening polymerization, which may proceed via the nucleophilic attack of a thiocarbonyl sulfur to a cationic initiator, followed by the nucleophilic attack of a thiocarbonyl sulfur to α-position of an ether oxygen in a cyclic carbenium ion.

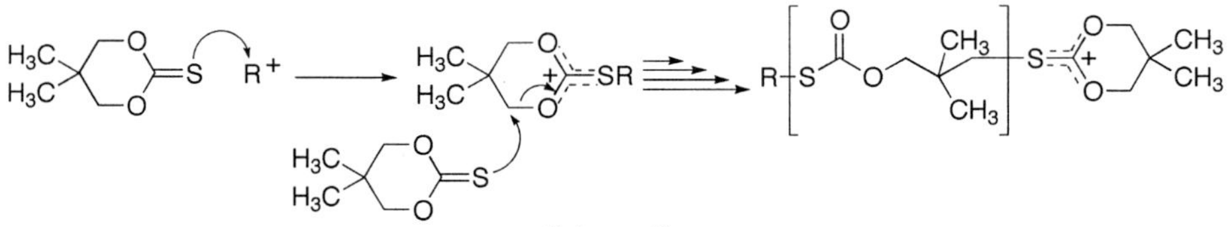

Scheme 2.

The relationship between the M_n of the polymer and feed ratio of [**1**]/[Et_3OBF_4] was linear, and the M_w/M_n was around or less than 1.1 as shown in Figure 2. The living character of the present polymerization was confirmed by postpolymerization. After the first stage polymerization of **1** (20 equiv. to Et_3OBF_4) for 12 h at 0 °C, the second monomer (40 equiv. to Et_3OBF_4) was fed to react continuously for an additional 12 h at 30 °C. The M_n of the obtained polymer increased by the second monomer feed as shown in Figure 3, though a small part of the prepolymer chains did not completely grow, supporting quasi-living character of the polymerization.

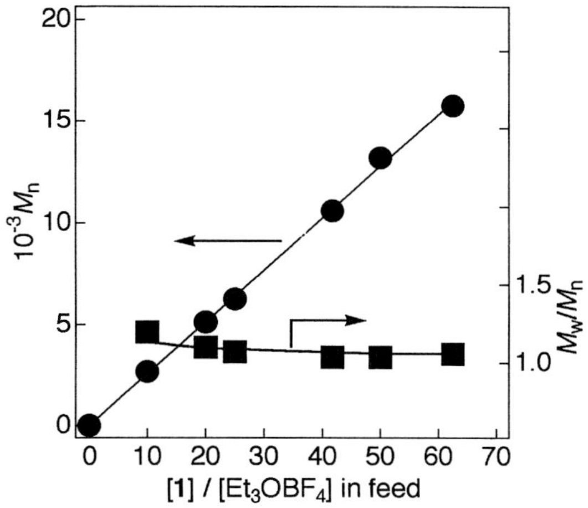

Figure 2. Relationships between M_n, M_w/M_n and feed ratio of [**1**]/[Et_3OBF_4]. Conditions: monomer, 0.25 mmol; solvent, 0.25 mL of CH_2Cl_2; temperature, 30 °C.

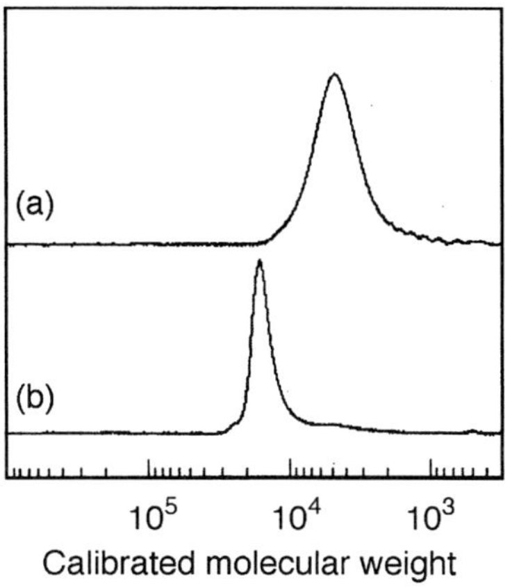

Figure 3. GPC profiles of the obtained polythiocarbonate in post polymerization: (a) after prepolymerization of **1** (20 equiv to Et_3OBF_4), M_n = 4200 (M_w/M_n = 1.16), (b) after postpolymerization of **1** (40 equiv to Et_3OBF_4), M_n = 12400 (M_w/M_n = 1.15).

Conclusions

The polymerization of 5,5-dimethyl-1,3-dioxane-2-thione (**1**) by several cationic initiators afforded polythiocarbonate with good solubility in common organic solvents and narrow molecular-weight-distribution. The molecular weight of the obtained polymer could be controlled by the feed ratio of the monomer to initiator and increased when the second monomer was added to the polymerization mixture after quantitative consumption of **1** in the first stage, supporting that the cationic ring-opening polymerization of **1** proceeded via a quasi-living process.

[1] (a) H. Höcker, H. Keul, S. Kühling, W. Hovestadt, A. Müller, B. Wurm, *Makromol. Chem., Macromol. Symp.* **1993**, *73*, 1. (b) H. Höcker, H. Keul, *Macromol. Symp.* **1994**, *85*, 211. (c) K. D. Weilandt, H. Keul, H. Höcker, *Macromol. Chem. Phys.* **1996**, *197*, 2539. (d) F. Schmitz, H. Keul, H. Höcker, *Macromol. Rapid Commun.* **1997**, *18*, 699. (e) H. R. Kricheldorf, B. Weegen-Schulz, J. Jenssen, *Makromol. Chem., Macromol. Symp.* **1992**, *60*, 119. (f) H. R. Kricheldorf, B. Weegen-Schulz, *Makromol. Chem., Rapid Commun.* **1993**, *14*, 405. (g) H. R. Kricheldorf, A. Mahler, S.-R. Lee, *New Polym. Mater.* **1996**, *5*, 25. (h) H. R. Kricheldorf, S.-R. Lee, B. Weegen-Schulz, *Macromol. Chem. Phys.* **1996**, *197*, 1043. (i) M. Hayakawa, M. Mitani, T. Yamada, T. Mukaiyama, *Macromol. Rapid Commun.* **1996**, *17*, 865. (j) M. Hayakawa, M. Mitani, T. Yamada, T. Mukaiyama, *Macromol. Chem. Phys.* **1997**, *198*, 1305.

[2] (a) T. Takata, T. Endo, *Prog. Polym. Sci.* **1993**, 18, 839. (b) Takata, T.; Endo, T. In *Expanding Monomers: Synthesis, Characterization, and Applications*; R. S. Sadhir, R. M. Luck, Eds., CRC Press: Boca Raton, FL, 1992; Chapter 3, p 63. (c) T. Takata, K. Amachi, K. Kitazawa, T. Endo, *Macromolecules* **1989**, *22*, 3188. (d) T. Takata, F. Sanda, T. Ariga, H. Nemoto, T. Endo, *Macromol. Rapid Commun.* **1997**, 18, 461.

[3] (a) Y. Imai, M. Ueda, M. Ii, *Makromol. Chem.* **1978**, *179*, 2085. (b) Y. Imai, M. Ueda, M. Ii, *J. Polym. Sci., Polym. Lett. Ed.* **1979**, *17*, 85. (c) Y. Imai, A. Kato, M. Ii, M. Ueda, *J. Polym. Sci., Polym. Lett. Ed.* **1979**, *79*, 579.

[4] W. Podkoscielny, S. Szubinska, *J. Appl. Polym. Sci.* **1988**, *35*, 85.

[5] F. Sanda, J. Kamatani, T. Endo, *Macromolecules* **1999**, *32*, 5715.

[6] H. R. Kricheldorf, D.-O. Damrau, *Macromol. Chem. Phys.* **1998**, *199*, 2589.

Synthesis of Branched Polymers by Means of Anionic Polymerization. 11. Anionic Synthesis of Densely Branched Polystyrenes Carrying Double Branches in Each Repeating Unit

Akira Hirao, Sang Wong Ryu*

Polymeric and Organic Materials Department, Graduate School of Science and Technology, Tokyo Institute of Technology, 2-12-1, Ohokayama, Meguro-ku, Tokyo 152-8552, Japan

Summary: We have demonstrated the first successful synthesis of well-defined densely branched polystyrenes carrying double branches in each of all repeating unit of backbone polymer by the coupling reaction of 1,1-diphenylethylene-end-capped polystyryllithiums with novel polystyrene derivatives having two benzyl bromide moieties in each monomer unit. The coupling reaction efficiently and quantitatively proceeded without any steric limitations to introduce two polystyrene branches in each repeating unit under certain conditions in THF at -40°C. Thus, a series of 36.4- and 181-arm densely branched polystyrenes were synthesized (M_w = 365 ~ 2 000 kg/mol, M_w/M_n = 1.01 ~ 1.03). In the 36.4-arm branched polystyrenes thus synthesized, the maximum M_w value of the introduced polystyrene as a branch segment was 55.1 kg/mol. The experimental g' values of the 36.4-arm branched polystyrenes deterimined in the range of 0.14 ~ 0.15 were very close to the value of 0.13 calculated from the empirical equation previously reported by Roovers.

Introduction

Branched polymers belong to one of the most important class of specially shaped polymeric materials and exhibit interesting and unique properties originated from their inherent branch structures. Therefore, these polymers have been extensively studied from synthetic as well as theoretical points of view. [1-3] For their synthesis, "grafting-onto method" is one of the most generally established and practical synthetic procedures. It has however long been believed that the maximum number of branch to be introduced into backbone chain is definitely limited by steric hindrance arising from the crowding of branch.[3]

Recently, Deffieux and Schappacher have first demonstrated that one branch chain can be introduced into each of all repeating unit of backbone polymer by means of a coupling reaction based on the grafting-onto method.[4,5] Their method involves the synthesis of poly(2-chloroethyl vinyl ether) by the living cationic polymerization and the subsequent

 CCC 1022-1360/00/$ 17,50+.50/0

coupling reaction of the resulting polymer with polystyryllithium. Surprisingly, the coupling reaction proceeded quantitatively in benzene in the presence of N,N,N',N'-tetramethylethylenediamine. Thus, a series of densely branched polymers having single branch in each repeating unit were successfully synthesized. Moreover, their branch and backbone chains were precisely controlled in molecular weight and molecular weight distribution, since they were obtained by the living anionic and cationic polymerizations.

More recently, we have successfully synthesized similar well-defined densely branched polystyrenes and graft copolymers by the coupling reaction of living anionic polymers of styrene, isoprene, 2-vinylpyridine, and *tert*-butyl methacrylate with precisely controlled poly(3-halomethylstyrene)s in chain length prepared *via* living anionic polymerization.[6-8] In the coupling reaction with use of the poly(3-bromomethylstyrene) (DP_w = 30) as a backbone chain under the conditions in THF at -40°C, a high molecular weight polystyrene branch with the M_w value of 68.8 kg/mol (DP_w = 662) could quantitatively be introduced. Almost at the same time, Hadjichristidis et al. have demonstrated that the poly(4-chloromethylstyrene)s obtained by the TEMPO-mediated living radical polymerization quantitatively reacted with DPE-end-capped either polystyryllithiums or polyisoprenyllithiums to afford various densely branched polymers with well-defined architectures.[9] In general, the above-mentioned coupling reactions efficiently and quantitatively proceed without any steric limitations as predicted when the polymerization degrees of the backbone chains are less than around 100, although the coupling efficiency seems to gradually decrease with increasing the polymerization degree of the backbone polymer. In the coupling reaction of poly(2-chloroethyl vinyl ether) with polystyryllithium, however, a very high efficiency of 85 % was still attained in the reaction between a high molecular weight poly(2-chloroethyl vinyl ether) (M_n = 111 kg/mol, DP_n = 1042) and polystyryllithium (M_n = 3.00 kg/mol).

During our synthetic studies, we found a very interesting and suggestive paper on the synthesis of graft copolymers previously reported by Roovers et al.[10] In this paper, they reported that the coupling reaction of poly(1,3-butadienyl)lithium with the hydrosilylated poly(1,3-butadiene) having $SiCl_2$ group underwent essentially to completion to introduce two poly(1,3-butadiene) segments at all of the available $SiCl_2$ groups. Although the number of the reaction sites that were hydrosilylated was not quantitative, the maximum average number of the introduced poly(1,3-butadiene) branch reached to *ca.* 1.6 branches in each repeating unit. This strongly suggests the possible synthesis of more densely branched polymers by means of the similar coupling reactions based on grafting-onto approach.

In this contribution, we report on the first successful synthesis of densely branched polystyrenes carrying double branches in each repeating unit by the coupling reaction of living anionic polystyrenes with novel polystyrene derivatives having two benzyl bromide moieties in each monomer unit.

Experimental Part

Purification of all the solvents and reagents (purchased from Tokyo Kasei, Japan) used in this study was carried out according to our previous papers.[6-8]

4-[3-(4-*Tert*-butyldimethylsilyloxymethylphenyl)propyl]styrene (1)

The title compound, **1**, was synthesized in 50 % yield by the Li_2CuCl_4-mediated coupling reaction of 4-(3-bromopropyl)styrene with the Grignard reagent prepared from 1-bromo-4-*tert*-butyldimethylsilyloxymethylbenzene and magnesium in THF by modifying the method previously reported [11]: ^{1}H NMR (300 MHz, $CDCl_3$): δ 0.10 (s, 6H, Si-CH_3), 0.93 (s, 9H, Si-C-CH_3), 1.95 (m, 2H, $CH_2\mathit{CH_2}CH_2$), 2.60 (t, 4H, $\mathit{CH_2}CH_2\mathit{CH_2}$), 4.71 (s, 2H, O-$CH_2$), 5.20 (d, 1H, CH_2=), 5.70 (d, 1H, CH_2=), 6.70 (q, 1H, CH=), 7.12 ~ 7.34 (m, 8H, ArH). ^{13}C NMR (75 MHz, $CDCl_3$): δ -5.12 ($SiCH_3$), 18.5 (Si-C), 26.1 ($SiCCH_3$), 33.0, 35.2 ($CH_2CH_2CH_2$), 65.0 (CH_2OSi), 112.9 (CH_2=), 136.8 (CH=), 126.3, 128.4, 128.7, 135.2, 138.9, 140.9, 142.1 (Ar C).

Anionic Polymerization of 1

The anionic polymerization of **1** was carried with *sec*-BuLi in THF at -78 °C for 0.2 ~ 0.5 h under high vacuum conditions (10^{-6} torr) using break-seal technique. The reaction mixture always showed a characteristic red color, which remained unchanged during the course of the polymerization. The polymerization reaction was terminated with degassed methanol and the polymer was precipitated in methanol. Yields of polymers were always quantitative under the conditions. They were reprecipitated twice from THF to methanol and freeze-dried from their benzene solutions. The results are summarized in Tab. 1. SEC profiles of the resulting polymers showed sharp monomodal distributions, M_w/M_n values being less than 1.06. The observed M_n values agreed well with those predicted. These results clearly indicate the living character of the polymerization of **1**. In the case for a longer reaction time to 2 h or at a higher [M] to [I] ratio of 200, a small amount of high molecular weight shoulder (< 10 %) was often formed. These high molecular weight reactions if formed were removed by SEC fractionation and used in the next reaction.

Table 1. Anionic polymerization of **1** with *sec*-BuLi in THF at -78 °C for 0.2 - 0.5 h[a)]

[1] / [*sec*-BuLi]	M_n (kg/mol)			M_w / M_n [b)]
	cacld	SEC	^{1}H NMR	
19.1	7.08	6.00	6.98	1.03
56.0	20.5	15.5	19.4[c)]	1.02
95.3	34.9	26.1	33.2[c)]	1.06

a) Yields of polymers were 100 % in all cases.
b) Determined by SEC.
c) M_w (kg/mol) determined by SLS.

Synthesis of Polystyrene Having One Benzyl Bromide Moiety in Each Repeating Unit, PS-1Br

The synthesis of **PS-1Br** was carried out by the transformation reaction of the poly(**1**) by treatment with a 50-fold excess of $(CH_3)_3SiCl$-LiBr in a mixture of acetonitrile and chloroform (3/5, *v/v*) at 40 °C for 24 h according to our previous procedure.[11] The transformation reaction proceeded cleanly and quantitatively to afford **PS-1Br**: ^{1}H NMR (300 MHz, $CDCl_3$): δ 0.67 (m, CH_3), 1.29 ~ 2.30 (m, CH_2-CH- and $CH_2CH_2CH_2$), 2.56 (t, $CH_2CH_2CH_2$), 4.47 (s, CH_2Br), 6.51 ~ 7.29 (m, ArH). IR (thin film on silicon wafer): 604, 1220cm^{-1} (CH_2-Br), 780, 1200, 1507cm^{-1} (Ar-H). Elemental analysis: $(C_{18}H_{19}Br)_n$ calculated; C 68.58, H 6.07, Br 25.35, observed; C 68.54, H 6.26, Br 25.93.

Synthesis of Polystyrene Having Two Benzyl Bromide Moieties in Each Monomer Unit, PS-2Br

The title polymer, **PS-2Br**, was synthesized by the coupling reaction of **PS-1Br** with a 1.2-fold excess of the functionalized anion in THF at -78 °C for 72 h prepared from 1,1-bis(3-*tert*-butyldimethylsilyloxymethylphenyl)ethylene (**2**) and *sec*-BuLi, followed by treatment with a 50-fold excess of $(CH_3)_3SiCl$-LiBr under the same conditions as mentioned before.[12] Both the reactions proceeded cleanly and quantitatively to afford **PS-2Br** as desired: ^{1}H NMR (300

MHz, $CDCl_3$): δ 0.48 ~ 0.68 (m, CH_3), 0.95 ~ 1.95 (m, CH_2-CH- and $CH_2\mathit{CH_2}CH_2$), 2.46 (t, $\mathit{CH_2}CH_2\mathit{CH_2}$), 4.36 (s, CH_2Br), 6.37 ~ 7.17 (m, ArH). IR (thin film on silicon wafer): 573 cm^{-1} (CH_2-Br), 794, 1210, 1511cm^{-1} (Ar-H). Elemental analysis: $(C_{38}H_{42}Br_2)_n$ calculated; C 69.30, H 6.43, Br 24.27, observed; C 69.41, H 6.52, Br 24.12.

Synthesis of Densely Branched Polymers Carrying Double Branches in Each Repeating Unit of Backbone Chain

The title densely branched polymer was synthesized by the coupling reaction of **PS-2Br** with DPE-end-capped polystyryllithium. The **PS-2Br** dissolved in THF (*ca.* 5 wt %) was added to the living polymer solution at -78 °C and the mixture was allowed to stand at -40 °C for 24 h. The living anionic polymer was usually used in a 1.5-fold excess to the benzyl bromide moiety. The reaction was quenched with degassed methanol and the polymer mixtures were precipitated in methanol. The resulting branched polymer was then isolated by fractional precipitation at 5 °C using a mixture of cyclohexane and hexane (5/1, *v/v*) and freeze-dried from its benzene solution.

Measurements

Both ^1H and ^{13}C NMR spectra were recorded on a BRUKER DPX (300 MHz for ^1H and 75 MHz for ^{13}C) in $CDCl_3$. Chemical shits were reported in ppm downfield relative to $(CH_3)_4Si$ (δ 0) for ^1H NMR and $CDCl_3$ (δ 77.1) for ^{13}C NMR spectra as standard. Size-exclusion chromatography (SEC) was obtained at 40 °C with a TOSOH HLC 8020 instrument with UV (254 nm) or refractive index detection. THF was used as a carrier solvent at a flow rate of 1.0 mL/min. Three polystyrene gel columns ($TSK_{gel}G4000H_{XL}$ + $G3000H_{XL}$ + $G2000H_{XL}$ or $G5000H_{XL}$ + $G4000H_{XL}$ + $G3000H_{XL}$) were used. Measurable molecular weight ranges in these columns are from 10^3 to 4 x 10^6 g/mol. A calibration curve was made to determine M_n and M_w/M_n values with standard polystyrene samples. Fractionation by SEC was performed at 40 °C using a TOSOH HLC 8020 fully automatic instrument equipped with a TSK-4000H_{HR} column (measurable molecular weight range: 10^3 ~ 5 x 10^5 g/mol). All runs for fractionation were made with THF as an eluent. The concentration of the polymer solution was adjusted to 10 ~ 20 *w/v* %, depending on the molecular weight of the sample. Static light scattering (SLS) measurements were performed with an Ootsuka Electronics Photal SLS-600R instrument (He-Ne laser, 633nm) in THF or benzene. The refractive index increments (dn/dc) of the polymers have been measured by an Ootsuka Electronics Photal DRM-1020. Intrinsic viscosity was measured by Ubbelohde-type capillary viscometer in toluene at 35 °C. FT-IR spectra were run with a JIR-AQS20M, JEOL. A THF solution (0.2g/mL) of polymer was spin-coated on the well-washed silicon wafer and dried for 20 min under an atmosphere of nitrogen prior to measurement.

Results and Discussion

Synthesis of Novel Well-Defined Polystyrenes Having Two Benzyl Bromide Moieties in Each Monomer Units, PS-2Br

As illustrated in Scheme 1, densely branched polystyrenes carrying double branches in each repeating unit were synthesized by the coupling reaction of **PS-2Br** with living anionic polymers of styrene. For this synthesis, **PS-2Br** used as a reactive backbone chain was prepared as illustrated in Scheme 2. At first, the precisely controlled poly(**1**) in chain length was obtained by the living anionic polymerization of **1**. The resulting polymer was quantitatively transformed into a polystyrene having one benzyl bromide moiety in each monomer unit, **PS-1Br**, by treatment with LiBr-$(CH_3)_3SiCl$. It was then reacted with the functionalized anion prepared from **2** and *sec*-BuLi. The two *tert*-butyldimethylsilyloxymethylphenyl groups thus introduced were again quantitatively transformed into two benzyl bromide moieties by treatment with LiBr-$(CH_3)_3SiCl$, thus forming a polystyrene having two benzyl bromide moiety in each monomer unit, **PS-2Br**. Two **PS-2Br** samples having different molecular weights were prepared in this study. The characterization results are summarized in Tab. 2.

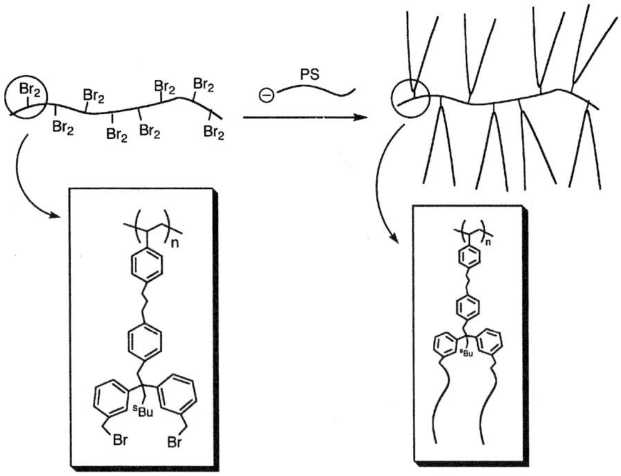

Scheme 1

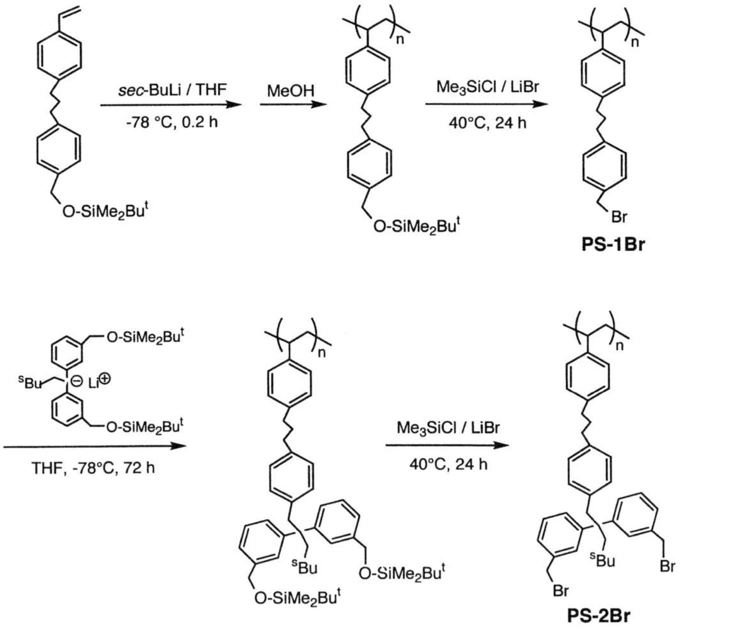

Scheme 2

Table 2. Characterization of **PS-2Br**

M_n (kg/mol)			M_w (kg/mol) (DP_w)		M_w / M_n [a)]
cacld	SEC	^{1}H NMR	cacld	SLS	
12.0	7.36	11.9	12.1	12.0 (18.2)	1.01
56.4	26.8	56.6	57.7	59.6 (90.5)	1.02

[a)] Determined by SEC.

The expected structure of **PS-2Br** was clearly confirmed by IR, ^{1}H and ^{13}C NMR measurements, and elemental analysis. SEC profiles of the resulting polymers show symmetrical monomodal distributions (see Fig. 1). Their molecular weight distributions were very narrow, the M_w/M_n values being less than 1.02.

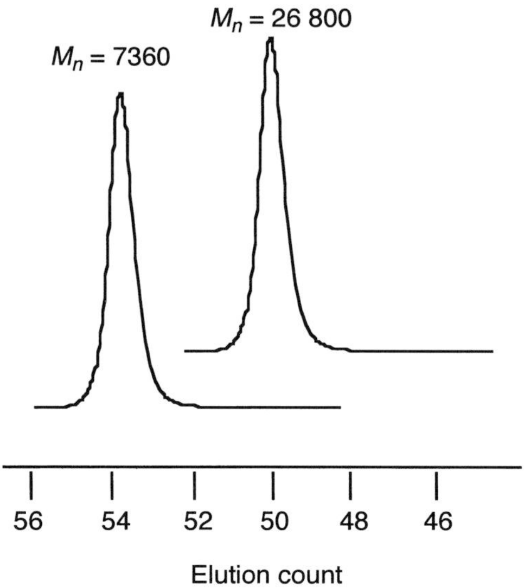

Figure 1. SEC profiles of **PS-2Br** samples

Although the M_n values estimated from SEC using polystyrene standard samples were always much smaller than those calculated, the molecular weights determined by ^{1}H NMR and SLS were in good agreement with those calculated. Thus, two polymers were precisely controlled in chain length and well defined in chemical structure as designed. Needless to say, the molecular weight of **PS-2Br** can readily be controlled by the stage of the living anionic polymerizations of **1**. The results also clearly indicate that all of the reactions used in the synthesis of **PS-2Br** cleanly and quantitatively proceed.

Synthesis of Densely Branched Polystyrenes Carrying Two Branches in Each Repeating Unit with Well-Defined Architecture

The coupling reaction of **PS-2Br** with DPE-end-capped polystyryllithium was carried out in THF at -40 ° C. The reaction time of 24 h was set in each case. The end-capping of polystyryllithium with DPE was essential in the coupling reaction to suppress presumable side reactions such as Li-Br exchange and single-electron transfer reactions occurred in the analogous reaction of poly(3-bromomethylstyrene) and polystyryllithium previously reported. Two **PS-2Br** samples (M_w = 12.0 and 59.6 kg/mol) and three DPE-end-capped polystyryllithiums (M_w = 9.40, 22.9, and 55.1 kg/mol) were used. The living polymer was always used in a 1.5-fold molar excess relative to the benzyl bromide moiety. After terminating the reaction with degassed methanol, the polymer mixtures were precipitated in methanol. The objective branched polymers were then isolated nearly quantitatively by fractional precipitation.

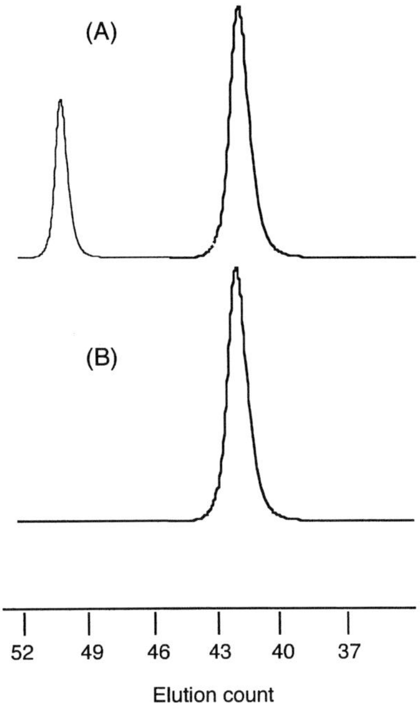

Figure 2. SEC profiles of the polymers before (A) and after (B) fractionation

Figs. 2(A) and (B) show typical SEC profiles of the polymers before and after fractionation. As you can see in Fig 2(A), there are two sharp single SEC peaks corresponding to the branched polymer and unreacted polystyrene used in excess. The branched polymer isolated by fractionation shows a sharp single SEC peak, indicating that this polymer is pure and completely free of its prepolymers (see Fig 2(B)). The characterization results are summarized in Tab. 3.

The M_w values of the resulting polymers estimated by SEC were always much smaller than those calculated as expected from their densely branched structures. On the other hand, the absolute M_w values determined by SLS agreed quite well with those calculated in all polymers obtained by the reactions of **PS-2Br** (M_w = 12.0 kg/mol, DP_w = 18.2 x 2) with DPE-end-capped polystyryllithiums (M_w = 9.40 ~ 55.1 kg/mol, DP_w = 90 ~ 529). This clearly indicates that all of the available benzyl bromide moieties react quantitatively with living polystyrene to

Table 3. Synthesis of densely branched polystyrenes carrying double branches in each repeating unit[a)]

PS-2Br DP_w	PSDLi M_w (kg/mol) (DP_w)	M_w (kg/mol) cacld	SEC	SLS	M_w / M_n [b)]	Coupling efficiency %
18.2 x 2	9.40 (90)	351	128	365	1.03	103
18.2 x 2	22.9 (219)	843	293	859	1.02	102
18.2 x 2	55.1 (529)	2015	628	2000	1.03	99.5
90.5 x 2	9.40 (90)	1744	291	1740	1.03	99.6

a) Yields of polymers were 100 % in all cases.
b) Determined by SEC.

introduce two branches in each repeating unit of the backbone polymer in each reaction. The coupling reaction of a high molecular weight **PS-2Br** (M_w =59.6 kg/mol, DP_w = 90.5 x 2) with DPE-end-capped polystyryllithium (M_w = 9.40 kg/mol, DP_w = 90) also quantitatively underwent under the same conditions. Accordingly, under the conditions employed here the steric hindering effect arising from the branch chain crowding as predicted seems to be not present in the coupling reaction.

The small deviation of individual coupling efficiency can be ascribed to errors in the determination of M_w values of **PS-2Br**, living polystyrenes, and the resulting branched polymers. Therefore, the coupling efficiencies of the four reactions are considered to be virtually quantitative within the analytical limits. Thus, we could successfully synthesize a series of densely branched polystyrenes carrying double branches in each repeating unit. Their M_w values were in the range from 365 to 2 000 kg/mol and the total average branch numbers were 36.4 and 181, respectively. Any of the branched polymers possessed both precisely controlled branch and backbone chains and well-defined architectures.

It was observed that the coupling reaction was unexpectedly rapid under the conditions in THF at -40 °C. For example, the reaction of **PS-2Br** (M_w = 12.0 kg/mol) with DPE-end-capped polystyryllithium (M_w = 9.40 kg/mol) proceeded with 70 and 92 % efficiencies only after 3 and 10 min and was complete within 1 h. Very surprisingly, the reaction rate appears to be almost comparable to that between poly(3-bromomethylstyrene) and DPE-end-capped polystyryllithium under the same conditions.

Intrinsic viscosities of the three 36.4-arm branched polystyrenes measured in toluene at 35 °C were much smaller than those of linear polystyrenes having the same molecular weights obtained from the established equation, [13] as listed in Tab. 4. In these polymers, degrees of polymerization of the branch chains (DP_w = 90, 219, and 529) were much larger than that of the backbone polymer (DP_w = 18.2 x 2). It can be therefore assumed that they adopt spherical starlike structures rather than comblike molecular brushes in solution. The branched architecture of star-branched polymer is evidenced by the extremely low g' value defined as the ratio of $[\eta]_{star}/[\eta]_{linear}$ where $[\eta]_{star}$ and $[\eta]_{linear}$ are the intrinsic viscosities of star and the corresponding linear polymer with the same molecular weight, respectively.[14] The overall experimental g' values thus determined were in the range of 0.14 ~ 0.15 (see Tab. 4). These values were very close to the calculated value of 0.13 from the empirical equation [f (arm number) = 36.4] based on the experimental results of star-branched polymers in good solvents proposed by Roovers.[15]

$$\log g' = 0.36 - 0.80 \log f$$

This also provides a good evidence for the starlike architecture of the 36.4-arm densely branched polystyrenes herein synthesized.

Table 4. Intrinsic viscosities and g' values of branched polystyrenes carrying double branches[a)]

DP_w		M_w (kg/mol)	$[\eta]_{branch}$	$[\eta]_{linear}$ [b)]	$g' = [\eta]_{branch}/[\eta]_{linear}$	
backbone	branch	SLS	(dl/g)	(dl/g)	experimental	equation[c)]
18.2 x 2	90	365	0.16	1.12	0.15	0.13
18.2 x 2	219	859	0.30	2.06	0.14	0.13
18.2 x 2	529	2000	0.52	3.75	0.14	0.13

a) Measured in toluene at 35 °C.

b) $[\eta] = 1.26 \times 10^{-4} M_w^{0.71}$.

c) $\log g' = 0.36 - 0.80 \log f$

Conclusions

We have successfully synthesized well-defined densely branched polystyrenes carrying double branches in each of all repeating units by the coupling reaction of DPE-end-capped polystyryllithiums with novel polystyrenes having two benzyl bromide moieties in each monomer unit, **PS-2Br**. The coupling reaction of either **PS-2Br** (M_w = 12.0 kg/mol, DP_w = 18.2 x 2) with living polystyrene (M_w = 9.40, 22.9, or 55.1 kg/mol, DP_w = 90, 219, or 529) or **PS-2Br** (M_w = 59.6 kg/mol, DP_w = 90.5 x 2) with living polystyrene (M_w = 9.40 kg/mol, DP_w = 90) quantitatively proceeded in THF at -40 °C without any steric limitations to afford 36.4 and 181-arm branched polystyrenes with well-defined architectures. As expected from branched architecture of the resulting 36.4-arm branched polystyrenes, their intrinsic viscosities measured in toluene at 35 °C were much smaller than those of linear polystyrenes having the same molecular weights. Their experimental g' values obtained in the range of 0.14 ~ 0.15 were very close to the calculated value of 0.13 by the empirical equation based on the previous experimental results of star-branched polymers in good solvents.

[1] D. J. Meier, "Block and Graft Coplymers" J. J. Burke and V. Weiss, Eds., Syracuse Unversity Press, Syracuse, New York, 1973, p105.
[2] P. Dreyfuss, R. P. Quirk., "Encyclopedia of Polymer Science and Engineering" J. I. Kroschwitz, Ed., Wiley, New York, 1987, vol 7, p551.
[3] H. L. Hsieh, R. P. Quirk., "Anionic Polymerization. Principles and Practical Applications" Marcel Dekker, Inc., New York, Basel, Hong Kong 1996, p.369.
[4] M. Schappacher, A. Deffieux, *Macromol. Chem. Phys.* **1997**, *198*, 3953.
[5] M. Schappacher, C. Billaud, C. Paulo, A. Deffieux, *Macromol. Chem. Phys.* **1999**, *200*, 2377.
[6] S-W. Ryu, A. Hirao, *Macromolecules* **2000**, *33*, 4765.
[7] A. Hirao, H. Kawano, S-W. Ryu, *Polym. Adv. Tech. Technol.* In press.
[8] S-W. Ryu, A. Hirao, *Macromol. Chem. Phys.* **2001**, *202*, 1727.
[9] T. Tsoukatos, S. Pispas, N. Hadjichristidis, *Macromolecules* **2000**, *33*, 9504.
[10] J. Roovers, P. Toporowski, J. Martin, *Macromolecules* **1989**, *22*, 1897.
[11] M. Hayashi, A. Hirao, *Macromolecules* **1999**, *32*, 6450.
[12] M. Hayashi, A. Hirao, *Macromol. Chem. Phys.* **2001**, *202*, 1717.
[13] M. Nakata, *Makromol. Chem.* **1971**, *149*, 99.
[14] B. J. Bauer, L. J. Fetters, Rubber Chem. Tech. **1978**, *51*, 406.
[15] J. Roovers, "Star and Hyperbranched Polymers" M. K. Mishra, S. Kobayashi, Eds., Marcel. Dekker Inc. New York, 1998, p308.

Macromol. Symp. **192**, 43–61 (2003)

Synthesis, Properties and Molecular Modeling of Functional Hyperbranched Polymers and Dendrimers

S. Fomine, L. Fomina, P. Guadarrama*

Instituto de Investigaciones en Materiales, Universidad Nacional Autónoma de México, Apartado Postal 70-360, CU, Coyoacán, México DF 04510, México
E-mail: fomine@servidor.unam.mx

Summary: The design, synthesis, properties and molecular modeling of fully conjugated dendritic molecules and conjugated hyperbranched polymers are described. It has been shown that conjugated hyperbranched molecules are much more soluble than their linear analogues while maintaining all the properties characteristic of conjugated polymers. It was found that the use of polymeric solid support in hyperbranched polymerization allows to control molecular weight and degree of branching (DB). The molecular modeling of hyperbranched conjugated molecules reveals that hyperbranched structure of conjugated molecules affects significantly neither their stability nor the conjugation. On the other hand the terminal groups affect appreciably the electronic structure of conjugated hyperbranched molecules

Introduction

In the last decade much attention has been paid to the synthesis and detailed characterization of macromolecules with the architectures differed from conventional linear and comb-like structures.

Among others are calixarenes[1], fullerene-containing polymers[2,3] and, highly branched structures: dendrimers and hyperbranched polymers[4]. The interest in those types of macromolecules is due to the unique properties imparting to the polymeric materials derived from unusual molecular shape.

Hyperbranched macromolecules can be conveniently divided into two major groups. The first is the perfect dendritic macromolecules prepared by stepwise synthetic approaches[5,6,7]. Their properties are easy to control but they are obtained only in limited quantities and after much efforts. The second is hyperbranched polymers obtained by direct polymerization of AB_n monomers[8,9]. Hyperbranched polymers generally have less perfect branching, show polydispersity and their properties are not so easy tailored; however, they are readily available by simply one-step polymerization. The most attractive features of hyperbranched structures are their excellent solubility, low solution viscosity and high concentration of terminal groups even for high polymers.

 CCC 1022-1360/00/$ 17,50+.50/0

It is also well known that the conjugated polymers, which show conductivity and non-linear properties, are mostly insoluble unless they contain long aliphatic spacers. Those spacers deteriorate the polymer properties related with conjugation and interchain interactions. Fully and partially conjugated hyperbranched polymers could be interesting alternative to conventional conjugated polymers combining good tractability of hyperbranched structure with high concentration of conjugated units.

Although the chemistry of hyperbranched molecules in general is experiencing now explosive development, the conjugated dendritic molecules have been paid little attention and only a few papers have been published to date, probably due to rather sophisticated chemistry involved in the preparation of conjugated dendrimers. Among them is the synthesis of a conjugated monodendron having a smooth gradient in electronic excitation energy from the periphery to the core reported by Moore[10]. Miller and Neenan[11] published their efforts on all-hydrocarbon polyphenylene dendrimers. Recently the synthesis of dendrimers up to third generation with stilbenoid chromophores into dendritic structure has been described [12], considered as potential materials for light-emitting diodes, non linear optics (NLO) and optical imaging storage. A simple orthogonal approach to poly(phenylenevinylene) has been developed[13]. Other interesting synthetic approach was reported by Klaus Müllen et al. to obtain spherical polyphenylene dendrimer via Diels-Alder reactions. Important result was that all dendrimers prepared were soluble in common organic solvents[14].

In general, a clear understanding of the structure-properties relationship is crucial for the directional design of hyperbranched conjugated structures. The appropriate way to get this kind of information deals with the comparative analysis of the electronic structure of linear and hyperbranched conjugated structures, however, little has been published date.

This paper is an attempt to fill this gap. It briefly describes the advances in design, synthesis, characterization and electronic structure modeling made in our group in the last several years.

Design, Synthesis and Molecular Modeling of Fully Conjugated Dendritic Molecules

For the synthesis of fully conjugated hyperbranched polymers and dendrimers β,β-diethynylphenyl building block has been chosen for two reasons[15]: easiness of preparation and possibility of functionalization. The synthetic route to the monomer and its polymerization is shown in Scheme 1. A simple Pd catalyzed polymerization of AB_2 monomer **A** produced hyperbranched polymer **POLYA**. During the polymerization of monomer **A** the reaction solution turned yellow and then orange, indicative of the formation

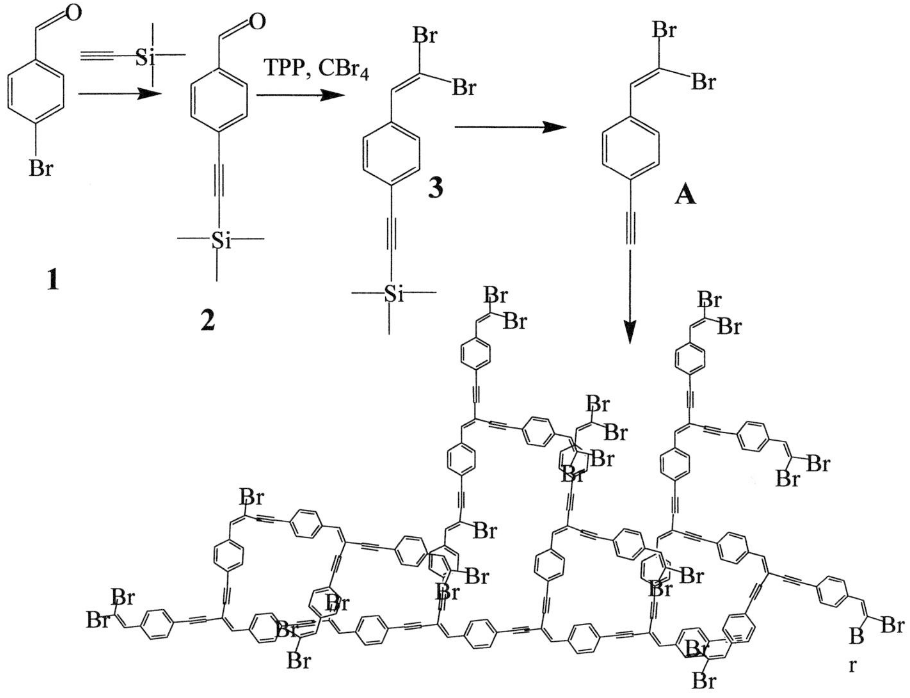

Scheme 1 Synthetic route to **POLYA**

of a polyconjugated structure. The resulting polymer was partially soluble in chloroform and toluene. The M_w of **POLYA** reached 70000 with polydispersity of 1.76. The bromine contents found by elemental analysis also suggested high molecular weight of the polymer. FTIR and NMR analysis confirmed the chemical structure of **POLYA**, however, the degree of branching could not be measured. Thus, absorption bands at 2108 and 3275 cm^{-1} corresponding to HC≡C and H-C≡ stretches in monomer **A** are completely disappeared in polymer and a new peak at 2190 cm^{-1} appeared in polymer due to stretching of conjugated internal triple bonds. It is interesting to note that while the long wave absorption maximum of monomer **A** is located at 286 nm ($CHCl_3$), a significant bathochromic shift of long wave absorption maximum observed for **POLYA** (90 nm) was attributed to the formation of conjugated hyperbranched structure.

Similar to linear polyphenylenevinylenes[16], hyperbranched polymer **POLYA** showed intense photoluminescence in the visible region. Thus, the luminescence spectrum of **POLYA** presents two maxims (at 500 and 460 nm) when excited at 420 nm.

To gain better insight in structure – properties relationship of fully conjugated hyperbranched polymers, several well defined oligomers of different generations of **POLYA** have been synthesized and characterized[17]. Convergent approach to the oligomer synthesis was applied (Scheme 2). The synthesis consisted of two repeating steps: The Pd-catalyzed cross-coupling of β,β-dibromovinylstyrene derivative (**1**) with 4-ethynylbenzaldehyde to generate formyl-terminated oligomers followed by the Wittig reaction of formyl-terminated oligomer with triphenyldibromomethylphosphorane generated from triphenylphosphine and carbon tetrabromide. After removing the protective trimethylsilyl group under mild basic conditions the prepared building block is reacted with **3** to produce next generation conjugated dendritic molecule. Similar to hyperbranched polymer **POLYA** the chemical structure of synthesized dendritic oligomers was proved by FTIR and NMR spectroscopy. Important information about conjugation in unsaturated dendritic structures can be obtained from comparison of long wave absorption in **POLYA** and well-defined oligomers of **POLYA**. Thus, the first generation oligomers **4** and **8** absorb at 377 and 367 nm which is very close to that of **POLYA** (367 nm). The second generation dendrimers (molecules **6**,**7**) showed their long wave absorption maxims at 399 and 383 nm, respectively.

This means that either steric hindrances in **POLYA** causes that the effective conjugation length does not exceed a repeating unit or **POLYA** has rather defective structure with low DB affecting the conjugation. It is seen that long wave absorption maxims of Br-terminated oligomers (**7**,**8**) are blue shifted compared to those of formyl terminated ones probably due to bulky bromine terminal atoms disturbing conjugation of Br-terminated oligomers.

Theoretical approaches have been used to understand the effect of steric hindrances and terminal groups on conjugation and electronic structure in general in hyperbranched conjugated molecules. Although a few theoretical studies of dendritic molecules have recently appeared[18,19,20,21] little is known about electronic structure of conjugated polymers. As simplest representative of conjugated hyperbranched polymers: 1→2 hyperbranched polyacetylene (PA) was used for this study[22]. Five generations of acetylenic dendrimers shown in Fig. 1 were modeled as idealized hyperbranched PA molecules at different stages of growth.

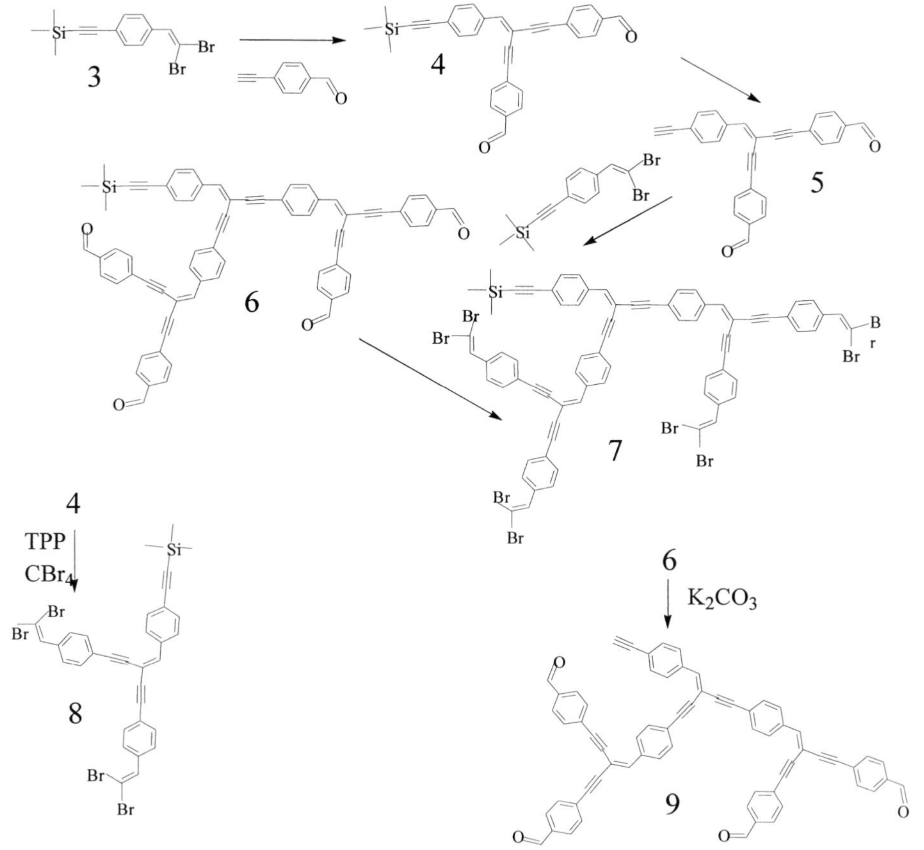

Scheme 2 Synthetic route well defined hyperbranched oligomers

The respective bromine-terminated acetylenic dendrimers up to fourth generation were also studied to clarify the effect of bulky and polarizable bromine terminal groups on geometry and electronic properties of hyperbranched PA. For comparison proposes linear trans-PA oligomers with the same number of atoms were also modeled. PM3[23] semiempirical hamiltonian was used for geometry optimization while Becke three parameter hybrid (B3) exchange functional in combination with the Lee-Yang-Parr (LYP) correlation functional (B3LYP) with split valence basis set (3-21G) was used for single point energy calculations to achieve the electron correlation. The vertical ionization potentials (IP_v) was taken as negative HOMO energy from RHF/3-21G single point calculations according to Koopman's theorem[24]. Good agreement was found between calculated and experimentally observed IP_v and energy gap (E_g) available for **Lin1**, **Lin2** and trans-PA, showing that the level of theory chosen for modeling was adequate.

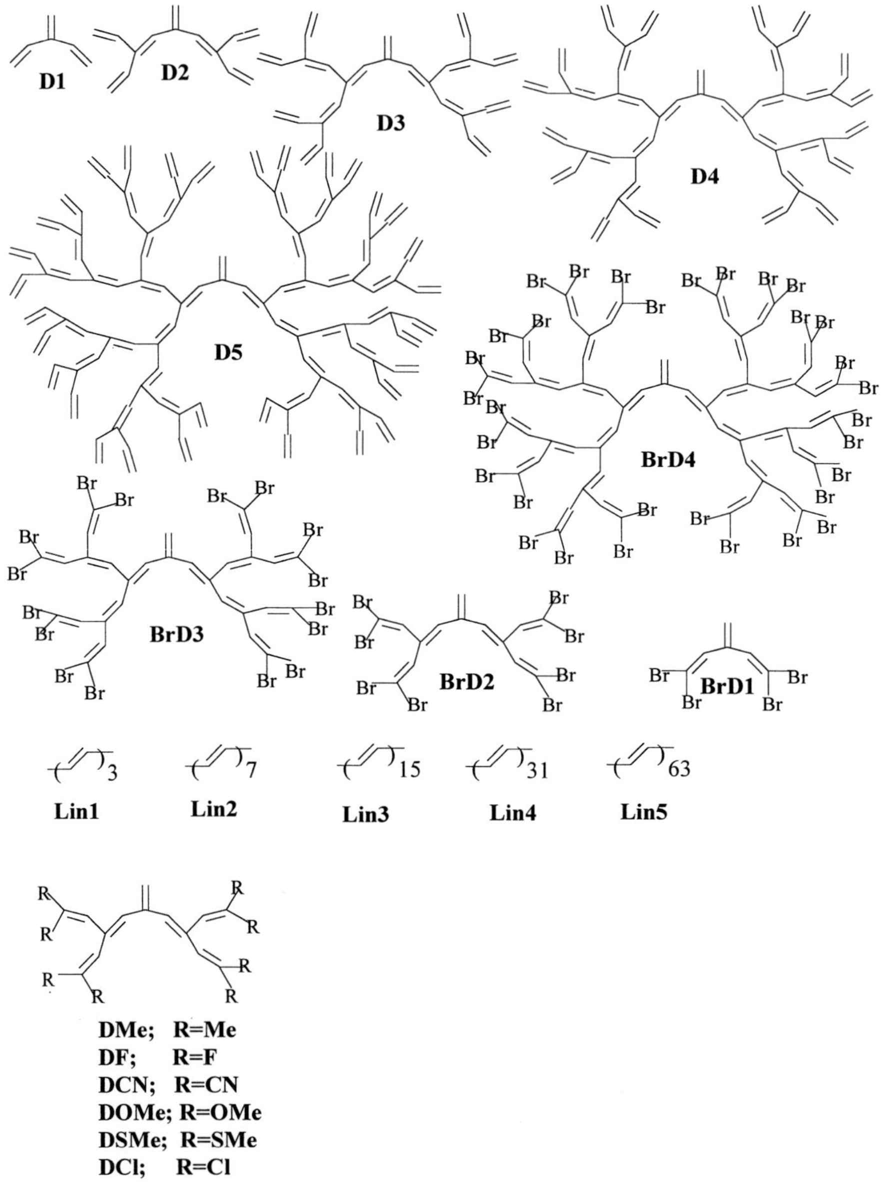

Fig.1 Idealized hyperbranched PA molecules at different stages of growth.

The relative stability of dendritic PA compared to linear trans-PA was estimated at B3LYP/3-

21G//PM3 and HF/3-21G//PM3 levels of theory as the total energy difference between dendrimer and the respective linear oligomer normalized by number of carbons in molecule. Both models present similar trends. Hyperbranched structure is less stable than linear one due to steric hindrances and destabilization increases with molecular weight. The destabilization of hyperbranched structure is, however, rather small. Both models gave the energy difference not to exceed 3.5 kcal/mol atom. Moreover, this difference seems tends to a limit with molecular weight being almost the same for **D3**, **D4** and **D5** molecules.

Both oligomer series, **D1-D5** and **BrD1-BrD4** follow the same trends as linear trans-PA oligomers (**Lin1-Lin5**). Both IP's and E_g's decrease with molecular weight due to expanding the conjugated system tending to a limit for high molecular weight dendrimers. Experimentally found E_g for trans PA (1.9 eV)[25] agrees well with that calculated for **Lin5** (1.94 eV) allowing us to extrapolate E_g's for dendrimers of **D** and **BrD** series with infinite molecular weight to 3.30-3.25 and 3.55-3.50 eV, respectively. Calculated E_g's and IP_v's of **D** and especially **BrD** series dendrimers are constantly higher than those of trans-PA oligomers. This reflects the twisting of conjugated double bonds in dendrimers especially notable for BrD series due to bulky Br terminal groups. Twisting of the double bonds decreases the conjugation in hyperbranched PA compared to linear trans-PA. However, the conjugation though less effective than in trans-PA is extended up to fourth or fifth generation.

While in linear polymers the numbers of terminal groups per repeat unit tends to zero with molecular weight not affecting properties of high polymers, the situation is completely different for hyperbranched polymers. The polymerization of AB_2 type monomer leads to a hyperbranched polymer similar to dendrimers of **BrD** series where the number of terminal group per repeat unit approaches 1 not 0 with molecular weight. Bromine terminal groups strongly affect the electronic structure of hyperbranched PA. On the one hand the bulky bromine atoms decrease even more the conjugation in bromine-terminated acetylene dendrimers due to steric hindrances as followed from calculated E_g's and PI_v constantly higher for **BrD** series. On the other hand highly polarizable bromine atoms reduces significantly the adiabatic PI's to be very close to that for trans-PA oligomers. Such a strong impact of terminal groups on the electronic structure of hyperbranched polymers is a particular feature of hyperbranched architecture and very unusual for linear polymers. Therefore, the correct choice of terminal groups is extremely important to impart desirable properties to a conjugated hyperbranched polymer.

To clarify the impact of the terminal groups on such important properties of hyperbranched polyacetylene as IP, E_g and electron affinity (EA) second generation dendritic PA oligomer (**D2**) bearing different terminal groups was modeled[26]. The end groups have been chosen in such a way to cover all possible combination of electronic effects. The results were compared with the corresponding linear oligomer **Lin2**. Higher level of theory was used in this study: all geometry's were minimized at HF/3-21G (d) level of theory while for single point energy calculations B3LYP hybrid functional was applied with 6-31G (d) basis set. The calculation showed that the geometry of dendrimer is strongly affected by the nature of terminal group. There can be considered two main factors affecting the molecular geometry, steric and electronic. While the steric hindrances caused by terminal groups tend to deteriorate conjugation, the lonely pairs and vacant orbitals of terminal groups involved in conjugation with the rest of the molecule lead to flattening of molecule. Similar effect of planarization of molecular geometry produces ionization or addition of an electron to dendritic molecule. This effect is easy to understand taking into account energy gain produced by delocalization of excessive charge on planarization. It is noteworthy that the lone electron pairs and π-electrons of terminal groups participate in the stabilization of positive charge in dendrimers as followed from the shortening R-C bond in cation radicals compared to neutral molecules in the case of R bearing lone pairs or π-electrons.

Table 1. Calculated ionization potentials (IP), electron affinity (EA) and band-gap (E_g) for the dendrimers with different terminal groups (in eV)

Compound	Formula	IP_v a)	IP_a b)	IP_v-IP_a	EA c)	E_{g1} d)	E_{g2} e)	σ_{para} f)
$\mathbf{D_{OMe}}$	$C_{22}H_{32}O_8$	5.47	5.09	0.41	0.17	3.93	3.90	-0.28
$\mathbf{D_{SMe}}$	$C_{22}H_{32}S_8$	6.61	5.74	0.87	-1.30	4.31	4.63	-
$\mathbf{D_{Me}}$	$C_{22}H_{32}$	6.76	5.91	0.85	-0.39	4.55	4.66	-0.14
$\mathbf{D_{H}}$	$C_{14}H_{16}$	7.24	6.73	0.51	-0.63	4.41	4.51	0.00
$\mathbf{D_{F}}$	$C_{14}H_8F_8$	7.30	6.92	0.38	-0.70	4.13	4.20	0.15
$\mathbf{D_{Cl}}$	$C_{14}H_8Cl_8$	7.63	7.07	0.56	-1.59	4.14	2.74	0.24
$\mathbf{D_{CN}}$	$C_{22}H_8N_8$	9.16	8.74	0.42	-3.70	3.66	3.90	0.70
Lin2	$C_{14}H_{16}$	6.39	5.89	0.50	-1.15	3.20 (3.18)g)	3.62	-

a). Adiabatic ionization potentials were calculated as the total energy difference between the neutral molecule and the corresponding cation-radical with molecular geometry optimized at HF/3-21G (d) level and single point energy calculations at B3LYP/6-31G (d) level.
b). Vertical ionization potentials were obtained as B3LYP/6-31G (d) energies difference between cation-radical and neutral molecule calculated on HF/3-21G (d) optimized geometries.
c). Electron affinities were calculated as the total energy difference between the neutral molecule and the corresponding anion-radical with molecular geometry optimized at HF/3-21G (d) level and single point energy calculations at B3LYP/6-31G (d) level.
d). Band gap calculated as HOMO-LUMO energy difference taken from a single point energy calculations at B3LYP/6-31G (d)//HF/3-21G (d) level.
e). Band gap taken from ZINDO calculations as the first singled excited state.
f). Hammett constants of the corresponding terminal groups
g) Experimental band gap for **Lin2** (Ref. 30)

The Table 1 shows calculated IP's, EA's and E_g's of the dendrimers. As seen the electronic structure of hyperbranched polyacetylene is strongly affected by the nature of the terminal groups unlike linear conjugated polymers. IP's show clear correlation with Hammett constants (σ_{para}) of the corresponding terminal groups[27]. Thus, the lowest IP's showing by **DOMe** molecule correspond to the lowest negative σ_{para} value and the highest IP's (**DCN**) are in agreement with the highest positive σ_{para} for CN group. This is a clear indication of the participation of terminal groups in the stabilization of excessive positive charge. The lower σ_{para} the better electrondonating properties of terminal groups, are and therefore, the lower the IP's of the corresponding dendrimer. The dendrimers bearing strong electrondonor terminal groups such as OMe and SMe show IP_a even lower than that of linear oligomer **Lin2**. A similar correlation holds also for EA's. The larger σ_{para} the greater the EA's. However, **DSMe** is eventually dropped out of the correlation. Even though σ_{para} is not available for SMe, the second lowest IP's among all dendrimers showed by **DSMe** would suggest strong electrondonating properties for this group and, therefore, low EA for **DSMe** molecule. Nevertheless, EA of **DSMe** is rather high reaching -1.30 eV. This apparent inconsistency is due to the participation of 3d orbitals of sulfur atom in the stabilization of the negative charge. Similar effect contributes to the higher EA of **DCl** molecule compared to **DF** one.

All other things being equal the IP_v reflects the conjugation in neutral molecule while the IP_a also accounts for the relaxation of molecular geometry and electron density distribution following the ionization process. Fast ionization produced by a photon impact when nucleus and electrons have no time to relax follows the vertical mechanism, whereas the ionization related with slow electron transfer (chemical oxidation) is adiabatic process and the difference between PI_v and PI_a reflects the energy gain due to the relaxation of molecular geometry. Generalizing the data presented in the Tables it can be said that the more voluminous the terminal group is the larger IP_v-IP_a difference. This trend is broken, however, when strong interaction between lonely pairs of a terminal groups and the rest of conjugated molecule takes place similar to that of **DF** and **DOMe** molecules.

As seen from the Table 1 the E_g's calculated by two different methods show similar trends. However, the LUMO-HOMO energy difference taken from a single point energy calculations at B3LYP/6-31G(d)//HF/3-21d(d) level reproduces best experimental E_g for **Lin2** molecule. Although there is no rigorous correlation between the volume of the terminal group and E_g's, the larger the terminal group the wider E_g. This reflects the twisting of conjugated double bonds in dendrimers, which is especially notable for large terminal groups. At the same time the electronic properties of the terminal groups strongly affected E_g values. Thus, all other things being equal the dendrimers bearing terminal groups with their electrons involved in conjugation show lower E_g's (**DF** compared to **DH**, **DOMe** and **DCN** compared to **DMe**).

To gain better insight in properties-structure relationship of synthesized dendritic oligomers of β,β-dibromo-4-ethynyl styrene those oligomers were modeled using quantum chemistry tools. In order to evaluate the accuracy of HF/3-21G and PM3 model chemistries in the geometry optimization, NMR shielding tensors of dendritic molecules were computed at HF/3-21G //HF/3-21G level of theory and the simulated ^{1}H-NMR spectra were compared with experimental data[28]. Computed NMR shielding tensors are known to be very sensitive to the molecular geometry[29,30]. The results of the calculations are presented in Fig. 2. As can be seen from the picture, the calculated proton chemical shifts agree reasonably well with experimentally observed. Only for aldehyde protons the difference is about 0.7 ppm while for all other protons the deviations from experimental values are not to exceed 0.35 ppm.

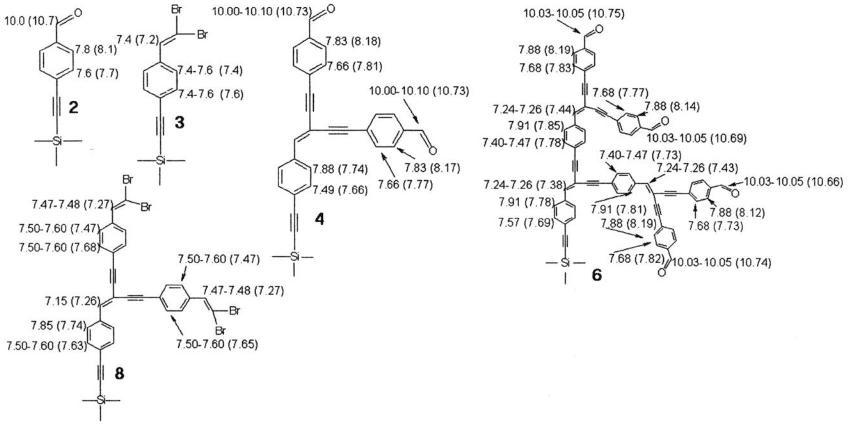

Fig. 2 Experimentally observed and theoretically calculated chemical shifts of some dendrimers.

Taking into account the above, HF/3-21G (d) level of theory is reliable enough for accurate geometry prediction of synthesized molecules. On the other hand when comparing HF/3-21G (d) Fig 2 and PM3 optimized structures the bond lengths and bond angles differed very little in the conjugated hydrocarbon part of molecules (0.01-0.02 A and 1-2° respectively).

However, C-Si and C-Br bonds lengths in PM3 optimized structures differed by 0.04-0.09 A and Br-C-Br angles even by 10° from these in HF/3-21G (d) optimized structures thus revealing certain deficiencies of PM3 method in the accurate treating of the atoms heavier than second row elements. Fortunately, both HF/3-21G (d) and PM3 optimized geometry produced similar molecular properties when using B3LYP/3-21G single point energy calculations as seen from the Table 1. Therefore, the most computationally demanding adiabatic ionization potential (IPa) calculations were run at PM3-optimized geometries.

As a rule IPa's is lower than IPv's, the energy difference accounts for the relaxation of molecular geometry and electron density distribution following the ionization process. This is also the case for the studied dendrimers. The difference between IP_v and IP_a is larger for bromine terminated molecules and tends to increase with generation number. Fig. 3 shows PM3 optimized molecular structures of the dendritic oligomers and the respective cation-radicals. As seen there is a significant difference in shape between neutral molecules and cation-radicals. Ionization causes flattening of molecular geometry. This behavior is similar to that observed for hyperbranched polyacetylene cation-radicals[22] and the explanation is that flattening of the molecule allows better stabilization of excessive positive charge by its delocalization. The larger the conjugated system the better stabilization of excessive charge can be achieved by the geometry adjusting which agrees with the increase of the difference between IP_v and IP_a with generation number. It is noteworthy that while **8+** is almost completely plane in **4+** molecule one of the benzene rings is almost perpendicular (at 81°) to the rest of the molecule. From our point of view this difference in molecular geometry is due to the influence of Br atoms. While in plane **8+** molecule the energy gain due to charge delocalization overcompensates the strain induced by the molecule flattening for polarizable bromine atoms participating in charge delocalization, a completely flat conformation for 4+ is not so favorable and the energy minimum is reached when one benzene ring is out of plane to decrease the steric hindrances with the rest of the molecule completely flat to achieve better delocalization of excessive positive charge.

Loose dendritic architecture of β,β-dibromo-4-ethynyl styrene oligomers contributes little to the instability and conjugation disruption compared to 1→2 branched polyacetylene. However, Br terminal atoms of dendrimers affect strongly the electronic structure of studied dendrimers. On the one hand the bulky bromine atoms decrease the conjugation in bromine-terminated dendrimers due to steric hindrances as followed from measured and calculated Eg's. On the other hand highly polarizable bromine atoms reduced significantly IPa's to be up to 1.5 eV lower than corresponding IPv.

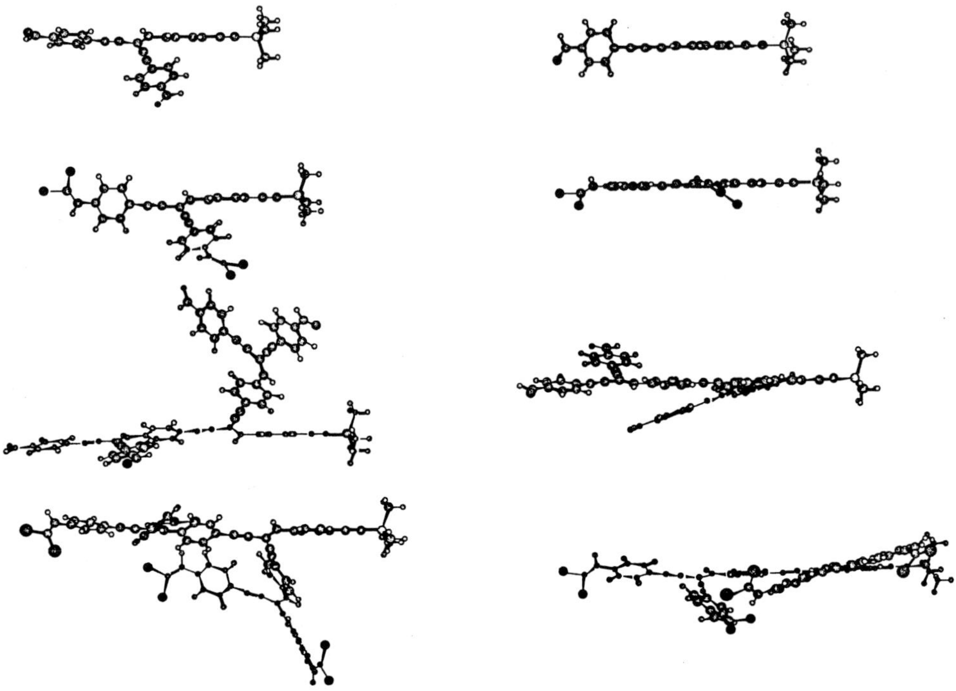

Fig. 3. PM3 optimized geometries of some conjugated dendrimers

Design, Synthesis and Molecular Modeling of Dendritic Molecules Containing Discrete Conjugated Units

In spite of much better solubility of hyperbranched fully conjugated polymers compared to their linear analogs they are still not soluble enough in low boiling solvents to form good quality films essentials for optical applications. Moreover, as it follows from electronic absorption spectra of **POLYA** and the results of molecular modeling of **POLYA** oligomers, the conjugation does not exceed a repeating unit in those types of hyperbranched polymers due to steric hindrances. On the other hand for many applications there is no need for very large conjugation system due to the fact that non-linear optical susceptibility and energy gap tends to a limit with effective conjugation length.

Therefore, a hyperbranched polymer having discrete conjugated groups connected by flexible spacers would combine the desirable electronic properties of fully conjugated polymer with improving tractability and film forming properties. Two approaches to hyperbranched polymers with discrete conjugated units have been studied; polymerization of AB_2 type monomer when discrete conjugated unit is formed during the polymerization and polymerization of AB_2 monomer where conjugated unit is part of monomer.

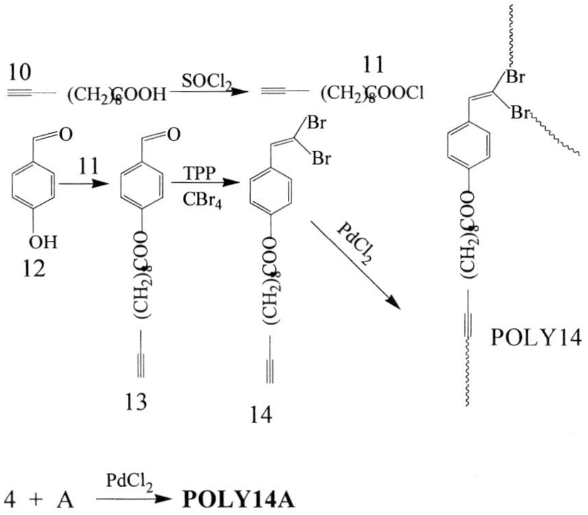

Scheme 3 Synthetic route to a hyperbranched polymer containing discrete conjugated units

The first approach is shown in Scheme 3. Monomer **14** having octamethylene spacer produces polymer **POLY14** under conditions similar to formation of **POLYA**. During the polymerization β,β-diethynylstyryl conjugated group is formed[31]. Hyperbranched polymer **POLY14** was completely soluble in aliphatic chlorinated hydrocarbons at room temperature and gave transparent and flexible films on casting from chloroform. GPC gave M_w of 11500 for **POLY14**. Although, the DB of **POLY14** could not be exactly determined, the data of ^{13}C-NMR spectroscopy suggest the presence of linear fragments of C≡CBrC=CH type where only one bromine reacts with terminal acetylenic group. UV-visible absorption spectra of **POLY14** clearly show the formation of discrete conjugated groups during the hyperbranched polymerization reflecting in bathochromic shift of long wave absorption maxim in **POLY14** compared to monomer **14** (from 226 to 300 nm). The relative amount of conjugated units in hyperbranched polymer can be increased by copolymerization of monomers **A** and **14** resulted in hyperbranched polymer **POLY14A**. When molar ratio of monomers is 1:1 **POLY14A** was almost completely soluble in hot chlorinated solvents like chlorobenzene and o-dichlorobenzene and showed MW of 11800. It is noteworthy that long wave absorption maxim of **POLY14A** (360 nm) is significantly red shifted compared to **POLY14** reflecting longer average conjugation in the former. Both hyperbranched polymers **POLY14** and **POLY14A** present intense photoluminescence in visible region. While **POLY14** emits blue light (maximum at 468 nm) **POLY14A** shows fluorescence in green region (maximum at 524 nm) when excited at 350 nm. Therefore, just by changing monomer ratio in copolymer it is possible to tune the emission spectrum of conjugated hyperbranched copolymer.

An undesirable feature of one-step AB_2 polymerization is loss of control over molecular weight accompanied by a broad molecular weight distribution. Recently it has been found that hyperbranched polymerization on an insoluble solid support allows better control over the structure of hyperbranched polymers[32]. Therefore this method was tested for the synthesis of hyperbranched polymers with discrete conjugated units[33]. Novel monomer and novel polymer supports have been synthesized (Scheme 4). AB_2 type monomer **18** was synthesized starting from 4-hydroxybenzaldehyde and 5-hexyne-1-ol (**15**) in three steps. The tosylation of **15** with TsCl produced tosylate **16**. The alkylation of 4-hydroxybenzaldehyde gave **16**, which is converted in monomer **18** by the Wittig reaction with dibromotriphenylphosphorane. Several new aromatic polyamides bearing aromatic bromine have been prepared and used as solid supports in hyperbranched polymerization. When monomer **18** is polymerized in the presence of a solid support **POLY18** is formed both in solution and on the support due to the Pd catalyzed cross-coupling reaction of terminal acetylene with aromatic bromine thus formed $sp\text{-}sp^2$ link. The polymer formed in solution is soluble in the reaction mixture and is separated from insoluble solid support by simple filtration. The support bounded hyperbranched polymer **POLY18-P** is cleaved by K_2CO_3 catalyzed transesterification in MeOH followed by the extraction with chloroform as shown in Scheme 4. The M_w found for **POLY18** formed in solution was of 17500 with polydispersity of 3.5 according to GPC. The DB, determined from ^{1}H-NMR spectra of protons *ortho* to vinyl group for this polymer was 30-35 %. Another part of polymer **POLY18** obtained after the detaching from the insoluble polymer support differed in properties from the polymer obtained in solution. **POLY18** synthesized using solid support showed lower molecular weight (Mn between 7300 and 12000) depending on polymeric support nature and lower polydispersity (about 1.8). It should be mentioned two important observations made for this type of hyperbranched polymerization. The first is that only the supports bearing bromine very active in Pd catalyzed cross-coupling reaction caused the formation of bounded hyperbranched polymer. Thus, dibromovinyl group is active enough to

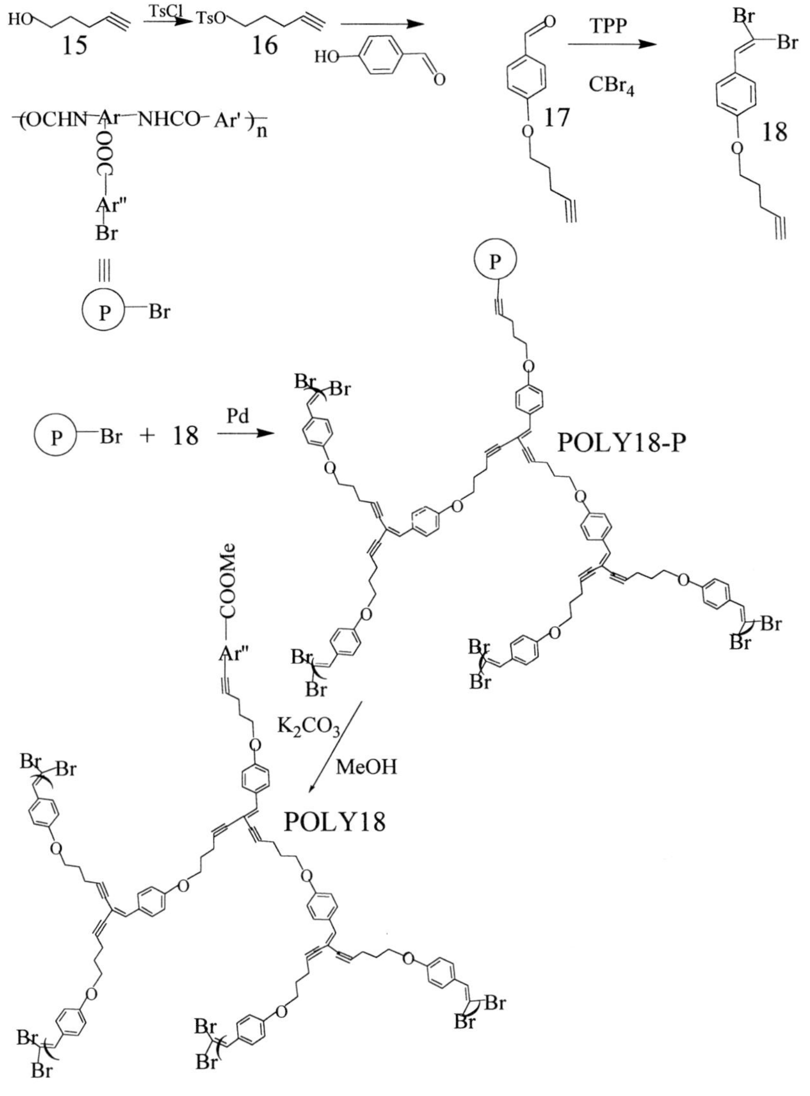

Scheme 4 Polymer supported hyperbranched polymer synthesis

cause the formation of bounded polymer while aromatic bromine of 4-bromo benzoyloxy group is not. Another interesting feature is that the molecular weight of detached polymer increased with the distance between active sites polymeric support. To gain better understanding of the nature of this phenomenon the process of solid supported hyperbranched polymerization was modeled using MM2 force field[34]. The polymeric supports were represented by oligomers of 7 and 14 repeating units while the hyperbranched polymer at different stages of growth were modeled by perfect dendritic molecules from first to sixth generation. The results showed that the decrease in molecular weight of detached polymer with decrease of the distance between active sites of polymeric support is due to steric hindrances between hyperbranched molecules attached to adjacent active sites of polymeric supports. It is noteworthy, that it has not been found any steric hindrances between the support and hyperbranched molecule attached to the support molecule.

Another approach to the synthesis of hyperbranched polymers with discrete conjugated units is shown in Scheme 5[35].

Coumarins have been recently paid attention for their fluorescent properties and they are widely used as laser dyes due to their excellent photostability and quantum yield of luminescence[36,37]. Coumarin-containing AB_2 monomer **25** was synthesized starting from 3-carboxycoumarin **19** in 6 steps. The nitration of coumarin **21** followed by the reduction of formed 6-nitrocoumarin-3-carboxylic acid **20** produced 6-aminocoumarin-3-carboxylic acid **21**. After esterification of carboxylic group of **21** the diazonium salt of **22** was reacted with benzoquinone to give substituted benzoquinone **23**. The reduction of quinone **23** with simultaneous hydrolysis of carbomethoxy group and acetylation of formed hydroquinone **24** with acetic anhydride leads directly to AB_2 coumarin containing monomer **25**.

High temperature transesterification of monomer **25** in melt results in hyperbranched coumarin containing polymer **POLY25** which is completely soluble in chloroform. The DB was found to be 1 using the ratio of the inner to outer protons in ^{1}H-NMR spectra which was also confirmed by the analysis of NMR signals of CH_3 protons of acetyl groups. The number average molecular weight calculated from ^{1}H-NMR spectrum by the end group analysis was of 27000, corresponding to degree of polymerization of 60. A series of copolymers of monomer **25** with AB monomer m-acetoxybenzoic acid have been prepared to correlate the polymer properties with the density of branching. Among all described here hyperbranched polymers **POLY25** and the respective copolymers were most thermostable which is even reflected by the method of preparation. The 10% weight loss temperatures were in the range of 370-400 °C. As a result of the presence of 6-phenylcoumarines chromophores in **POLY25** and copolymers were blue emitters, emitting in the range of 450-490 nm when excited at 400 nm. Single layer electroluminescent devices on the basis of **POLY25** have been made showing turn on voltage from 3 to 8 V.

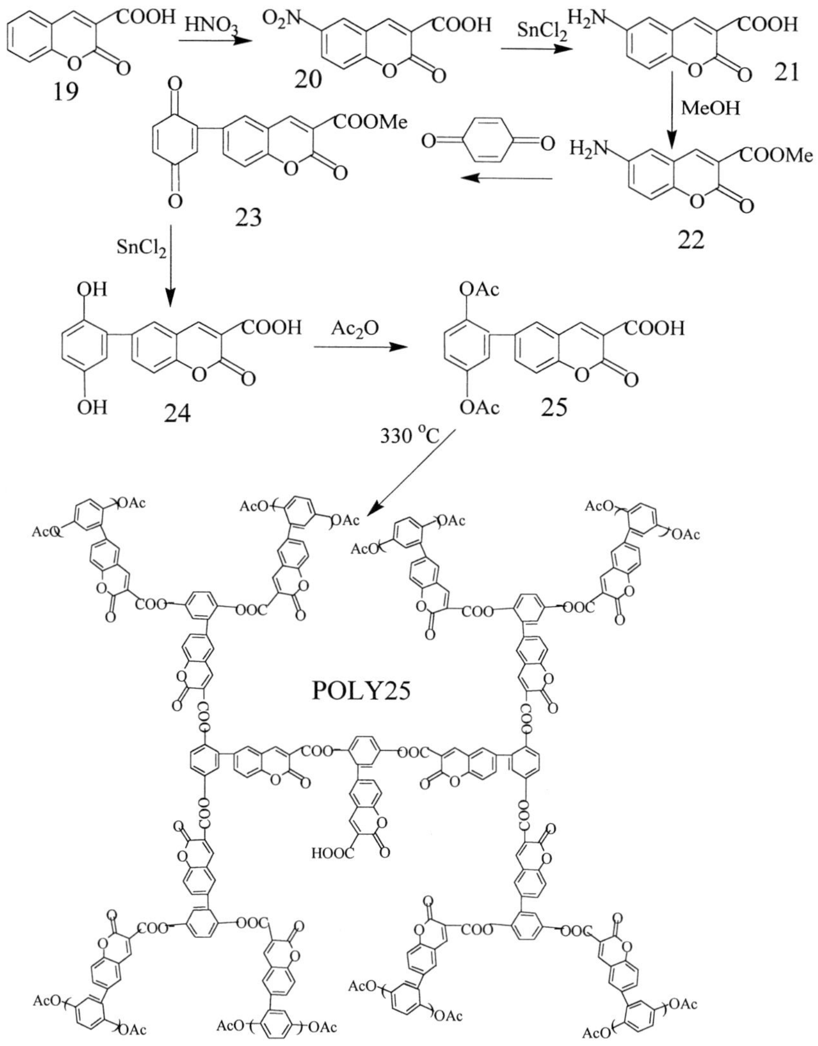

Scheme 5 Synthesis of coumarin containing hyperbranched polymer

Conclusions

Hyperbranched conjugated molecules have been synthesized both by polymerization of AB_2 type of monomers and stepwise synthesis. Fully conjugated hyperbranched polymers were found to be much more soluble than the respective linear polymers while maintaining all other properties characteristic of conjugated polymer. The introduction of flexible spacer between conjugated fragments increases even more the processability of hyperbranched polymers. In the case of hyperbranched polymerization of coumarin-containing monomer of AB_2 type novel hyperbranched coumarin-containing polymers and copolymers were prepared with DB close to 1. Those polymers combined excellent solubility and film forming properties with intense photo and electro-luminescence properties. To improve the control over the structure of hyperbranched molecules solid-supported synthesis has been tested. It was found that choosing appropriate solid support it was possible to control the molecular weight and DB of a hyperbranched polymer. The theoretical analysis of the model hyperbranched polyacetylene showed that although hyperbranched structures were less stable than the respective linear analogs the difference is not excessively large and tends to a limit with molecular weight of hyperbranched molecule while the conjugation in conjugated dendritic molecule extends up to fourth or fifth generation. It was shown that the electronic properties of terminal groups were of importance and allow to tailor the electronic properties of hyperbranched molecule.

Acknowledgments

This work was supported by a Grant from CONACyT with contract 32560-E

[1] C. David Gutsche *Calixarenes* The Royal Society of Chemistry 1989.
[2] K. M. Kadish, R. S. Ruoff Eds *Fullerenes* Volume 4, The Electrochemical Society Inc. 1997.
[3] R. Taylor Ed, *Chemistry of Fullerenes* World Scientific Publishing Co Pte. Ltd.
[4] G. R. Newkome, C. N. Moorefielf, F. Vögtle, *Dendritic molecules*, VCH Verlagsgesellschaft mbH, Weinheim, 1996.
[5] D. A. Tomalia, A. M. Naylor, and W. A. Goddard, *Angew. Chem. Int. Ed. Engl.*, **1990,** *29*, 138
[6] C.J. Hawker and J.M.J. Frechet, *J. Am Chem. Soc*, **1992**, *114*, 8405.
[7] C.J. Hawker and J.M.J. Frechet, *J. Am Chem. Soc*, **1990**, *112*, 7638.
[8] Y. Kim and O.W. Webster, *Macromolecules*, **1992**, *25*, 5561.
[9] Y. Kim and O.W. Webster, *Polym. Prepr.* **1988**, *29*, 310.
[10] Z. Xu, J. Moore, *Acta Polymerica*, **1994**, *45*, 83.
[11] T. Miller, T. Neenan, *Chem. Mater.* **1990**, *2* 346.
[12] H. Meier, M. Lehmann, *Angew, Chem. Int. Ed.* **1998**, *37*, 643.
[13] S. K. Deb, T. M. Maddux and L. Yu, *J. Am. Chem. Soc.*, **1997**, *119*, 9079.
[14] F. Morgenroth, A. J. Berresheim, M. Wagner and K. Müllen, *Chem. Commun.* **1998**, 1139.
[15] L. Fomina, R. Salcedo, *Polymer*, **1996**, *37*, 1723.
[16] A. Greiner, B. Bolle, P. Hesemann, J. Oberski, R. Sander, *Macromol. Chem. Phys.*, **1996**, *197*, 113.
[17] L. Fomina, P. Guadarrama, S. Fomine, R. Salcedo, and T. Ogawa, *Polymer*, **1998**, *39*, 2629.
[18] M. Mansfield, L. Klushin, *Macromolecules*, **1993**, *26*, 4262.
[19] A. Naylor, W. Goddard, G. Kiefer, A. Tomalia, *J. Am. Chem. Soc.*, **1989**, *111*, 2339.
[20] R. Lescanec, M. Muthukumar, *Macromolecules* **1990**, *23*, 2280.
[21] J. Scheutjens, G. Fleer, *Macromolecules* **1985**, *18*, 1882.
[22] S. Fomine, L. Fomina, and P. Guadarrama, *Macromolecular Theory and Simulations* **1999**, *8*, 54.
[23] J.J.P. Stewart, *J. Comp. Chem.* **1989**, *10*, 209.
[24] T. Koopmans, *Physica*, **1933**, *1*, 104.
[25] T. Chung, F. Morales, J. Flood, A. Heeger, *Phys. Rev.* **1984**, *B29*, 2341.

[26] L. Fomina, P. Ponce, P. Guadarrama and S. Fomine. *Macromolecular Theory and Simulations* **1999**, *8*, 403.
[27] J. March, *Advanced Organic Chemistry*, 3rd edition, J Wiley&Sons, New York 1985.
[28] S. Fomine, L. Fomina and P. Guadarrama, *Journal of Molecular Structure (TheoChem)*,**1999**, *488*, 207.
[29] J. Cheeseman, G. Truck, T. Keith, M. Frisch, *J. Chem. Phys.* **1996**, *104*, 5497.
[30] W. Kutzelnigg, *Einführung in die Theoretische Chemie*, Band 2 Verlag Chemie, Weinheim 1978.
[31] L. Fomina, S. Fomine, R. Salcedo, and T. Ogawa, *Polymer Journal,* **1996**, *28* 1071.
[32] P. Bharathi, and J. Moore *J. Am. Chem. Soc.*, **1997**, *119*, 3391.
[33] P. Guadarrama, L. Fomina, V. Pankov. W. Matus, and S. Fomine, *Polymer Journal* **1999**, **31**, 423.
[34] U. Burkert, N. L. Allinger, *Molecular Mechanics* ACS Monograph 177, American Chemical Society, Washington, DC 1982.
[35] S. Fomine, E. Rivera, L. Fomina, A. Ortiz and T. Ogawa, *Polymer*, **1998**, **39**, 3551.
[36] K. Drexhage ed., *Dye Lasers*, Mir Moscow, 1976
[37] M. Maeda ed., *Laser Dyes* OHM, Tokyo, 1984.

Amphiphilic and Mesomorphic Fullerene-Based Dendrimers

*Daniel Guillon**[a], *Jean-François Nierengarten*[a], *Jean-Louis Gallani*[a], *Jean-François Eckert*[a], *Yannick Rio*[a], *Maria del Pilar Carreon*[a], *Blaise Dardel*[b], *Robert Deschenaux*[b]

[a] Institut de Physique et Chimie des Matériaux de Strasbourg, Groupe des Matériaux Organiques, 23 rue du Loess, 67037 Strasbourg Cedex, France
[b] Institut de Chimie, Université de Neuchâtel, Avenue de Bellevaux 51, Case Postale 2, CH-2007 Neuchâtel, Switzerland

Summary: Growing attention is currently devoted to large dendritic structures for applications in nanotechnology and materials science. In this respect, the incorporation of such compounds into thin ordered films appears to be an important issue. One of the most widely pursued approaches to structurally ordered dendrimer assemblies has been the preparation of Langmuir films at the air-water interface. We report on the case of a diblock globular fullerene-based dendrimer and show that peripheral substitution of the dendrimer with hydrophobic chains on one hemisphere and hydrophilic groups on the other provides the required hydrophobic/hydrophilic balance allowing the formation of stable Langmuir films. A second approach has been to consider the case of fullerene containing dendrimers terminated by mesogenic groups such as cyanobiphenyl subunits. Whatever the generation is, up to the fourth one, all these compounds exhibit a well-defined liquid crystalline smectic A phase. The molecular organisation within the smectic layers is found to be monolayered or bilayered depending on the generation. For the smallest dendrimers, the organisation is mainly governed by the size of the fullerene moiety, whereas for the higher ones, it is governed by the interactions between the terminal mesogenic groups. These two approaches appear particularly interesting for functional groups such as fullerenes, which are not well adapted to be organised in nanoscale architectures. The present study shows that fullerenes can indeed be introduced into different types of ordered structure when they have been chemically adequately modified.

Introduction

Fullerenes and their derivatives have shown a wide range of chemical and physical properties that make them attractive candidates for a variety of interesting features in supramolecular chemistry and materials science[1]. An important issue for applications of this new carbon allotrope appears to be its incorporation into well-ordered structures such as Langmuir-

 CCC 1022-1360/00/$ 17,50+.50/0

Blodgett (LB) films or liquid crystals.

Monolayers of C_{60} itself at the air-water interface are difficult to achieve due to strong fullerene-fullerene interactions and three-dimensional aggregation and all attempts to create defined LB multilayers of fullerenes have failed.[2-4]. Whereas the functionalisation of the fullerene sphere with hydrophilic addends leads to significant improvements[4-5], fullerene derivatives with good spreading characteristics and reversible compression/expansion behaviour are quite rare[6-7]. In a collaborative work amongst the research groups of Diederich, Stoddart, Echegoyen and Leblanc, dendrimers with a fullerene core and peripheral acylated glucose units have been prepared and studied[6]. These fullerene derivatives have shown for the first time a reversible behaviour of fullerene monolayers in successive compression/decompression cycles, the dendritic part of the compounds preventing the irreversible aggregation of the fullerene moieties. We have recently shown that the encapsulation of the fullerene sphere in a cyclic addend is also able to prevent the aggregation usually observed for amphiphilic fullerene derivatives[7].

Fullerene-containing liquid crystals appear to be another important class of materials for technological applications. However, the liquid crystalline ordering of fullerenes is difficult since the C_{60} subunit does not behave as a mesogenic unit[8]. During the last few years, several approaches have been developed for the preparation of fullerene-containing liquid crystalline derivatives.

The first C_{60}-containing thermotropic liquid crystal was described in 1996.[9] The functionalization of C_{60} with a malonic ester bearing two mesogenic cholesterol subunits resulted in a fullerene derivative with liquid crystalline properties. A mesophase was observed from 146 to 190°C and identified as a monotropic smectic A phase. More recently, the same group has shown that a dendritic addend exhibits similar mesogenic properties[10]. In this case, the C_{60} core is buried in the middle of the dendritic structure and thus prevents unfavourable effects of the C_{60} unit such as aggregation or steric hindering.

In this paper, we will present recent results obtained in our laboratories regarding both ways of introducing fullerene groups into ordered structures. Langmuir-Blodgett films containing fullerene will be considered in a first part, whereas smectic liquid-crystalline structures containing C_{60} will be in a second one. In both cases, the nanostructures are obtained with fullerene-based dendrimers.

Langmuir-Blodgett Films of Amphiphilic Diblock Fullerene Based-Dendrimers

We have shown recently that fullerene functionalised dendrons with peripheral octyl chains and a carboxylic acid at the focal point are able to form Langmuir films at the air-water interface[11]. We also succeeded in forming Langmuir-Blodgett (LB) films by transferring the monolayers onto hydrophobic substrates. However, due to the difference in size between the the hydrophobic and hydrophilic groups, the preparation of multilayered films was found to be difficult and the transfer ration not too good (~~0.7-0.8). Thus, it was decided to consider the case of a diblock globular dendrimer (figure 1) where the peripheral substitution of the dendrimer is obtained with hydrophobic chains on one hemisphere and hydrophilic groups on the other, in order to provide the perfect hydrophobic/hydrophilic balance allowing the formation of stable Langmuir films[12].

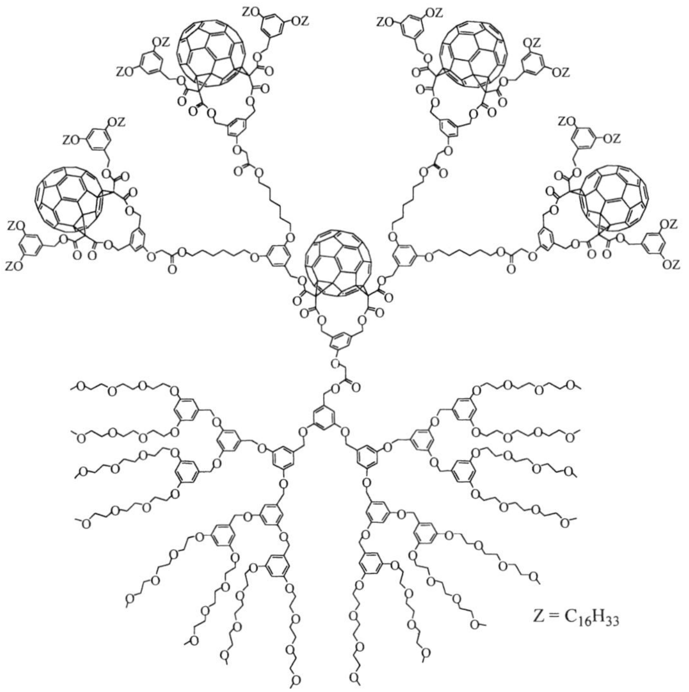

Figure 1. Chemical structure of the amphiphilic dendrimer considered in the present study.

This compound is able to form Langmuir films at the air-water interface. The molecular area extrapolated to zero pressure is 610 ± 30 Å^2 in good agreement with the value estimated by molecular modelling. The films are stable and a perfect reversibility has been observed in successive compression/decompression cycles (figure 2).

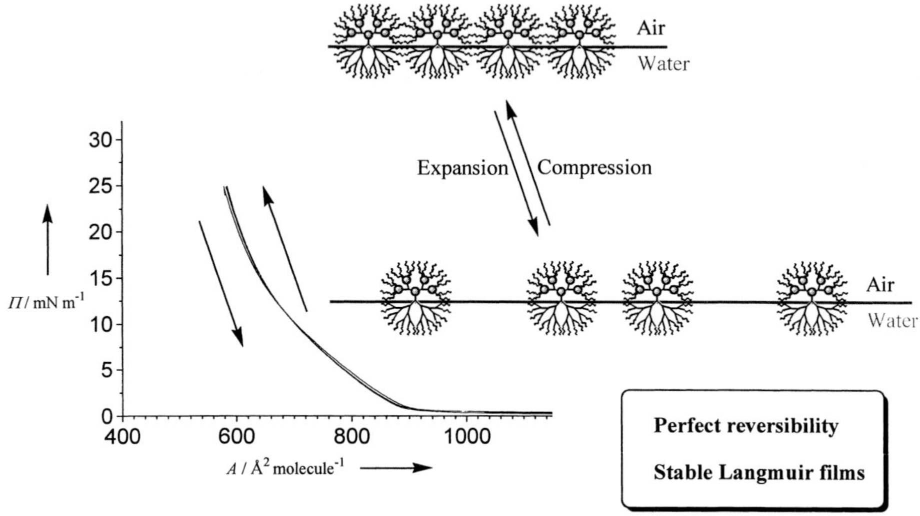

Figure 2. Four successive compression/decompression cycles with a monolayer of the amphiphilic dendrimer (25°C; compression/decompression speed, 5 mm mn^{-1}) showing the perfect reversibility of the process.

The onset of the surface being quite progressive, one could have expected the existence of some liquid condensed phases in the Langmuir film. Brewster angle microscopy (BAM) reveals that this is actually not the case and that the molecules already associate to form islands at large molecular areas, as seen in figure 3a. Still, these islands fuse together in a nice and reversible way and finally yield a very homogeneous film, as clearly shown in figure 3b. It is noteworthy that the five C_{60} units are buried in the middle of the dendritic structure, which is capable of providing a compact insulating layer around the carbon spheres, thus preventing the irreversible three-dimensional aggregation resulting from strong fullerene-fullerene interactions as usually observed with C_{60} derivatives[13]. The polar head-group causes an attractive interaction with the aqueous subphase, forcing the molecules to a two-dimensional arrangement on the water surface. In the case of the present dendrimer, it seems there exists a perfect hydrophilic/hydrophobic balance provided by the 16 triethylene glycol units in the diblock dendrimer, allowing the preparation of stable Langmuir films. Therefore, the functionalisation of the branching shell of a diblock dendrimer structure is an efficient strategy for the incorporation of functional groups in thin ordered films.

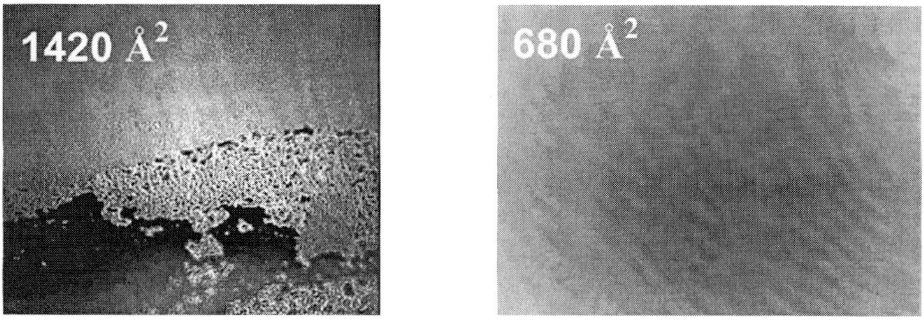

a **b**

Figure 3. Brewster angle microscopy images for the diblock dendrimer at large molecular areas (a) showing the formation of islands, and at smaller molecular areas (b) showing the formation of an homogeneous film.

Transfer experiments of the Langmuir films onto solid substrates and the preparation of LB films were investigated. The deposition of films of the diblock dendrimer occurred regularly on fused silica slides or silicon wafers with a transfer ratio of 1 ± 0.05. Once again, the

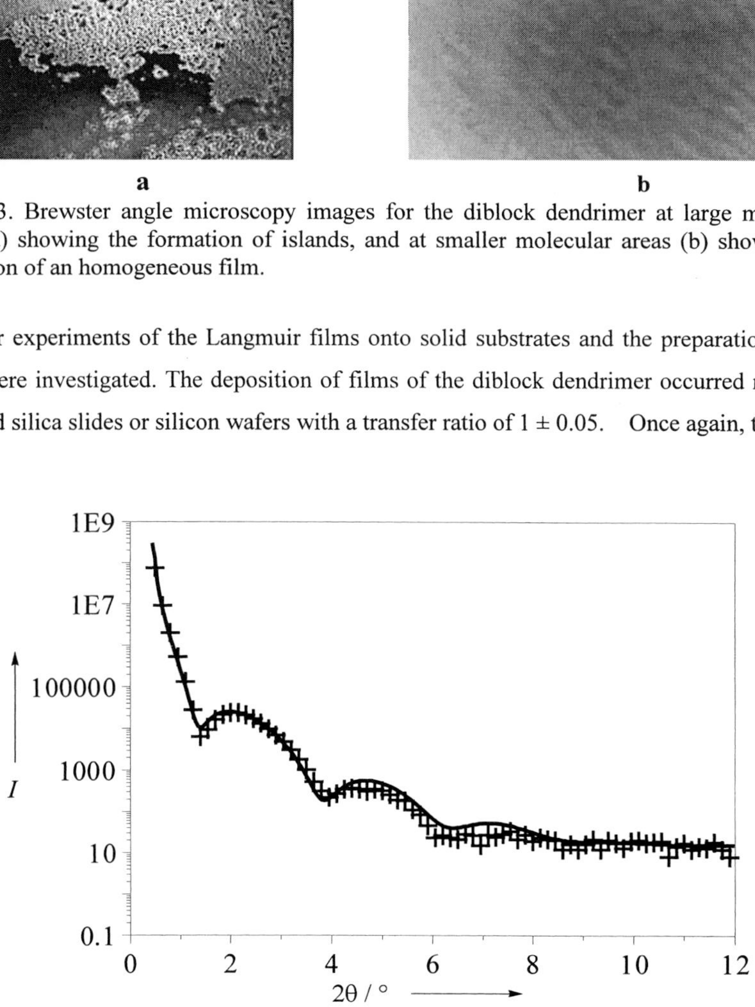

Figure 4. Grazing incidence X-ray pattern of a monolayer, together with the fit (continuous line) of the data.

perfect hydrophilic/hydrophobic balance also appeared crucial for efficient transfers of the Langmuir films in order to obtain well-ordered multilayered LB films. The structural quality of mono- and multilayer films was investigated by grazing X-ray diffraction. The presence of low-angle Kiessig fringes in the grazing X-ray patterns indicates that the overall quality of the films is good. The fit obtained for a monomolecular film gives a thickness of 36 ± 1 Å and a roughness of ~ 2 Å (figure 4). For the multilayer films, the average layer thickness was found to be ~ 36 Å, indicating no or little interpenetration of successive layers. The excellent quality of the LB films is also deduced from the plot of their UV/visible absorbance as a function of the layer number, which results in a straight line, indicating an efficient stacking of the layers (figure 5). It is worth stressing the quality of the stacking and, as a consequence, the quality of the multilayered films obtained with such a megamolecule.

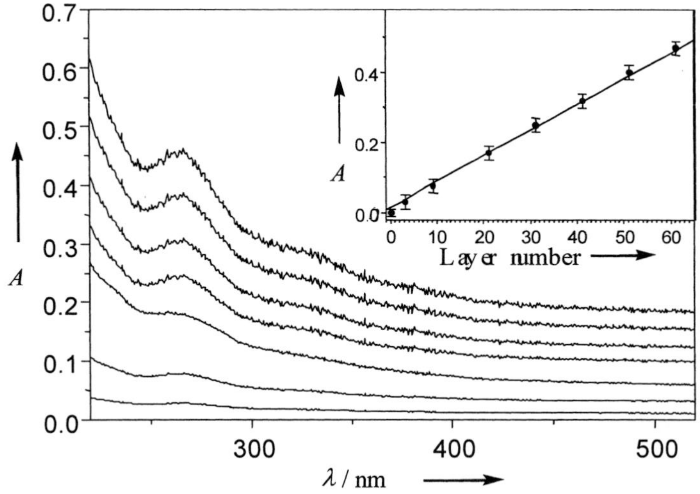

Figure 5. UV-visible spectra of the LB films of the diblock dendrimer (from bottom to top: 3, 9, 21, 31, 41, 51 and 61 layers). The inset shows the plot of the absorbance at 265 nm against the layers number.

Fullerene-Containing Liquid Crystalline Dendrimers

Two main objectives motivate the research in fullerene-containing thermotropic liquid crystals: 1) the understanding of how a sphere-like structure can be incorporated into liquid-crystalline phases (this study will lead to a better understanding of the structure-organization relationship for fullerene-based supramolecular assemblies), and 2) the design of new anisotropic materials.

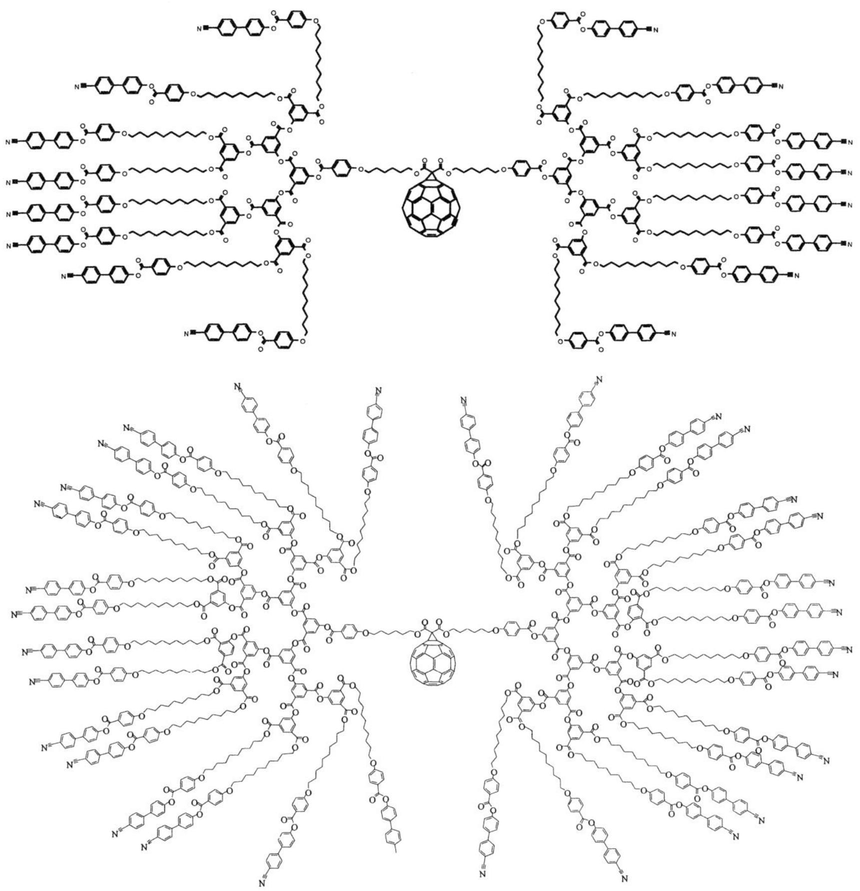

Figure 6. Chemical structures of the dendrimers of third and fourth generation containing 16 and 32 mesogenic cyanobiphenyl terminal groups.

In order to circumvent the size and the shape of C_{60}, which decrease naturally the liquid-crystalline potentiality of the corresponding materials, the use of dendritic addends could be an interesting means for the elaboration of fullerene-containing liquid crystals with tailor-made properties, which could be controlled *via* the dendrimer generation. This is the case of the two dendrimers presented in figure 6, where the dendritic part is ended by a large number of mesogenic groups.

Both compounds exhibit a smectic A phase as revealed by the X-ray diffraction patterns (figure 7) which contain, in the low angle region, two sharp reflections, the corresponding spacing being in the 1:2 ratio, and, in the wide angle region, a diffuse signal corresponding to the liquid-like arrangement of the mesogenic groups. The layer spacing is very similar for the two compounds and ranges between 55-56 Å. This value is rather small compared to the size of the dendrimers. In addition to the X-ray diffraction signals characteristic of a smectic A phase, one should mention the presence of diffuse bands corresponding to spacings of 22 and 29 Å respectively. It is also important to remark that the intensity of the Bragg reflections decreases as the generation number increases, and that these reflections are not as sharp as for the low generations.

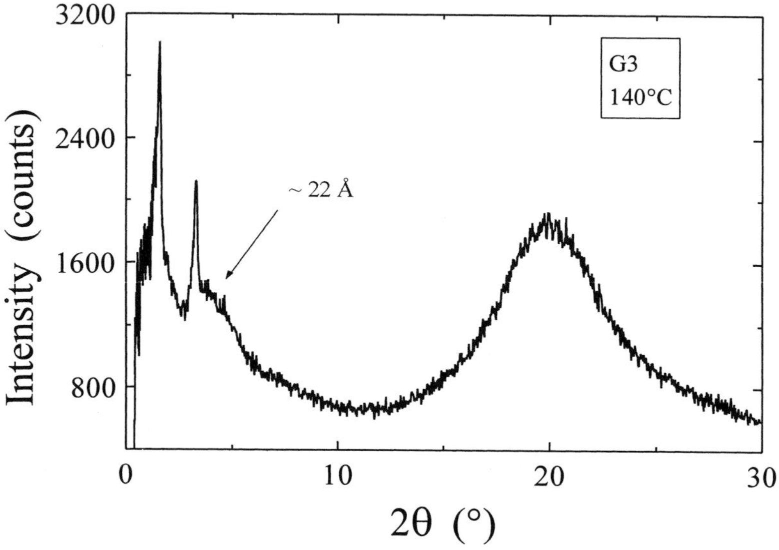

Figure 7. X-ray diffractogram pattern of the third generation dendrimer in the smectic A phase recorded at 150°C.

To describe the molecular organization, it is necessary to realize that these dendrimers are mainly constituted by the branching nodes and by the 16 and 32 mesogenic moieties including the aliphatic spacers; the fullerene moiety represents henceforth a minor part. It is reasonable to admit that the structure is governed essentially by the polar cyano mesogenic groups. The latter should tend to form an anisotropic micro-domain and tend to be oriented parallel to one another. However, due to the fact that these groups are attached to one of their extremity and due also to the fact that the branching nodes are rather voluminous, their parallel orientation is far from being optimized. As it is shown in a molecular simulation, free space is available between the terminal cyano groups[14]. Such a conformation is therefore very favorable for an inter-penetration of mesogenic groups of adjacent layers. Thus, the molecular organization can be schematized as illustrated in Figure 8 for the dendrimer of third generation (the molecular organization of the dendrimer of fourth generation is similar, the only difference lying in the number of mesogenic groups). The central part of the layer is constituted by the fullerene moiety embedded by the important branching part, and the layer interface is formed by the mesogenic cyano groups oriented in one main direction and partially interdigitated from one layer to the adjacent one. This interdigitation is in agreement with the decrease in intensity of the Bragg signals corresponding to the layering and to their simultaneous small broadening; this is similar to X-ray investigations of the smectic A to nematic transition where the interface between the layers becomes more and more diffuse as the transition temperature is approached.[15] Let us remark that such an organization and the good space filling are presumably made possible thanks to the presence of rather long aliphatic spacers and to a large number of ester groups which ensure the necessary flexibility to the dendrimer. Finally, in such a structure, the diffuse X-ray signals corresponding to spacings of 22 and 29 Å may be

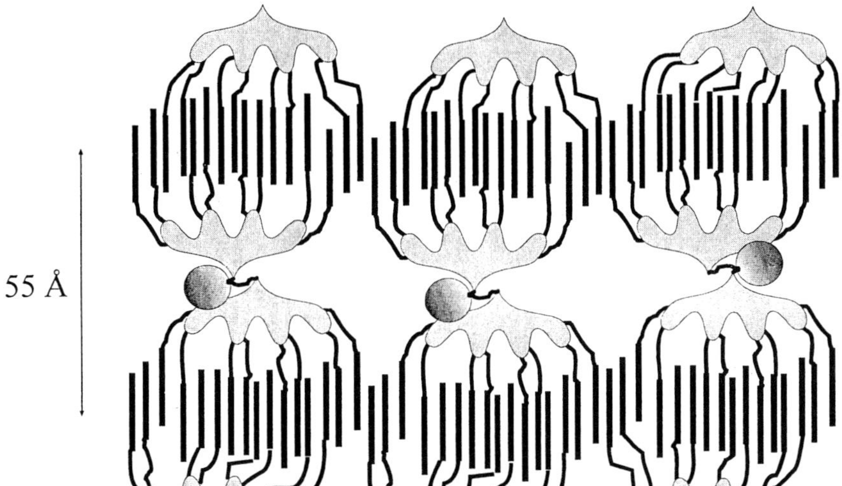

Figure 8. Schematic representation of the supramolecular organisation of the dendrimer of third generation within the smectic A layers. The full circles represent the C_{60} groups, the vertical full rectangles the mesogenic groups, the wavy lines the aliphatic spacers and the rest the dendritic part.

attributed to the mean average distance between the fullerene moieties within the smectic layers, this distance increasing as expected with the size of the dendrimer. This indicates that the expansion of the dendritic volume takes place in two dimensions essentially, as the dendrimer generation increases.

Conclusion

The above examples show how fullerene moieties can be incorporated into ordered nanostructures. In the case of LB films, the peripheral substitution of a globular dendrimer with hydrophobic chains on one hemisphere and hydrophilic groups on the other provides the perfect hydrophobic/hydrophilic balance allowing the formation of stable Langmuir films. On one hand, this approach shows some of the fundamental architecture requirements for obtaining stable films with amphiphilic dendrimers. On the other hand, functional groups not well adapted for the preparation of Langmuir and LB films such as fullerenes can be attached into the branching shell of the dendritic structure and, thus, efficiently incorporated in thin order films.

In the case of liquid-crystalline dendrimers, the fullerene moiety is embedded by an important branching part surrounded by a large number of mesogenic cyano groups. The latter ensure the formation of a mesomorphic phase, whereas the bulkiness of the branching part prevents the C_{60} groups from aggregation. Once more, functional groups such as fullerenes which are not intrinsically mesogenic can however be incorporated into ordered structures such as smectic mesophases and participate to the production of new mesomorphic materials with very interesting physical properties associated to the C_{60} units.

[1] F. Diederich and M. Gomez-Lopez, *Chem. Soc. Rev.*, **1999**, *28*, 263; M. Prato, *J. Mater. Chem.*, **1997**, *7*, 1097.
[2] C. A. Mirkin and W. B. Caldwell, *Tetrahedron*, **1996**, *52*, 5113, and references therein.
[3] Y. S. Obeng and A. J. Bard, *J. Am. Chem. Soc*, **1991**, *113*, 6279.
[4] U. Jonas, F. Cardullo, P. Belik, F. Diederich A. Gügel, E. Harth, A. Herrmann, L. Isaacs, K. Müllen, H. Ringsdorf, C. Thilgen, P. Uhlmann, A. Vasella, C. A. A. Waldraff and M. Walter, *Chem. Eur. J.*, **1995**, *1*, 243, and references therein.
[5] For examples, see: M. Maggini, A. Karisson, L. Pasimeni, G. Scorrano, M. Prato and L. Valli, *Tetrahedron Lett.*, **1994**, *35*, 2985; C. J. Hawker, P. M. Saville, J. W. White, *J. Org. Chem.*, **59**, 3503 (1994); S. Ravaine, F. Le Pecq, C. Mingotaud, P. Delhaes, J. C. Hummelen, F. Wudl and L. K. Patterson, *J. Phys. Chem.*, **99**, 9551 (1995).
[6] F. Cardullo, F. Diederich, L. Echegoyen, T. Habicher, N. Jayaraman, R. M. Leblanc, J. F. Stoddart, S. Wang, *Langmuir*, **1998**, *14*, 1955.
[7] J.-F. Nierengarten, C. Schall, J.-F. Nicoud, B. Heinrich, D. Guillon, *Tetrahedron Lett.*, **1998**, *39*, 5747.
[8] S. Ravaine, F. Vicentini, M. Mauzac and P. Delhaes, *New J. Chem.*, **19**, 1 (1995).
[9] T. Chuard and R. Deschenaux, *Helv. Chim. Acta*, **1996**, *79*, 736.
[10] B. Dardel, R. Deschenaux, M. Even and E. Serrano, *Macromolecules*, **1999**, *32*, 5193.
[11] D. Felder, J.L. Gallani, D. Guillon, B. Heinrich, J.F. Nicoud, J.F. Nierengarten, *Angew. Chem.,Int. Ed.*,

2000, *39*, 201.
[12] J.F. Nierengarten, J.F. Eckert, Y. Rio, M. del Pilar Carreon, J.L. Gallani, D. Guillon, *J. Am. Chem. Soc.*, **2001**, *123*, 9743.
[13] Y.S. Obeng, A.J. Bard, *J. Am. Chem. Soc.*, **1991**, *113*, 6279; T. Nakamura, H. Tachibana, M. Yumura, M. Matsumoto, R. Azumi, M. Tanaka, Y. Kawabata, *Langmuir*, **1992**, *8*, 4.
[14] B. Dardel, D. Guillon, B. Heinrich, R. Deschenaux, *J. Mater. Chem.*, **2001**, *11*, 2814.
[15] J.D. Lister, R.J. Birgeneau, M. Kaplan, C.R. Safinya, J.Als-Nielsen,, in "Ordering in Strongly Fluctuating Condensed Matter Systems", ed. T. Riste, Plenum Press, New York, 1980.

π–Conjugated Columnar Polyacetylenes Prepared with Rh Complex Catalysts

Tabata Masayoshi, Sone Takeyuki, Mawatari Yasuteru, Yonemoto Daigo, Miyasaka Atsushi, Fukushima Takashi, Sadahiro Yoshikazu*

Department of Molecular Chemistry, Graduate School of Engineering, Hokkaido University Sapporo 060-8628 Japan. E-mail: tabata@eng.hokudai.ac.jp

Summary: Highly stereospecific polymerization of monosubstituted acetylenes was carried out using the Rh complex, $[Rh(norbornadiene)Cl]_2$ catalysts. The resulting polyacetylenes were characterized in detail by ^{1}HNMR, ESR, laser Raman, diffuse reflective UV, and wide angle X-ray diffraction methods. The data showed that the Rh complex were the preferred catalyst to selectively yield the corresponding cis-transoid polymers even at room temperature when alcohol, triethylamine or water was used as the polymerization solvent. Additionally, the resulting cis polyacetylenes were found to have a helical form whose polymer is amorphous or composed of pseudohexagonal structures called π-conjugated columnar as self-assembly or super structure. Further compression of the amorphous cis polymers resulted in cis to trans isomerization at room temperature under vacuum, breaking rotationally the cis C=C bonds giving π-radicals called solitons as the origin of a polymer magnet. On the other hand, the π-conjugated columnar was also found to show an extremely longer wavelength absorption compared with that of the amorphous one, although the absorption maximum was shifted to a shorter wavelength when the columnar was destroyed by the compression. Therefore, the formation of the π-conjugated columnar can be considered as a new and quite useful control method concerning color of such conjugated polymers, i.e., a new concept concerning the color of conjugated polymers.

Keywords: polyacetylene; stereospecific polymers; self-assembly; catalysts; WAXS

Introduction

Scientific interest in monosubstituted polyacetylenes as π-conjugated polymer which has rapidly grown over the past years has focused recently on novel properties, in particular on photoconductivity[1-2], ferromagnetism[3-4], humidity sensor[5-6], non linear optical (NLO) materials[7-8], electroluminescence (EL) materials[9], oxygen permeability[10] and so on.

Preparation of the poly(phenylacetylene) (see **Scheme 1**) has been achieved by the use of various types of catalysts, e.g., Lutinger catalyst, $Co(NO_2)_2$-$NaBH_4$[11], Ziegler-Natta catalyst[12], Pt or Ni complex catalysts[13-14], and metathesis catalyst, e.g., WCl_6[15]. Especially

 CCC 1022-1360/00/$ 17,50+.50/0

metathesis catalyst developed by Masuda and Higashimura group[15] progressed the chemistry of not only monosubstituted but also disubstituted polyacetylenes which are soluble in ordinary solvents and mainly amorphous. It is conjectured that the cis-rich polyacetylene is

Scheme 1

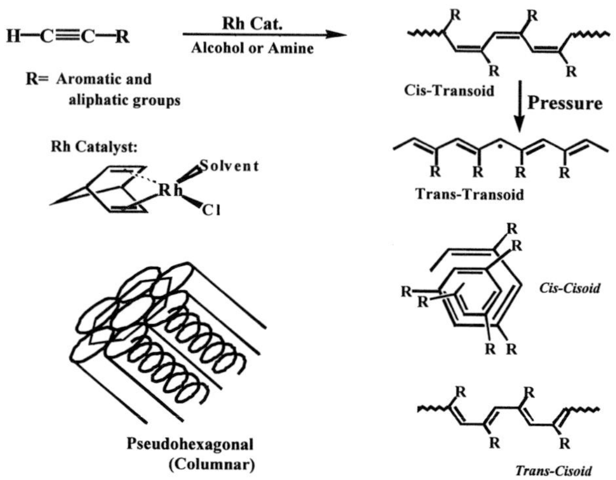

mainly produced by the use of $MoCl_5$ at 40°C while trans-rich polyacetylene is homogeneously produced by the WCl_6 catalyst in toluene solution[16]. Preliminary stereoregular polymerizations of phenylacetylene together with its living polymerization were first achieved using the Rh complex catalyst by Tabata's group[17-19], although the modified metathesis catalysts, Mo and W based complexes in the case of *ortho* substituted polyacetylene were used to give fairly narrow molecular weight distribution in high yields[16]. However, except for the precise control of the molecular weight and its dispersity there has been no clear and unequivocal assignment of the geometry of the resulting polymers with respect to the C=C bond in the main chain of the polymer (see **Scheme 1**), including selective formation of cis and/or trans monosubstituted polyacetylenes (**i**), copolymerization (**ii**), cis-to-trans isomerization (**iii**) at room temperature (**iv**), generation of conjugated π-radicals as solitons through the rotational scission of the cis C=C bond, and the role of the radicals to the conductivity and/or magnetism of the polymer (**v**).

In this report we show the important stereospecific polymerization of monosubstituted acetylenes to selectively yield the cis form (**a**), control of the secondary structure of the relevant π-conjugated polymer chain in the solid state, i.e., formation of a pseudohexagonal structure as columnar (**b**), pressure-induced cis-to-trans isomerizations (**c**), associated with generation of π-radicals through the rotational scission of the cis C=C bonds (**d**), together with detailed characterization of the resulting polymer structures using ^{1}HNMR, ESR, laser Raman, diffuse reflective UV-Vis (DRUV-Vis), and wide angle X-ray diffraction (XRD) methods. Previously, Kern (1968)[20] has showed that the so-called Wilkinson catalyst, $RhCl(PPh_3)_3$ is active for polymerization of phenylacetylene. Recently, analogs of the catalyst were extensively developed in order to selectively produce the polyphenylacetylene, **Ia** with cis-transoid form even at room temperature[17-19, 21-22a,b], although one should be reminded that the complete cis isomer of nonsubstitued polyacetylene can be only generated at lower temperature, *ca*. -78 °C in the case of Ziegler-Natta catalyst.

Results and discussion

Stereospecific Polymerization of Aromatic Acetylenes

Previously we revealed[23] that the so-called Wilkinson complex such as $[Rh(L\text{-}L)Cl]_2$, (L-L=ligand) is the useful stereoregular catalyst for not only aromatic acetylenes but also aliphatic acetylenes, e.g., propiolates called acetylene esters[24-25] to selectively produce a cis-

Table 1. Various derivatives (I) of PPA. 1)

	R_1	R_2	R_3	R_4	[Lit.]
a:	H	H	H	H	12-14,17-22
b:	H	CH_3	H	H	34,46-47
c:	*t*-Bu	OH	*t*-Bu	H	30
d:	H	O_nC_{2n+1} (n= 1~14)	H	H	23,30,34
e:	H	H	H	O_nC_{2n+1} (n=1 and 2)	32-33
f:	H	H	H	CF_3	41
g:	H	$OCHCH\text{-}CH(CH_3)_2$	H	H	27
h:	H	$NHC(O)C_nH_{2n+1}$ (n=1 , 3, 5, 7)	H	H	This work
i:	H	$N(CH_3)_2$	H	H	28
j:	H	NO_2	H	H	29
k:	H	—N(O•)—*t*-Bu	H	H	38
l:	*t*-Bu	—O•	*t*-Bu	H	30
m:	H	halogen	H	H	This work

1)

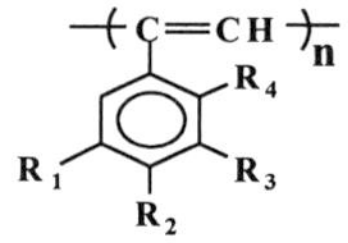

Table 2. Polyacetylenes (II) having various side chains. [1)]

	R	[Lit.]
a:	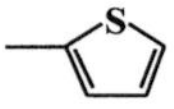	39a
b:	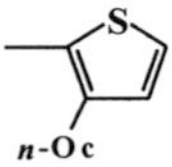	39b
c:	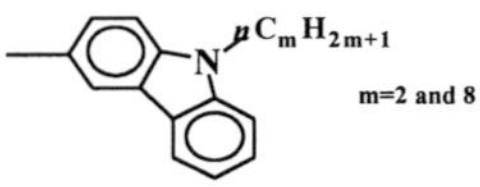	40
d:	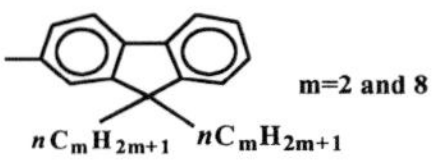	31
e:	$-COO\ nC_mH_{2m+1}$ (m=1~18)	24,48,52
f:	$-CH_2CH_2COO-$ 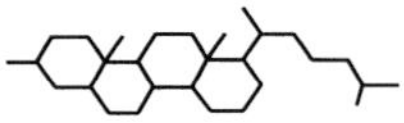	42

1) : $-(C(R)=CH)_n-$

transoid form when alcohol or amine was used under quite mild conditions (see **Tables 1, 2, 3, and 4**). We also showed that the catalytic activity in the $[Rh(L\text{-}L)Cl]_2$ complex is shown in **Scheme 2**. This shows that the NBD ligand is strongly coordinated to the Rh atom during the polymerization without replacement of the ligand with a donor molecule[23]. We also found that the molecular weight dispersity was not narrow but approximately Mn/Mw= 2-4.

Scheme 2

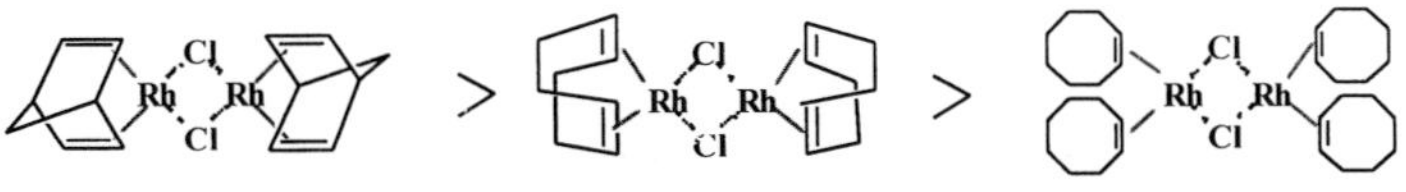

$[Rh(NBD)Cl]_2$	$[Rh(COD)Cl]_2$	$[Rh(bis\text{-}COT)Cl]_2$
NBD:norbornadiene	COD:cyclooctadiene	COT: cyclooctene

Solvent Effect

Previously, we reported that an important feature of the stereospecific polymerization using the Rh complex catalyst is that alcohols[18a] or water[26-27] as an aprotic solvent or a strong organic amine, e.g., triethylamine (TEA)[19,28] can be used as the polymerization solvent.

Scheme 3

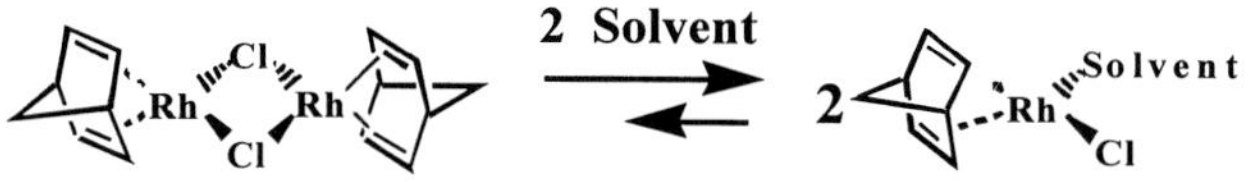

(2)

These solvents were proven to promote the dissociation of the bidentate Rh complex to a monomeric species as shown in **Scheme 3**. We also established that the order of the solvent activity[18a,b,27-29] as the cocatalyst is as follows.

$N(CH_2CH_3)_3$ > ROH (R=aliphatic group) ≒ H_2O

Table 3. Polymerization results of Ia and Ig monomers by $[Rh(NBD)Cl]_2$, (Rh I) and $[Rh(COD)Cl]_2$, (Rh II), and [Rh(bis-COD)Cl]$_2$, (Rh III) catalysts under various solvents.[a)]

Monomer	Rh cat.	Solvent	Yield in %	Mn X10[4b]	Mw/Mn[b)]	Cis in %
Ia	(Rh I)	TEA	100	16.0	1.76	100
Ia	(Rh I)	MeOH	90	7.31	2.13	100
Ia	(Rh I)	THF	84	6.85	3.43	85
Ia	(Rh II)	TEA	92	2.47	2.16	100
Ia	(Rh II)	MeOH	66	2.47	2.50	100
Ia	(Rh III)	MeOH	32	2.53	2.64	100
Ig[c)]	(Rh I)	EtOH[d)]	96	9.90	2.66	100
Ig[c)]	(Rh I)	Water[d)]	70	23.1	2.23	100
Ig[c)]	(Rh I)	TEA	80	7.20	2.92	100

[a)]Polymn. temp., 30℃.; [M] = 0.5 mol/l.; [Cat.] = 0.005 mol/l.
[b)]Measured with GPC using chloroform as an eluent, and polystyrene as a standard.
[c)][M] = 0.20 mol/l.
[d)]TEA was used as the cocatalyst, [Cocat.]/[Cat.] =100.

To the best of our knowledge, such solvents have not hitherto been used even in the polymerization of ordinary vinyl monomers since the start of polymer science. This indicates that the Rh catalyst is completely different from that of metathesis catalyst, such as the WCl_6[15] or Ziegler-Natta catalyst, i.e., $AlEt_3$-Ti(O*n*-Bu)$_4$[12], which are easily deactivated by such aprotic solvents because of strong Lewis acid. This chemical property can be effectively applied in the polymerization of monosubstituted acetylenes, because even acetylene monomers, those bearing hydroxy[30], N,N'-dimethylamino[28], amide[31], alkoxy[23,32,34-36], nitro[29], hydroxyl[37], nitroxide[38], thiophens[39], and carbazole[40] groups can be polymerized under quite mild condition. This also clearly means that a new and useful stereospecific polymerization catalyst for various monosubstituted acetylenes was born to generate useful advanced materials (see **Tables 1** and **2**).

Table 4. Polymerization results of poly(phenylacetylene), Ia and poly(*p*-methylphenylacetylene), Ib initiated with $[Rh(NBD)Cl]_2$ catalyst in various solvents.[a)]

Monomer	Solvent	Cat. conc. mM	Yield %	Mw $x10^4$	Color	Crystallinity
Ia	EtOH	5.00	81	30.0	Yellow	*ca.* 10
Ia	Toluene	5.00	100	-[c]	Red[d]	*ca.* 60
Ib	EtOH	5.00	54	14.0b	Yellow	*ca.* 3
Ib	Toluene	5.00	54	-[c]	Red[d]	*ca.* 53

[a]Polymn. Time, 20h; Polymn. temp., 30℃; [M] = 0.7 M.
[b]THF soluble part.
[c]Insoluble in ordinary solvents.
[d]The yellow-colored polymers of **Ia** and **Ib** were immersed in toluene at room temperature for 0.5hr.

Structural Differences of Polymers Prepared with Rh Complex and Metathesis Catalyst

^{1}HNMR

Figure 1 shows the ^{1}HNMR spectrum of poly((*p*-3-methylbutyl)phenylacetylene), (**Ig** in **Table 1**) prepared with the Rh complex catalyst and observed in chloroform-d_1 ($CDCl_3$) at room temperature[27]. It is clear that the spectrum is composed of extremely sharp lines,

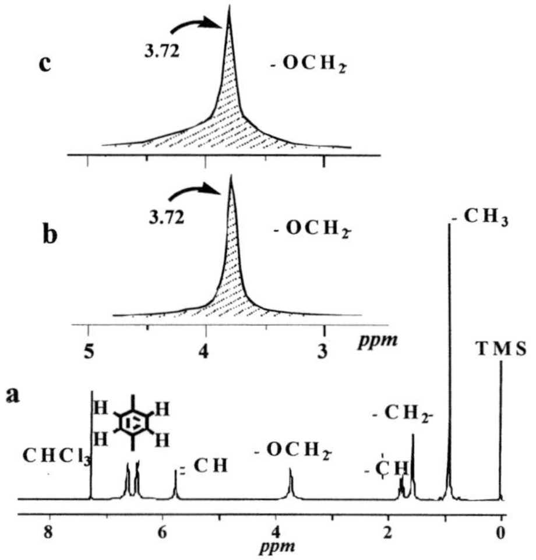

Fig. 1 ^{1}H NMR spectra of poly((*p*-3-methylbutyl)phenylacetylene))s, **Ig** (see **Scheme 2**) prepared with $[Rh(NBD)Cl]_2$ in the presence of triethylamine. (a) full scale spectrum of **Ig**, (b) expanded peak of $-OCH_2-$ protons before compression, and (c) expanded peak of $-OCH_2-$ protons after compression at 100kg/cm^2 under vacuum at room temperature.

suggesting highly stereospecific polymerization, i.e., selective formation of the cis-transoid isomer even if the molecular weight is very large, i.e., more than Mn = $5x10^5$.

Figure 2 shows the typical ^{1}HNMR spectrum of poly(2-octylthienylacetylene), **IIb**[39] observed at room temperature using $CDCl_3$ solution. This spectrum is composed of fairly broadened lines, although this polymer was also prepared with the same Rh complex catalyst using a mixed solvent: $CHCl_3$/TEA=100/1 (v/v) and observed in $CDCl_3$ solution at room temperature. However, in this case the cis ratio was estimated as ca. 70%, whose value was deduced with the aid of computer simulation (see **Figure 2**). This indicates that even if TEA as the cosolvent was used as the polymerization solvent no more the cis isomer was quantitatively obtained but the ratio is decreased to a large extent if a thiophene moiety is included in the monomer.

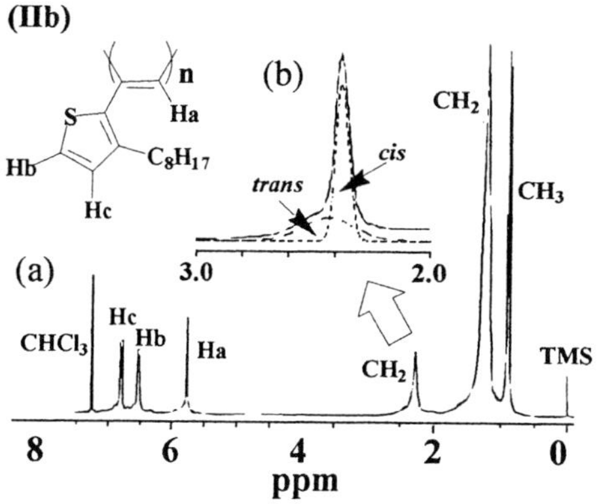

Fig. 2 ^{1}H NMR spectra of poly((2-*n*-octylthienyl)acetylene))s, **IIb** (see **Table 2**) prepared with $[Rh(NBD)Cl]_2$ in the presence of a mixed solvent: TEA/THF=1/100(v/v). (a) full scale spectrum of **IIb**, (b) expanded peak of the $-OCH_2-$ protons region in the side chain.

Figure 3 shows the ^{1}HNMR spectra of poly((*o*-trifluoromethyl)phenylacetylene), **If** which were prepared with not only the Rh complex but also the metathesis catalyst, $MoCl_5$, and observed in $CDCl_3$ solvent at room temperature[41]. It should be noted that in the case of the metathesis catalyst the spectral assignment is almost impossible unlike the case of the Rh complex catalyst as shown in **Figures 3a** and **b**. **Figure 4** shows the ^{19}FNMR spectra observed in order to deduce the possible geometrical structures [41]. The observed spectra were deconvoluted to determine, cis, trans, and boundary components connecting both isomers, together with random trans components like irregular bonds such as head-to-head

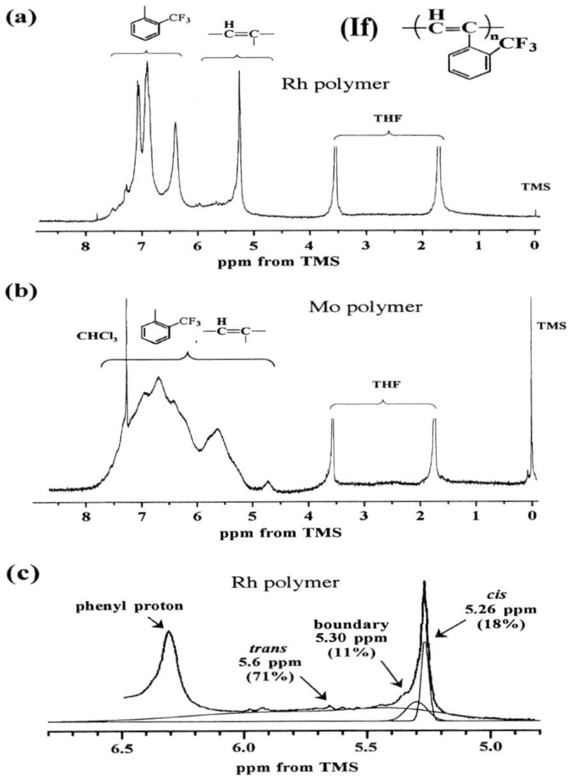

Fig. 3 ^{1}H NMR spectra of P((*o*-trifluoromethyl)phenylacetylene)), **If** polymers observed in THF solutions which contain a small amount of chloroform-d, 99.8 atom % with 1% v/v TMS at room temperature: (a) $[Rh(NBD)Cl]_2$ catalyst; (b) $MoOCl_4$-*n*-Bu_4Sn-EtOH catalyst; (c) expanded spectrum for $[Rh(NBD)Cl]_2$ catalyst.

and/or tail-to-tail using a computer simulation method. Thus, the data showed that the polymer (=Mo polymer) prepared with the $MoCl_5$ catalyst contains a large amount of the irregular structure as the random trans structure, i.e., head-to-head and/or tail-to-tail structures called ***disordered trans sequences***[41] together with boundary components **(see Figure 5b)**. On the other hand, the latter polymer (=Rh polymer) prepared with the Rh catalyst is inversely composed of regular and repeated head-to-tail structure called ***ordered cia amd trans sequences***[41] together with the boundary components (see **Figure 5a**). It seems, therefore, that the Rh polymer is very important as a new advanced material because there is no defect in terms of the cis-trans geometrical structure.

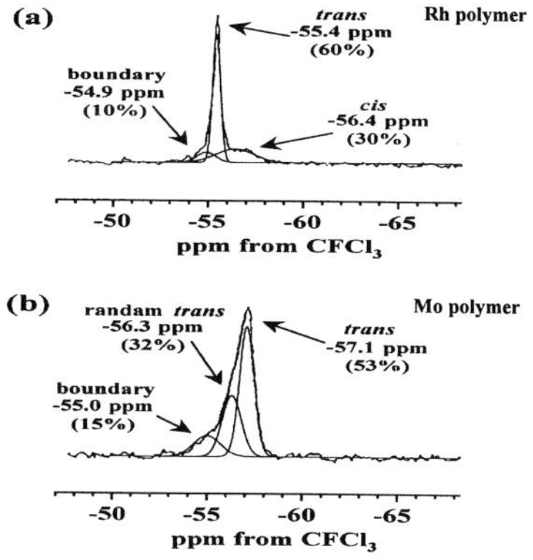

Fig. 4 ^{19}F NMR spectra of **If** polymers observed in THF-d_8 including a small amount of C_6F_6 as the internal standards at room temperature: (a) $[Rh(NBD)Cl]_2$ catalyst; (b) $MoOCl_4$-*n*-Bu_4Sn-EtOH catalyst.

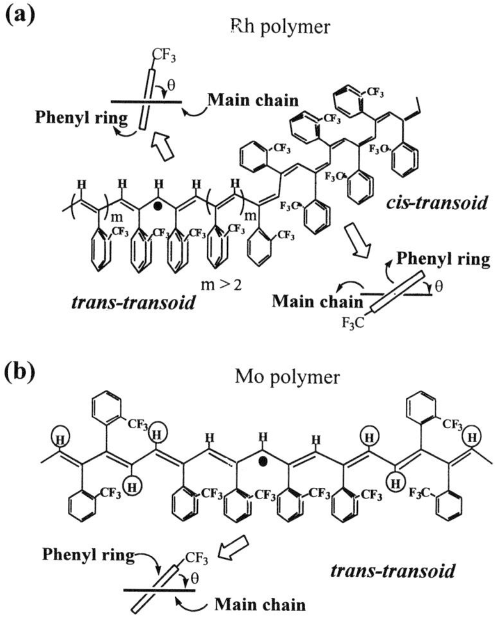

Fig. 5 Proposed conformations of **If** polymers prepared with (a) $[Rh(NBD)Cl]_2$ catalyst; (b) $MoOCl_4$-*n*-Bu_4Sn-EtOH catalyst.

Raman Study

Figure 6a shows the resonance Raman (RR) spectra of poly(dihydrocholesteryl- pentynoate)s, **IIf** prepared with the Rh complex catalyst uisng triethylamine[42]. It is clear that in the case of

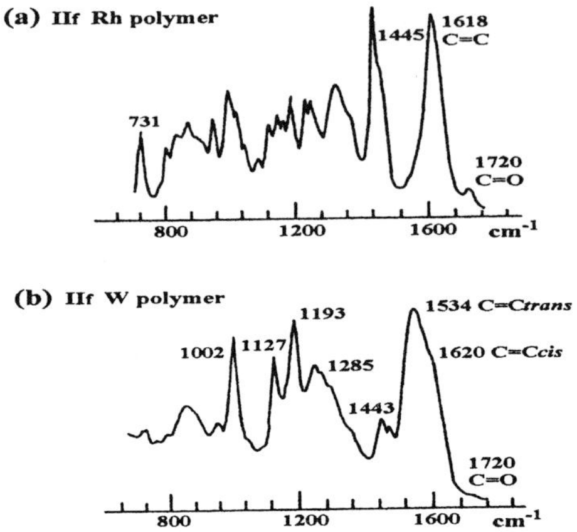

Fig. 6 Laser raman spectra of poly(dihydrocholesterylpentynoate), **IIf** polymer. (a) polymerized with the Rh complex catalyst in TEA. (b) Polymerized with WCl_6 catalyst in dioxane.

the Rh polymer a large peak at *ca.* 1618cm^{-1} is assigned to that of the cis main chain C=C bond, although a peak due to the carbonyl moiety in the ester group is quite small[43]. The large peak at 1445cm^{-1} may be assigned to the CH_3 or CH_2 groups which is characteristic for the aliphatic cis polyacetylenes[42a]. The peak at 731cm^{-1} was ascribed to that of the C-H bond in the cis main chain. On the other hand, **Figure 6b** shows the RR spectrum of **IIf** prepared with the so-called metathesis catalyst: WCl_6 at 40°C in dioxane[42a]. A peak at 1534cm^{-1} with a shoulder at ca. 1620cm^{-1} was assigned to that of the trans C=C bonds, and the shoulder at ca. 1620cm^{-1} can be ascribed to that of the remaining cis C=C bonds in the polymer. The peak at 1443cm^{-1} may also be due to CH_3 or CH_2 groups. The several peaks observed at 1400cm^{-1}~1000cm^{-1} may reflect the resulting ***disordered trans sequences*** as mentioned above[41-42]. The peak intensity at 731cm^{-1} due to the C-H in the cis isomer was decreased. This also means that the cis ratio is low in the W polymer compared with that of the Rh polymer. This is supported by the fact that the similar multiplet was also observed in the spectrum of the poly(*p*-chlorophenylacetylene), (**Im, R_2=Cl**) which was topochemically obtained by γ-ray irradiation of the monomer in the solid state[44].

Estimation of π-Conjugation Length

Previously, we showed[33-36] that the band ν_4 at around 1460~1640cm^{-1} in the RR spectra of

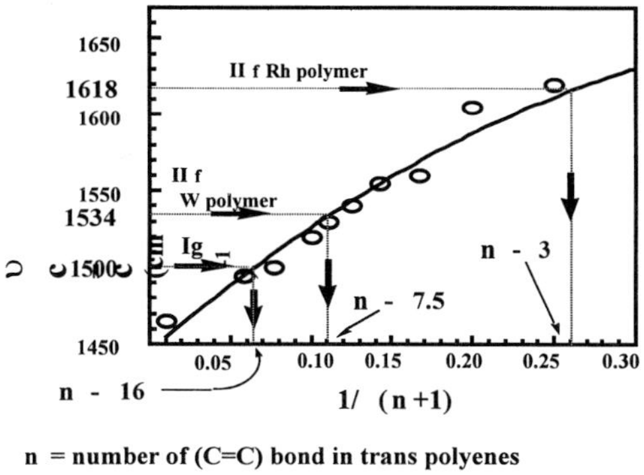

Fig. 7 Dependence of the trans conjugation length, (C=C)n, on the Raman peak of the red polymer of **Ig** polymer (**see Tables 1** and **2**) prepared with $[Rh(NBD)Cl]_2$ in the presence of triethylamine.

the trans polyenes was correlated with the number, $n_{C=C}$, of the trans conjugated C=C bonds. Therefore, the conjugated *trans* length can be estimated in the cases of **IIf** prepared with the Rh complex catalyst and the W catalyst, because the conjugation between the main-chain C=C bond and the ester group is completely decoupled by the ethylene, $-CH_2CH_2-$ moiety[42a]. Further, the two double C=C bonds in the *cis*-transoid form is *trans*-configurated to each other. On the other hand, in the cases of the trans-cisoid or cis-cisoid forms the configuration of the two C=C bonds is *cis*-configurated to each other as shown in **Scheme 1**, where such conjugation between the cis C-C bonds is not expected, if the dihedral angle between the double bonds is nearly zero degree[32,41]. Therefore, in the cases of the *trans-cisoid* and *cis-cisoid* polymers the estimation of the conjugation length is not allowed using the above relation. The trans conjugation length in the Rh polymer (**IIf**) was estimated as n=ca. 3 for 1618cm^{-1} using **Figure** 7.[42a] On the other hand, the trans conjugation length in the W polymer was also estimated as n = *ca.* 8[42a]. Thus, the geometrical configuration with respect to monosubstituted polyacetylenes can unequivocally be determined using the RR spectra. This assignment is extremely important because the possible four isomers (see **Scheme 1**) have not been unequivocally determined using a reliable and convenient spectroscopic method which is essential for development of new advanced materials.

Pressure–Induced cis-to-trans Isomerization

In our previous reports[29,32-36] we have shown that cis-to-trans isomerization of poly(phenylacetylene)s bearing *o*-alkoxy and *p*-alkoxy, and *p*-nitro groups as the side chain in the phenyl ring, can be induced by compression at *ca.* 100~200 kg/cm^2 under reduced pressure, ca. 10^{-2} Torr even at room temperature, when these polymers are stereoregularly prepared using a Rh complex catalyst in the presence of TEA or alcohol as the polymerization solvent.

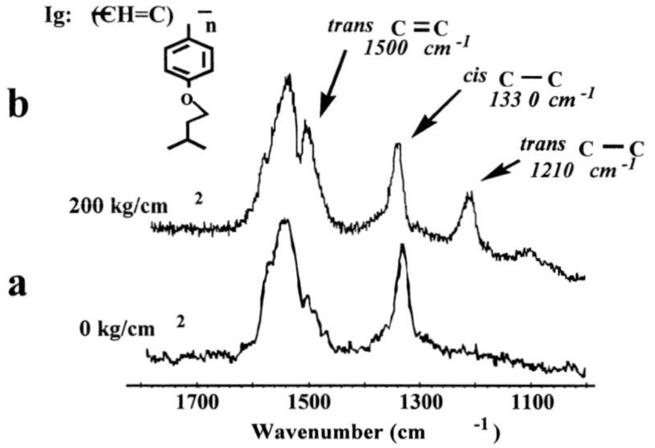

Fig. 8 Laser Raman spectra of **Ig** polymer (see **Table 1**) observed at room temperature. (a) before and (b) after compression at 200kg/cm^2 under vacuum at room temperature.

Figure 8 shows the laser Raman spectra of the red-colored polymer, **Ig**, which was prepared with the Rh catalyst using TEA solvent (see **Tables 1, 2,** and **3**) and compressed at 200kg/cm^2 under reduced pressure. It is clear that, after compression, new peaks at 1210 cm^{-1} and 1500 cm^{-1} appeared and the peak intensity at 960 cm^{-1} is decreased by the compression, though the peak at 960 cm^{-1} is not shown in **Figure 8**.

The two peaks at 1500cm^{-1} and at 1210 cm^{-1} are assigned to those of the C=C bond and C-C bond vibrations in the resulting trans form, respectively[43-44]. The peak at 1330 cm^{-1} is assigned to that of the C-C bond in the cis form which may be also superposed with that of the bond connecting the main chain with the side chain[43-44]. The conjugation length was estimated to be n≈16 for 1500 cm^{-1} as shown in **Figure 7**. Thus, it is suggested that compression of the red polymer has cis-transoid form, which undergoes cis-to-trans isomerization even at room temperature in order to create such a new trans π–conjugated sequences.

We also found that the pristine yellow **Ig** polymer showed almost the same Raman spectrum as that of the red polymer shown in **Figure 8a**. This strongly indicates that the yellow as well as the red-colored polymers have also the same cis-transoid form, suggesting a difference of the secondary structure.

Diffuse Reflective UV-Vis

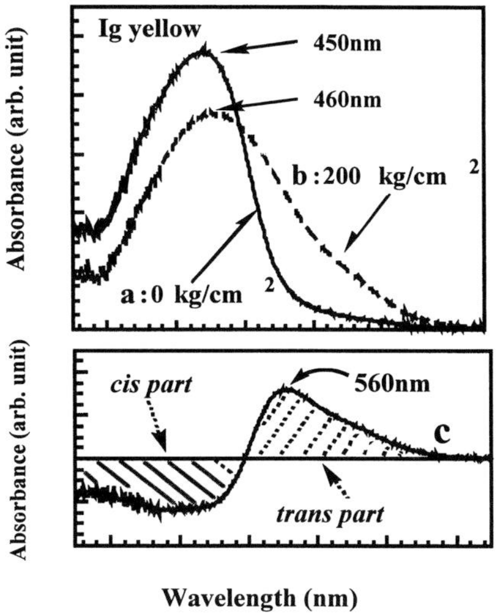

Fig. 9 Diffuse reflective UV-Vis spectra of the yellow polymer of **Ig** polymer (**see Tables 1** and **3**) observed at room temperature. (a) before and (b) after compression at 200kg/cm^2 under vacuum at room temperature, and (c) the subtraction spectrum.

Figure 9 shows the diffuse reflective UV-Vis (DRUV-Vis) spectra observed before and after compression of **Ig** with a yellow color under reduced pressure, ca. 10^{-1} Torr (see **Table 1**) at room temperature[27]. The yellow-colored polymer showed an absorption maximum, 450nm. We found that the absorption maximum in the pristine yellow polymer was shifted to a higher wavelength side when compressed, i.e., from λmax=450nm to λmax=460nm and the absorption intensity decreased. The difference spectrum obtained by subtraction of the absorption spectra observed before and after compression at room temperature is shown in

Figure 9c where a new absorption band is observed at higher wavelength, i.e., 560nm. The band observed at 560nm can be attributed to the *trans* conjugated sequences. Length of the conjugated *trans* sequences generated by the compression of the yellow polymer was also estimated, assuming that the trans conjugation length, (C=C)n, is related with the DRUV-Vis peak of the yellow **Ig** polymer as shown in **Figure 7** and **Table 5** [27].

Table 5. Absorption maxima and π–conjugation lengths, (C=C)n created by the compression of Ig polymers at room temperature.[a]

No.	Abs. max. (nm) before		Abs. max. (nm) of trans form	N in trans $(C=C)_n$	Color
1	450	460	560	29	Yellow
3	490	450	-	-	Red

[a]Polymers were compressed under vacuum at ca. 10^{-2} Torr for 10 min.

The *trans* conjugated lengths were estimated as *ca.* n=29 as shown in **Figure 10** and the data are shown in **Table 5**.

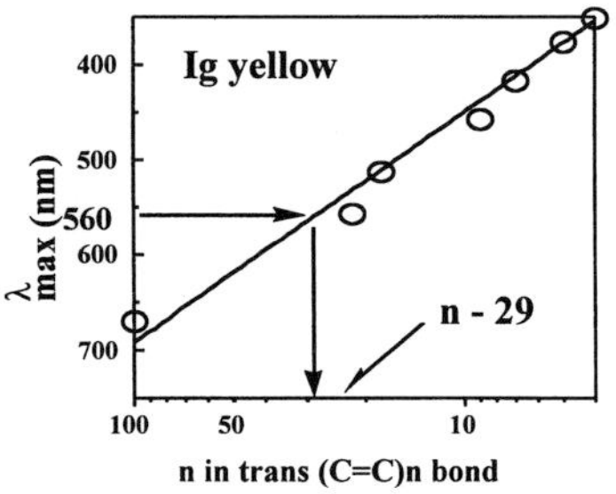

Fig. 10 Dependence of the trans conjugation length, (C=C)n, on the DRUV-Vis peak of the yellow polymer of **Ig** observed after compression at room temperature.

Figure 11 shows the DRUV-Vis spectra observed before and after compression of **Ig** having the red color under reduced pressure for 10 min. at room temperature (see **Table 1**). The red polymer showed a notably large absorption as an entire extending from ca, 300nm to ca. 840nm. However, such a long absorption tail has not been observed in other polymers which were prepared with the Rh complex catalyst. This observed chromophore may be newly explained in terms of formation of a pseudohexagonal structure called π-conjugated columnar, although this red polymer is composed of a cis-transoid form having regular head-to-tail structure as evidenced by the above Raman data. This indicates that the spectrum of the columnar polymer is attributed to that of specific π-conjugated self-assembly or self-organization which is fairly different from that of ordinary amorphous polymers, i.e., the yellow polymer, **Ig** as shown below[27]. A simple cis-transoid structure having sequences of one or two monomer units can not allow us to explain why the λmax must be observed at such

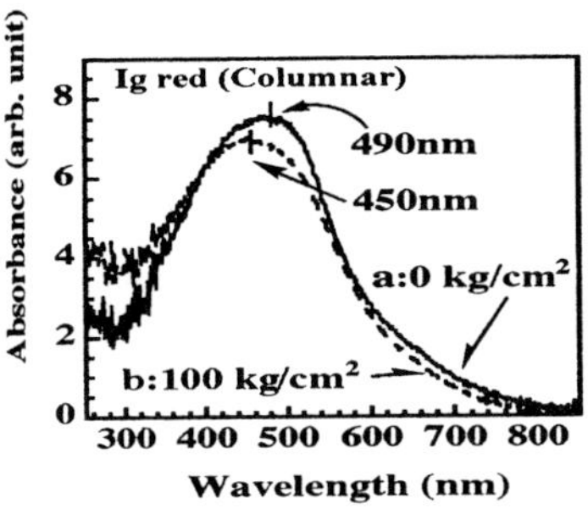

Fig. 11 Diffuse reflective UV-VIS spectra of the red polymer of **Ig** observed at room temperature. (a) before and (b) after compression at 200kg/cm^2 under vacuum at room temperature

a longer wave length region, i.e., at 490nm. Previously, we have shown that the amorphous yellow polymers, e.g., poly(*p*-methoxyphenylacetylene), (**Id**, n=1)[35] or poly(*p*-ethoxyphenylacetylene), (**Id**, n=2)[36] show the λmax at *ca.* 440nm and the absorption maximum shifted to a *longer wavelength, i.e., ca. 500~630 nm* by compression to produce the planar π-conjugated trans sequences, associated with a small peak due to that of *para* alkoxy styrene moiety[45]. On the other hand, in this experiment we also found that compression of **Ig** bearing the red color resulted in a large blue shift of the absorption maximum from 490nm to 450nm, though a shoulder appears at a shorter wavelength region of less than 300nm. Thus, compressions of the yellow and red polymers generate completely reverse phenomena regarding the shift of the absorption maximum, although the ^{1}HNMR spectra of the red and yellow polymer are completely the same in the $CDCl_3$ solution. Therefore, the observed large blue shift for the red polymer may be explained by destruction of the π-conjugated columnar as π-conjugated self-assembly, i.e., a specific aggregation of many polymer molecules in order to generate the trans π-conjugated sequences[27]. A small shoulder peak at less than 300nm is also ascribed to that of the *para*-3-methoxybutyl styrene moiety[45] in the resulting trans form as reported as before[45]. Appearance of this small peak also suggests that the *para*-3-

methylbutoxystyrene moiety plane is decoupled from the plane of the resulting trans zigzag conjugation sequences created by compression. This explanation will be strongly supported by the ESR parameters, g values observed before and after compression of the red polymer as mentioned below.

Formation of Columnar Polymer

The polymers, **Ia** and **Ib** were obtained as fine yellow powders when TEA or ethanol, EtOH was used as the polymerization solvent. On the other hand, orange colored polymers were also obtained when cyclohexanol, c-HexOH was employed. Further red colored polymers were obtained when the yellow colored polymers were immersed in toluene. We reported[46-47] that the yellow-colored polymers can be changed to the red-colored polymers by treatment with toluene (**see Ia in Table 4**). These red-colored polymers of **Ia** and **Ib** were actually insoluble in ordinary solvents such as tetrahydrofuran (THF), acetone, chloroform, and benzene unlike of **Ig**. Therefore, wide angle X-ray diffraction (XRD) patterns were observed to determine whether their polymers are amorphous or not. **Figure 12** shows the XRD patterns of **Ib** observed before and after treatment with toluene. The yellow polymer showed peaks at d=11.6Å and 4.7Å and the red colored polymer showed d=14.24Å, 7.12Å, 5.38Å, 4.75Å, 3.95Å, and 3.56Å. The latter red polymer peaks were indexed, assuming that the polymer is composed of a pseudohexagonal structure called ***columnar***[24,48-49] as molecular self-assembly or super structure where helical main-chains with cis-transoid form are packed, although the peak at ca. 4.75Å is due to amorphous halos. The indexed parameters are shown together with the columnar diameter estimated from the (100) reflection peak (see **Table 6** and **Figure 12c**). The crystalline size was also calculated as 124.4Å and 147.3Å before and after toluene treatment using the half width of the (100) reflection combined with Scherrer's equation[50]. The XRD patterns of **Ia** observed before and after the toluene treatment were nearly the same as that of **Ib**. This also indicates that **Ia** similarly forms such columnar polymers[24,27,32-36], although the columnar diameter is smaller than that of **Ib**, about 2.56 Å.

Interestingly, the peak observed at d= 3.56Å may be assigned to the helical pitch width due to the helical main-chain[48]. It is noteworthy that even the pristine yellow (**Ib**) polymer contains a small hexagonal component, i.e., *ca.* 2.5 % in the amorphous phase when ethanol was used in the polymerization, and that this content was greatly increased to *ca.* 52.5 % when the yellow polymer was immersed in toluene at room temperature. We also examined the wide angle XRD patterns observed before and after compression of the red polymer, **Ig** which was prepared using TEA solvent (see **Figure 12**)[27]. The peaks observed at 17.1Å, ~9.4Å, ~8.6Å, 6.5Å, and 4.5Å were analyzed, also assuming formation of a pseudohexagonal structure, i.e., columnar, though the (110) and (200) peaks were not clearly separated[27]. Those peaks were also attributed to the reflections of (100), (110), (200), (120), and amorphous halo, respectively. We found that the compression of the red polymer decreased the intensity of the peak at (100) reflection to a great extent. This indicates that the columnar content is decreased by the compression to make the polymer more amorphous. The columnar contents of the pristine red polymer and compressed polymer were estimated as *ca.* 53% and 34%, respectively[27]. The content of the columnar of the yellow polymer (**Ig** in **Table 1**) was also estimated as *ca.* 38%. However, the content was not found to be decreased by the compression unlike the case of the red-colored polymer.

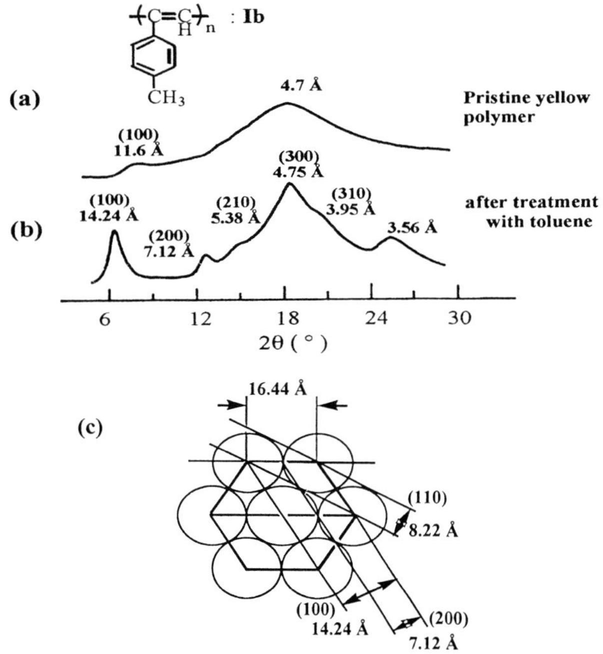

Fig. 12 X-ray diffraction patterns of **Ib** polymers observed at room temperature using Cu*K*α radiation, together with assigned indeces: (a) yellow polymer and (b) red polymer. The insert figure (c) is a pseudohexagonal structure generated by toluene treatment.

It is clearly concluded, therefore, that the columnar content of substituted polyacetylenes prepared with the Rh complex catalyst is correlated with the color, i.e., λmax (see **Figures 9** and **11**). In other word, formation of such a columnar is a preferred method to shift the original absorption of the cis polymer to longer wave length compared with that of the amorphous polymer without redesigning the polymer structure, although the crystallinity of the columnar is related with the degree of aggregation states called π–conjugated self-assembly as evidenced by the data shown above.

Table 6. Intensities and Bragg distances (Å) calculated and observed in the X-ray powder diffraction spectra and crystallite sizes (Å) calculated from (100) reflection peaks of Ib polymer synthesized with $[Rh(NBD)Cl]_2$ in triethylamine before (A) and after (B) toluene treatment.

(A)					(B)				
hkl	d*calc.*	d*obs.*	I*obs*[a].	Size	*hkl*	d*calc.*	d*obs.*	I*obs*[a].	Size
(100)	11.62	11.6	W	124.4	(100)	14.24	14.20	VS	147.3
(110)	6.71	-	VW		(110)	8.22	-	VW	
(200)	5.81	-	VW		(200)	7.12	7.12	S	
(210)	4.39	-	VW		(210)	5.38	5.38	W	
(300)	3.87	-	VW		(300)	4.75	4.75	W	

[a]VW, Very weak; W, Weak; S, Strong; VS, Very strong.

Stable Conformations

The most energetically stable configurations of **Ia** and **Ib** were deduced using the so-called *semi-empirical quantum chemical calculation method*, AM1, where 20 monomers were used for the computer simulation[51]. **Figure 13** shows the calculated stable conformation for **Ib**.

The data showed that in the case of **Ia** and **Ib** the cis-transoid isomer takes a helical structure, approximately 8/3 helix as the most energetically stable conformation. The calculation gave rise to *ca.* 14.65Å as the diameter of the helix chain. This diameter was nearly the same as 14.65Å obtained by the XRD method. In the case of **Ia** the calculated diameter was around 12.6Å which was also very near the observed diameter, 13.34Å (see **Figure 13**).

Recently, we found that *n*-alkyl propiolate, **IIc** called acetylene ester can be also stereoregularly polymerized using the $[Rh(NBD)Cl]_2$ catalyst and the resulting polymers are not oxidized in air at all, even after more than two years and the **IIc** polymers produce also columnar which can be used as an oxygen-permeable membrane[48,52]. Therefore, this finding clearly indicates that *tri*-methylsilyl moieties as the side chains of polyacetylene are not needed for an oxygen enriched polymer unlike the case of the Masuda group[10,53].

Ib polymer

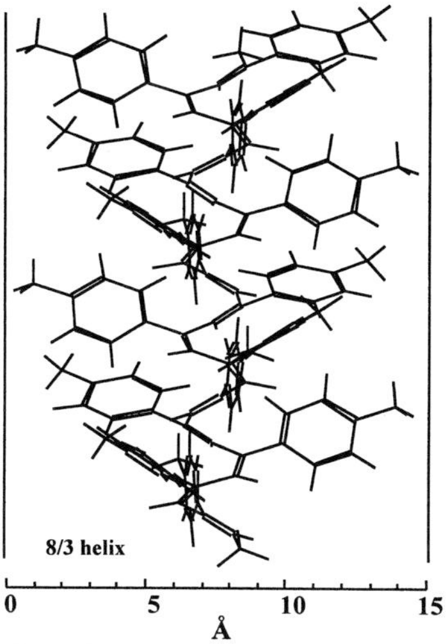

Fig. 13 Proposed helical conformation of **Ib** polymer obtained using **AM1** method.

ESR Spectra

Previously, we have shown[24,48] that the g value of not only aromatic but also aliphatic polyacetylenes such as poly(alkyl)propiolates called polyacetylene esters can be used to deduce the exact geometrical conformations, i.e., cis form or trans form, when one hetero atom such O, N, or halogen is substituted within the side chain of the polyacetylene molecules. Such a hetero atom has a relatively large spin orbit coupling constant, δ[54-55] compared with the carbon, and the hetero atom can make the observed g value shift to lower magnetic field, especially in the cis form, though such a shift in the case of the trans form is extremely small due to decoupling of the magnetic interaction between the side chain and the planar trans main-chain. This indicates that the side chain and planar trans-conjugation plane is nearly perpendicular where there is no strong magnetic interaction between the hetero atom in the side chain and the unpaired electrons. The ESR spectra were observed to determine the geometrical structure of the red polymers obtained before and after compression under vacuum for 10 min (see Figure 14)[27]. The observed parameters are shown in Table 7. It is clear that compression of the pristine red polymer decreased the g value from 2.0039 to 2.0034 and line width, ΔH_{msl} from 9.8Gauss to 9.1Gauss together with a large increase of the spin concentration from 1.1×10^{16} spins/gram to 7.3×10^{16} spins/gram, strongly supporting the compression-induced isomerization from cis form to trans form[27]. The compression of the polymer under reduced pressure can break the cis C=C double bonds to produce two radical

Table 7. ESR parameters observed before and after compression of poly(3MBOPA) prepared in the presence of TEA.

Pressure (kg/cm^2)	g-Value	ΔH / (Gauss) Room temp.	ΔH / (Gauss) 77K	Spin (g) $x10^{16}$
0	2.0039	9.8	10.4	1.1
100	2.0034	9.1	11.9	7.3

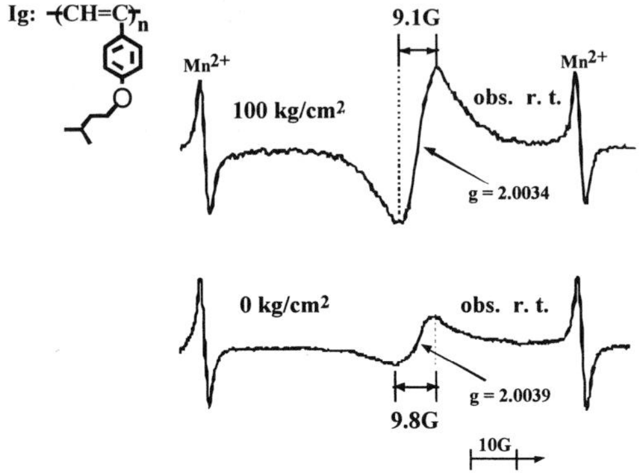

Fig. 14 ESR spectra of **Ig** polymer. (a) before and (b) after compression at $100kg/cm^2$ under vacuum at room temperature. Mn^{2+} shown in this Figure is used as a standard signal of magnetic field.

spins as biradicals, and the resulting spins can stabilize in the resulting planar trans sequences as mobile unpaired electrons called solitons[56-59] which should show motional narrowing effects in the line width, i.e., the change in the line width reversibly depends on temperature change, e.g., 9.1G at room temperature and 11.9 G at 77K just as shown in **Table 7**. Thus, these data clearly indicate that the pristine polymer can be assigned to that of the cis polymer as proven by not only Raman data but also DRUV-Vis data, although the red polymer as well as the yellow polymer with a cis form can be isomerized by the compression to the trans form polymer.

Origin of Color of π-Conjugated Polymer

Thus, the results on the Raman and DRUV-Vis observed before and after compression clearly agree with those of the ESR data without any inconsistency. The red color of **Ig** is neither simply explained in terms of chromophore of the trans structure nor the cis-transoid structure as the primary structure such as repeated double-single units where the side chain moiety partly cooperated with the π-conjugated main chain as previously reported by us[27]. Therefore, it is concluded that a new chromophore as the entire peak bearing λmax at 490nm is attributed to that of a pseudohexagonal structure called columnar as π–conjugated self-assembly or self-organization. This assembly may reflect the aggregation of π-conjugated polymers which are similar to that of the so-called *J* aggregation of LB molecules[60-61]. This is a new and important concept because preparation of π–conjugated polymers having columnar structure is a new method to control the color of the conjugated polymer which can be used as new and promising advanced materials such as NLO[62] or EL materials[63] in the solid state.

Conclusion

Stereoregular polymerization of monosubstituted acetylene was successfully performed using the $[Rh(NBD)Cl]_2$ catalyst in the presence of various solvents at around room temperature to selectively produce the corresponding cis-transoid polymer which can make a pseudohexagonal structure called columnar as self-assembly in high yields. The poly(phenylacetylene) derivatives obtained before and after not only solvent treatment but also compression under vacuum were characterized in detail using ^{1}H NMR, Raman, ESR, diffuse reflective UV-Vis, and X-ray diffraction methods. Consequently, the pristine polymer prepared using triethylamine manifested that the red color arises from that of the columnar as the π–conjugated self-assembly whose content can be also decreased with compression associated with a large blue shift of the absorption maximum in the UV-Vis spectrum of the original polymer. Thus, formation and destruction of the columnar as π–conjugated self-assembly of a conjugated polymer such as **Ia** or **Ig** is a very important concept for the color design of π-conjugated polymers, i.e., absorption maximum which may be related with ionization potential, *IP* as well as the Stokes shift on LED materials. Therefore, the columnar formation becomes a new method to control the color of the conjugated polymer even in the cis form without formation of trans polymers.

[1] (a) T. A. Skotheim (Ed.), *Handbook of Conducting Polymers*, Dekker, New York, **1986**, *Vol. 1-2*. (b) J. R. Ferraro and J. M. Williams, Eds., *Introduction to Synthetic Electrical Conductors*, Acad. Press. Inc., New York, **1987**. (c) J. C. Salomone (Ed.), *Polymeric Materials*, CRC Press, New York, **1996**, *Vol. 8*, 6481.

[2] B. Z. Tang, W. H. Poon, S. M. Lueng, W. H. Leung and H. Peng, *Macromolecules*, **1997**, *30*, 2209.

[3] Y. V. Korshak, T. V. Medvedeva, A. A. Ovchinnikov and V. Spector, *Nature*, **1987**, *32*, 370.

[4] (a) M. Tabata, Y. Nozaki, W. Yang, K. Yokota and Y. Tazuke, *Proc. Jpn. Acad. Ser.*, **1995**, *B71*, 219. (b) M. Tabata, T. Sone, Y. Sadahiro, W. Yang, S. Kobayashi, Y. Inaba and K. Yokota, *Kobunshi Ronbunsyu (Comprehensive Papers in Japanese)*, **1997**, *54*, 863. (c) M. Tabata, Y. Sadahiro, T. Sone, Y. Inaba and K.

Yokota, *Kobunshi Ronbunsyu (Comprehensive Papers in Japanese)*, **1999**, *56*, 350. (d) M. Tabata, T. Sone, M. Nakamura, Y. Mawatari, D. Yonemoto, K. Yokota and Y. Sadahiro, *Polym. Appl. (Comprehensive Papers in Japanese)*, **2001**, *50*, 211.
[5] A. Asdenete, A. Ottoboni, A. Furlani and M. V. Russo, *Chemtronics*, **1991**, *5*, 75.
[6] A. Furlani, G. Iucci, M. V. Russo, A. Bearzotti and R. D'Amato, *Sens. Actuators, B*, **1992**, *B7*, 447.
[7] D. Neher, A. Wolf, C. Bubeck and G. Wegner, *Chem. Phys. Lett.*, **1989**, *163*, 116.
[8] M. Tabata, T. Sone, K. Yokota, T. Wada and H. Sasabe, *Nonlinear Optics*, **1999**, *22*, 341.
[9] K. Tada, H. Sawada, J. Kyokane and K. Yoshino, *Jpn. J. Appl. Phys.*, **1995**, *34*, L1083.
[10] T. Masuda and T. Higashimura, *Adv. Polym. Sci.*, **1986**, *81*, 121.
[11] (a) F. D. Kleist and N. R. Byrd, *J. Polym. Sci., Polym. Chem. Ed.*, **1969**, *7*, 3419. (b) L. B. Luttinger, *J. Org. Chem.*, **1962**, *27*, 1591.
[12] (a) C. I. Simionescu, V. Percec and S. Dumitrescu, *J. Polym. Sci., Polym. Chem. Ed.*, **1977**, *15*, 2497. (b) N. Y. Nguyen, S. Amdur, P. Ehrlich and R. D. Allendoerfer, *J. Polym. Sci., Polym. Syn.*, **1978**, *65*, 63. (c) A. C. Chiang, P. F. Waters and M. H. Aldrdge, *J. Polym. Sci., Polym. Chem. Ed.*, **1982**, *20*, 1807.
[13] Furlani, I. Collamati and G. Sartori, *J. Organomet. Chem.*, **1969**, *17*, 463.
[14] Furlani, P. Biev, M. V. Russo and M. Fiorentio, *Gazz. Chim. Ital.*, **1987**, *107*, 373.
[15] (a) T. Masuda, N. Sasaki and T. Higashimura, *Macromolecules,* **1974**, *7*, 728. (b) T. Masuda and T. Higashimura, *Adv. Polym. Sci.*, **1986**, *81*, 121.
[16] Y. Fujita, Y. Misumi, M. Tabata and T. Masuda, *J. Polym. Sci., Part A, Polym. Chem.*, **1998**, *36*, 3157.
[17] (a) M. Tabata, W. Yang and K. Yokota, *Polym. Prepr. Jpn.*, **1987**, *36*, 2090. (b) M. Tabata, *Nikkei New Materials*, **1988**, *50*, 82.
[18] (a) M. Tabata, W. Yang and K. Yokota, *Polym. J.*, **1990**, *22*, 1105. (b) EP 420699 A2 910403, EP 90-310708 900928 (1990), invs.: M. Tabata, K. Yokota, Y. Yoshinaga, F. Kishi, N. Kaneko and N. Kushibiki, *Chem. Abstr.*, **1990**, *115*, 125436.
[19] W. Yang, M. Tabata, K. Yokota and A. Shimizu, *Polym. J.*, **1991**, *23*, 1135.
[20] R. J. Kern, *J. Polym. Sci., Polym. Chem. Ed.*, **1969**, *7*, 621.
[21] G. Mestromi, A. Camus, A. Furlani and M. V. Russo, *Gazz. Chem. Ital.*, **1982**, *112*, 1435.
[22] (a) A. Furlani, S. Licoccia, M. V. Russo, A. M. Camus and N. Marsichi, *J. Polym. Sci., Part A: Polym. Chem.*, **1986**, *24*, 991. (b) A. Furlani, C. Napoletano, M. V. Russo and W. J. Feast, *Polym. Bull. (Berlin)*, **1986**, *16*, 311.
[23] M. Tabata, W. Yang and K. Yokota, *J. Polym. Sci., Part A: Polym. Chem.*, **1994**, *32*, 1113.
[24] M. Tabata, Y. Inaba, K. Yokota and Y. Nozaki, *J. Macromol. Sci. Pure Appl. Chem.*, **1994**, *A31*, 465.
[25] H. Nakako, R. Nomura, M. Tabata and T. Masuda, *Macromolecules*, **1994**, *32*, 2061.
[26] X. Kong, J. K. Y. Lam and B. Z. Tang, *Macromolecules*, **1999**, *32*, 1772.
[27] Y. Mawatari, M. Tabata, T. Sone, K. Ito and Y. Sadahiro, *Macromolecules*, **2001**, *34(11)*, 3776.
[28] M. Lindgren, H-S. Lee, W. Yang, M. Tabata and K. Yokota, *Polymer*, **1990**, *32*, 1531.
[29] R. D'Amato, T. Sone, M. Tabata, M. V. Russo and A. Furlani, *Macromolecules*, **1998**, *31*, 8660.
[30] N. Yoshioka, H. Nishide, T. Kaneko, H. Yoshiki and E. Tsuchida, *Macromolecules*, **1992**, *25*, 3838.
[31] D. Yonemoto, M. Tabata, T. Sone and Y. Sadahiro, *Submitted for Macromolecules*, **2002**, *35*.
[32] M. Tabata, K. Takamura, K. Yokota, Y. Nozaki, T. Hoshina, H. Minakawa and K. Kodaira, *Macromolecules*, **1994**, *27*, 6234.
[33] M. Tabata, Y. Tanaka, Y. Sadahiro, T. Sone, K. Yokota and I. Miura, *Macromolecules*, **1997**, *30*, 5200.
[34] M. Tabata, T. Sone and Y. Sadahiro, *Macromol. Chem. Phys.*, **1999**, *200*, 265.
[35] M. Tabata, T. Sone, Y. Sadahiro, K. Yokota and Y. Nozaki, *J. Polym. Sci., Part A. Polym. Chem.*, **1998**, *36*, 217.
[36] M. Tabata, T. Sone, Y. Sadahiro and K. Yokota, *Macromol. Chem. Phys.*, **1998**, *199*, 1161.
[37] Y. Togo, N. Nakamura and H. Iwamura, *Chem. Lett.*, **1991**, 1201
[38] H. Iwamura, *Mol. Cryst. Liq. Cryst.*, **1989**, *176*, 33.
[39] M. Nakamura, M. Tabata, T. Sone and Y. Mawatari, *Macromolecules*, **2002**, *35*, in press. (b) A. Miyasaka, M. Tabata and T. Sone, *Submitted for Macromolecules*, **2002**.
[40] T. Fukushima, M. Tabata, T. Sone and Y. Sdahiro, *Submitted for Macromolecules*, **2002**, *34*.
[41] (a) T. Sone, R. Asako, T. Masuda, M. Tabata, T. Wada and H. Sasabe, *Macromolecules*, **2001**, *34*, 1586. (b) M. Tabata, K. Yokota and M. Namioka, *Macromol. Chem. Phys.*, **1995**, *196*, 2969. (c) M. Tabata, M. Namioka, K. Yokota and H. Minakawa, *Polymer*, **1996**, *37*, 1969.
[42] H. Minakawa, M. Tabata and K. Yokota, *J. Macromol. Sci., Pure Appl. Chem.*, **1996**, *A33*, 291.
[43] T. Harada, M. Tasumi and H. Shirakawa, *Chem. Lett.*, **1978**, 1411.
[44] M. Tabata, Y. Nozaki and K. Yokota, *J. Photopolym. Sci. Technol.*, **1993**, *6*, 1.
[45] H. H. Yaffe and M. Orchin, Ed., *Application of Ultraviolets Spectroscopy*, Wiley, New York, **1964**, 257.
[46] (a) M. Tabata, S. Kobayashi, Y. Sadahiro, Y. Nozaki, K. Yokota and W. Yang, *J. Macromo. Sci., Pure*

Appl. Chem., **1997**, *A34*, 505. (b) M. Tabata, Y. Sadahiro, Y. Nozaki, Y. Inaba and K. Yokota, *Jpn. J. Appl. Phy.*, **1996**, *35*, 5411.
[47] M. Tabata, Y. Sadahiro, K. Yokota and S. Kobayashi, *Jpn. J. Appl. Phys.*, **1996**, *35*, 5411.
[48] M. Tabata, Y. Sadahiro, Y. Nozaki, Y. Inaba and K. Yokota, *Macromolecules*, **1996**, *29*, 6673.
[49] G. Unger, *Polymer*, **1992**, *34*, 2050.
[50] P. Scherrer, *Göttinger Nachr.*, **1918**, 98.
[51] (a) J. J. P. Stewart and F. J. Seiler, Research Laboratory, U.S. Air Force Academy, Colorado Springs, Colorado 80840-6528, U. S. A. (b) MOPAC version 5.0, QCPE#455. (c) J. J. P. Stewart, *J. Comput. Chem.*, **1989**, *10*, 221.
[52] M. Tabata, Y. Sadahiro, T. Sone, K. Yokota and Y. Ishikawa, *J. Polym. Sci., Part A, Polym. Chem.*, **1998**, *36*, 2457.
[53] T. Masuda, E. Isobe, T. Higashimura and K. Takada, *J. Am. Chem. Soc.*, **1983**, *105*, 7473.
[54] S. Y. Pshezhetskii, A. G. Kotov, V. K. Milincuk, V. A. Roginskii and V. I. Tupikov, *ESR of Free Radicals in Radiation Chemistry*, Wiley, New York, **1972**, 22.
[55] M. Tabata and A. Lund, *Chem. Phys.*, **1983**, *75*, 379.
[56] (a) B. R. Weinberger, E. Ehrenfreud, A. Pron, A. J. Heeger and A. G. MacDiarmid, *J. Chem. Phys.*, **1980**, *72*, 4749. (b) S. Murakami, M. Tabata and M. Hatano, *J. Appl. Polym. Sci.*, **1984**, *29*, 3445. (c) J. Sohma and M. Tabata, *Polym. Degrad. Stab.*, **1987**, *17*, 5. (d) M. Tabata, T. Matsuura, S. Okawa, W. Yang, K. Yokota and J. Sohma, *Synth. Metal*, **1987**, *17*, 577.
[57] H. Thomann, H. Jin and G. L. Baker, *Phys. Rev. Lett.*, **1987**, *59*, 509.
[58] E. J. Hustedt, H. Thoman and B. H. Robinson, *J. Chem. Phys.*, **1990**, *92*, 978.
[59] M. Mehring, H. Seidel, G. Müller and G. Wegner, *Solid State Commun.*, **1983**, *45*, 1072.
[60] R. Steiger, R. Kitzing and P. Junod, *J. Photogr. Sci.*, **1973**, *21*, 107.
[61] Reicher, *Photogr. Sci. Eng.*, **1974**, *18*, 335.
[62] T. Wada, L. Wang, H. Okawa, T. Masuda, M. Tabata, M. Wan, M.-A. Kakimoto, Y. Imai and H. Sasabe, *Mol. Cryst. Liq. Cryst.*, **1997**, *294*, 254.
[63] M. Tabata, V. Boucard, D. Adés, A. Siove, T. Sone, T. Seino and Y. Mawatari, *Macromolecules*, **2001**, *34*, 8101.

Approaches to Conjugated Polymers via Solid State Polymerization

Daniel J. Sandman, Ik-Bum Kim, Jeffrey M. Njus, Dong-Chan Lee, Ashok L. Cholli, Sangrama Sahoo*

Summary: Topochemical solid state polymerization is a method of obtaining fully crystalline macromolecules. Herein, approaches to conjugated polymers via new solid state polymerizations and the properties of materials, especially polydiacetylenes(PDA), obtained via solid state polymerization are discussed. Our attempts to polymerize cyanoalkynes and to define the thermal reaction product of a dicyanodiazepine are presented. Certain PDA exhibit absorption, likely charge-transfer, at energies below the exciton. Nanocrystals of PDA-DCH react with liquid bromine to give the mixed polyacetylene PDA-$DCHBr_6$ whose visible spectrum exhibits structure not observed in spectra of bulk crystals of PDA- $DCHBr_6$. The thermochromic PDA of the dialkylurethanes of 5,7-dodecadiyn-1,12-diol were studied by variable temperature solid state ^{13}C nmr. The upfield shift of the acetylene carbons as the temperature is raised above that of the thermochromic phase transition indicates the role of mechanical strains in achieving thermochromism.

Introduction

Conjugated polymers have attracted considerable interdisciplinary scientific interest for over thirty years. The 2000 Nobel Prize in Chemistry[1-3], awarded to H. Shirakawa, A.G. MacDiarmid, and A.J. Heeger for "doping", via electron transfer chemistry, of polyacetylene, $(CH)_x$, to a metallic state may be taken as a recognition at the highest level of the scientific, technological, and commercial results of an extremely large body of work. Nevertheless, much of this work has been carried out on materials that are partially crystalline at best, and hence many properties, especially electronic transport, are limited and do not reach their intrinsic limit. Hence, it will be argued herein that there are still many important challenges remaining in the area of conjugated polymers, and materials with very high degrees of three dimensional order are key to several desirable advances. It is in this context that we note that G. Wegner's studies[4] of the topochemical and topotactic diacetylene (DA, **1**) polymerization predate the "doping" of $(CH)_x$.

As examples of fully ordered macroscopic polymer single crystals, polydiacetylenes (PDA, **2**) have structural and physical properties that provide far more detail and insight than those of

 CCC 1022-1360/00/$ 17,50+.50/0

less ordered systems. Scheme 1 illustrates the DA polymerization and gives the structures of the side group substituents under discussion herein. Let us provide two examples to illustrate that point chosen from the areas of odd order non-linear optics (NLO) and studies of exciton binding energy. The third order NLO coefficient, $X^{(3)}$, is related to the light intensity-dependent refractive index, the key parameter for all-optical signal processing. The studies[5] of $X^{(3)}$ in both PDA single crystals and related thin films

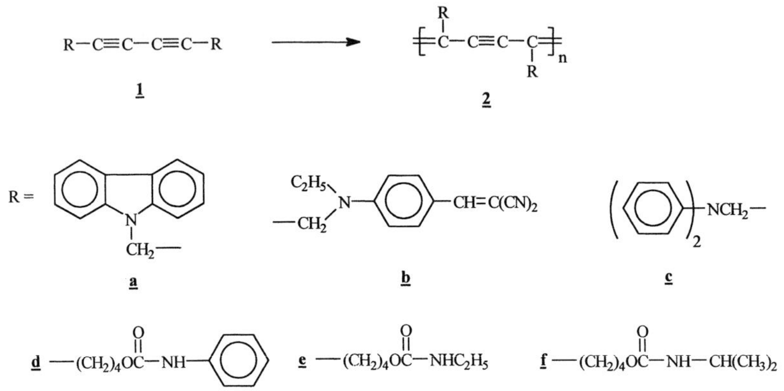

Scheme 1. Diacetylene polymerization and the structures of side groups R.

have revealed magnitudes of $X^{(3)}$ comparable to the best inorganic semiconductors with temporal responses in the transparent spectral region in the subpicosecond time frame. Recently, the dispersions of $X^{(5)}$ and $X^{(7)}$, higher order NLO coefficients, have been obtained[6] from electroabsorption spectroscopy using spin-cast thin films of soluble PDA. Hopefully, the extension of these studies to single crystal PDA will occur in the near future. Exciton binding energy (E_b) is the energy that must be exceeded for electron-hole formation, and subsequently carrier transport, to occur. Carrier transport is especially critical in polymeric light emitting diodes (LEDs). PDAs have a large E_b[7], approximately 1 eV, and this value is significantly larger than other conjugated polymers, such as polyphenylenevinylenes (PPVs), where the E_b is estimated to range from close to zero to 0.9 eV. An extensive literature[8] discussing experimental and theoretical aspects of this topic exists. Recently, the E_b in PPV was determined[9] to be 0.060 eV from studies of photoconductivity excitation profile spectroscopy as a function of light polarization, applied electric field, and temperature.This contrast between PDA and other less ordered conjugated polymers is illustrated[10] in the difficulty of observing photoinduced electron transfer from PDA to C_{60} versus the ease of the process with PPV.

PDA exhibit electronic carrier mobilities (μ) greater than 1 cm^2/volt-sec[11], slightly larger than molecular crystals at room temperature. Values of μ greater than 10^3 cm^2/volt-sec are an important aspect of the functioning of crystalline semiconductors such as Si and GaAs. The room temperature mobilities of crystalline PDA are 3-4 orders of magnitude larger than those of partially crystalline or amorphous conjugated polymers. Recent experiments by H. Schön and coworkers at Lucent Technologies Bell Laboratories reveal[12] carrier mobilities in molecular crystals in excess of 10^4 cm^2/volt-sec, at liquid helium temperatures in a field effect transistor (FET) device configuration. As a consequence of such large values of μ, electrically pumped laser action[13] and superconductivity[14]have been observed in molecular crystals such as anthracene, tetracene, and pentacene, and PPV oligomers[15]. In addition, the FET device configurations have been used to observe[16] superconductivity in a specimen of partially crystalline regioregular poly-3-hexylthiophene (P3HT) with T_c approx. 2°K. It is most likely that a more crystalline specimen of P3HT would exhibit a higher value of T_c. It is very likely that a full assessment of the scientific and technological potential of the phenomena dependent on large values of μ will require fully crystalline rigorously purified samples of conjugated polymers beginning with PDA.

The literature of approaches to conjugated polymers via solid state polymerization is overwhelmingly dominated by PDA[7,17-24] and closely related systems. Lauher, Fowler, and coworkers have recently reported[25] the topochemical and topotactic 1,6-addition polymerization of a triacetylene where the required monomer crystal structure is achieved by means of a host-guest strategy involving hydrogen bonding. There are a number of reports of studies of solid state reactivity and polymerization, especially of monoacetylenes. B.M. Foxman and coworkers have studied[26] the crystallography and solid state reactivity of numerous metal propiolates, $HC{\equiv}C\text{-}CO_2^-M^+$. The crystallographic information provides a compelling case that these polymerizations are topochemically initiated. Nevertheless, to date, the solid state chemistry of the overall reaction is single crystal monomer-to-amorphous polymer. Other acetylenes such as 10-undecynamide and the cyclohexyl urethane of 5-hexyn-1-ol display[27] intermolecular contacts between potentially reactive acetylenic carbons less than 3.70Å, yet these melt stable compounds appear to be unreactive to ionizing radiation due to a facile fluorescence process. In the interest of enhancing the reactivity of a crystalline monoacetylene by means of a heavy atom effect, 11-Bromo-10-undecynamide was synthesized.[28] This compound is reactive to ultraviolet light and ^{60}Co γ-radiation; the latter converts the crystalline monomer to an amorphous polymer.

As a result of these studies on monoacetylenes and other work it should be apparent to the casual reader and beginning researcher that solid state reactivity and polymerization can be very complex subjects. Not only must potentially reactive molecules have atoms that will form covalent bonds within about 4.20 Å of each other in accord with the topochemical postulate[17], but there must also be viable reaction mechanisms for thermal, ultraviolet light, and ionizing radiation to initiate and propagate a chain whose growth must also be compatible with the initial lattice. Just as William Shakespeare tells us in *The Merchant of Venice* that "all that glisters is not gold", so every colorless organic compound that becomes a dark reflective solid in response to thermal, uv light, and ionizing radiation stimuli is not necessarily experiencing processes analogous to the optimal DA-PDA conversion.

Herein, we summarize recent initial studies of the solid state reactivity of cyanoalkynes, compounds iso-π-electronic to DA, the definition of the initiation process and structure of the black reflective solid product of the solid state reaction of 2,3-dicyano-5,7-dimethyl-6H-1,4-diazepine (**3**), our initial observation of charge-transfer transitions in PDA, our observation of vibrational structure in the brominated product of PDA-DCH (**2a**) nanocrystals, and our deduction of the role of mechanical strains in thermochromic PDA via solid-state ^{13}C cross polarization magic angle spinning (CPMAS) NMR.

Cyanoalkyne Reactivity

Pursuant to the identification of other solid-state reactions that might be related to DA polymerization, we identified cyanoalkynes because they are iso-π-electronic to DA. Just as DA reactivity involves diradical and dicarbenes as intermediates, qualitatively, in principle, cyanoalkynes should be able to form analogous intermediates, as shown in Figure 1. Just as the observed en-yne form of a PDA chain is lower in energy than the butatriene form, so the imine-yne product of 1,4-addition polymerization of a cyanoalkyne is calculated to be lower in energy than a cumulene form (Scheme 2). The calculations were performed on pentamer oligomeric structures of both PDA and polymerized cyanoalkyne chains using the AM1 semiempirical method in the Hyperchem suite of programs. As shown in Scheme 2, reaction at either the acetylene or cyano groups of a cyanoalkyne is conceivable.

The first cyanoalkyne that we synthesized was N-(3-cyanoprop-2-ynyl)carbazole, **4**. This compound was found[29] to be relatively unreactive to uv light and ^{60}Co γ-radiation. Compound **4** exhibits m.p. 91-93°C, and it was heated at 80°C for an extended period of time.

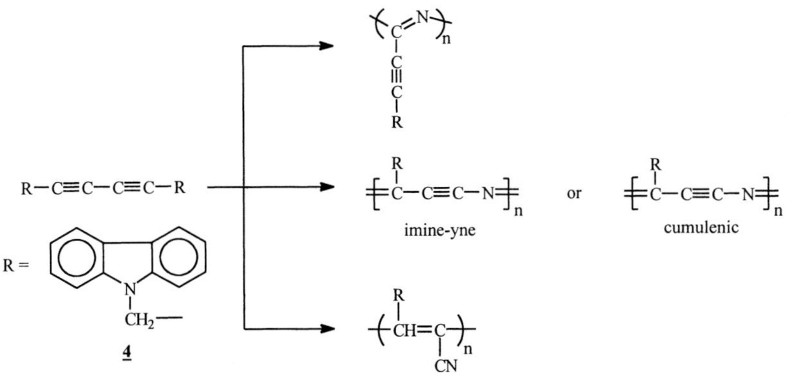

Scheme 2. Possible reaction pathways for reactions of cyanoalkynes.

The initially colorless crystals become light brown on heating. The crystal structure[29] reveals relatively short contacts between cyano groups with a N----C intermolecular distance as short as 3.339Å, but this structure is not one that allows the desired 1,4-addition polymerization. The characterization[29] of the thermal product reveals only reactivity between the cyano groups. Our further studies of cyanoalkynes will involve systems designed to bring about 1,4-addition.

Solid State Reaction of 2,3-Dicyano-5,7-dimethyl-6*H*-1,4-diazepine

We began the study of 2,3-dicyano-5,7-dimethyl-6H-1,4-diazepine (**3**) in connection with our studies of polymerization[30] and cyclotetramerization[31] of dicyanoalkenes and –arenes initiated by unprotected sugar reagents. This compound was hoped to be a cyclic analog of *bis* (dimethylamino) maleonitrile which reacts[32] with alkaline glucose to give a low yield of cyclotetramer. The original synthesis of **3** reported[33] melting at 199-200°C with decomposition. Of the dicyanoalkenes and –arenes that we have studied to date, **3** is the only compound that does not exhibit reversible melting behavior. The process of decomposition attracted our attention, and we observed that heating the colorless **3** at 150°C for one week resulted in formation of a black reflective solid (45% yield) accompanied by sublimation of **3** out of the reacting crystal. The solid product was amorphous, and its IR spectrum revealed that its average chain structure involved formation of a conjugated polymer via reaction at one of the two cyano groups, similar to that found in solution polymerization[30] involving sugars. The solid state reaction of **3** is illustrated in Scheme 3.

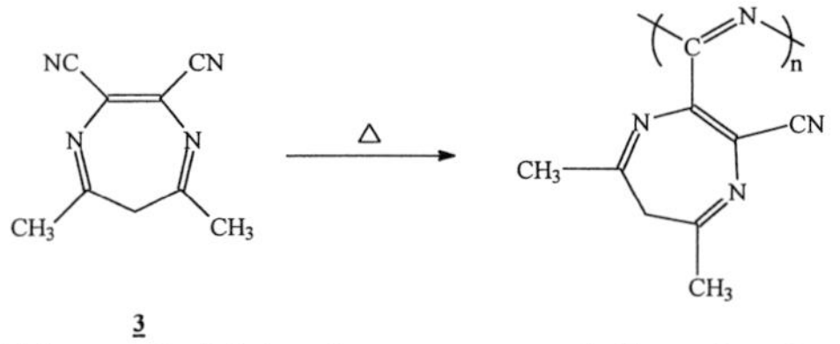

Scheme 3. Molecular structure of diazepine **3** and a possible structure of the product of its thermal solid state reaction.

As noted above, the key to understanding solid state reactivity is the reactant crystal structure and the intermolecular relationships of monomer molecules. Examination of the crystal structure[34,35] of **3** did not reveal a linear pathway for polymerization. The structure also exhibits an intermolecular N---C contact of 3.425Å between –CN groups. This experimental observation coupled with lack of other thermally reactive dicyano compounds raises the question as to how the thermal reaction initiates and how the chain grows. Our current understanding of these issues comes largely from solid state nmr studies of ^{13}C, ^{15}N, and ^{1}H nuclei.

The ^{15}N spectrum exhibits resonances in regions that allow assignment to tertiary amine and quaternary ammonium (R_3NH^+). The ^{1}H spectrum of partially reacted monomer reveals resonances assignable to olefinic H atoms. These observations may imply the involvement of tautomeric forms of the monomer. Some tautomers of **3** are illustrated in Figure 1. If reaction initiates from a tautomer such as **3b** by addition of the –NH group to a cyano group in an adjacent molecule, this process would involve a defect site because the NH is not found in the crystal structure[34,35] and its creation would involve formation

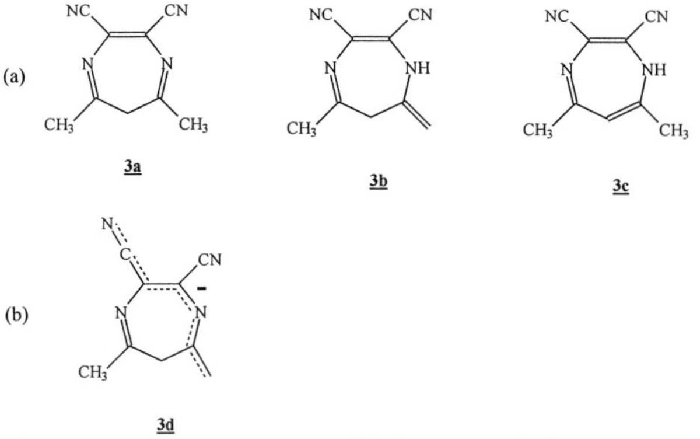

Figure 1. (a) Some tautomers of **3**. Tautomer **3c** is an 8π antiaromatic system. (b) **3d** is a delocalized anion derived from **3**.

of the NH bond which would have a length of approximately 1Å. This situation describes creation of a defect site, and the initiation may be topochemically controlled. It is apparent that the lattice does not control the chain growth process. If **3** can lose a proton in the solid state, the anion **3d** (or an equivalent group of resonance structures), also illustrated in Figure 1, would form. The presence of excess negative charge on a cyano nitrogen[30] in **3d** offers a rationalization as to why one cyano group is unreactive in this solid state reaction.

Polydiacetylenes: Electronic Transitions Below the Exciton

In the absence of oxidation[36] of a PDA backbone, PDA crystals are typically transparent at energies lower that of the excitonic absorption. We synthesized[37] a series of 2,4-hexadiyne DA monomers that have the characteristic that they absorb light at longer wavelengths than carbazole such that, on polymerization, they would provide an environment that would shift the excitonic absorption of the PDA to longer wavelengths than found in **2a**. The first monomer to give suitable single crystals for crystallographic study was 1,6-diethyl-(1,6-bis-*p*-benzylidenemalononitrile)-2,4-hexadiyndiamine, **1b**. This monomer is relatively inert to ionizing radiation and was thermally polymerized in a topotactic process[38] to **2b**. Polarized specular reflection spectra were obtained from the (001) face of crystals of **2b**. The reflection peak is found near 571 nm for pseudo-long-axis polarization, and significant anisotropy is apparent from the spectrum in the perpendicular direction. This reflection peak is in a similar position as that found[39] in the PDA of 1,1,6,6-tetraphenylhexadiyndiamine (THD, **2c**). In addition, heretofore unobserved structure for PDAs is observed in both principal directions from approximately 833 to 625 nm. The observation of a band at lower energy than the usual PDA absorption, its width, lack of vibronic structure, and apparent polarization suggest[38] that the band arises from one or more charge-transfer transitions. The donor would be the conjugated backbone of one polymer chain with the side group of an adjacent chain serving as acceptor. Such features to the red of PDA excitonic absorption had not been previously observed. Figure 2 displays the low energy portion of the reflection spectrum of **2b**. Subsequently, another PDA was reported[40] to have such absorption.

Polydiacetylenes: Bromination of PDA-DCH Nanocrystals

Single crystals of the PDA of 1,6-di-N-carbazolyl-2,4-hexadiyne (**2a**) react[41] with bromine in a crystal-to-crystal transformation that allows the material to gain 3-9 Br atoms per repeat unit. A particularly interesting member of this family is the material that has gained 6Br atoms per repeat unit. Here, bromine adds to each carbazole unit at the 3 and 6 positions and then proceeds to add to the triple bond. The brassy appearance of **2a** is changed to a copper bronze. Thus, the PDA backbone structure is converted largely to a mixed polyacetylene structure, as initially[42] deduced from ^{13}C CPMAS NMR studies. Evidence[43] supporting the mixed polyacetylene structure also came from Raman spectra. While the material is intensely colored, its visible spectrum was recorded[44] by diffuse reflectance. The reflectance was not particularly informative as a distinct visible maximum associated with the conjugated backbone was not observed.

In 1993 Nakanishi and coworkers described[45] the preparation of nanometer sized crystals of a variety of organic molecular compounds, some of which were DA monomers. Of particular interest to the present work, nanocrystals of **2a** obtained by uv photopolymerization have been extensively studied.[46] Specimens of **2a** in programmed[47] electrostatic assemblies formed by layer-by-layer deposition with polydiallyldimethyl- ammonum chloride (PDADMAC) as polycation were assesssed[48] for their third order NLO properties.

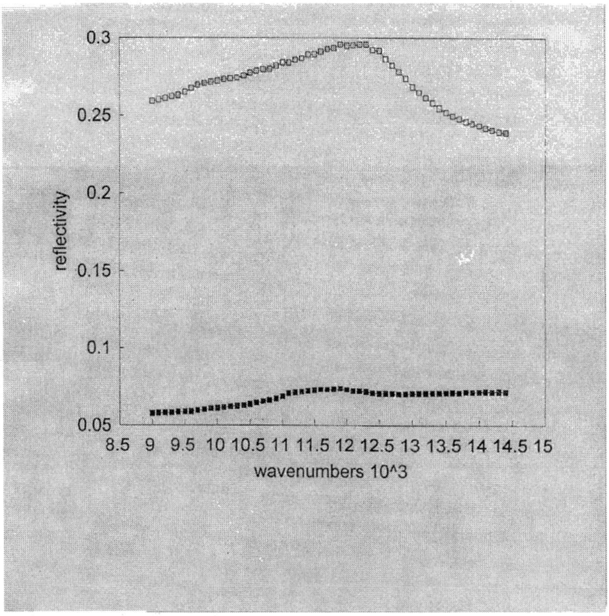

Figure 2. Polarized specular reflection spectra of **2b** in the infrared region. The upper spectrum is obtained with light polarized along the chain axis, and the lower spectrum is taken with light polarized perpendicular to the upper spectrum.

In our studies, nanocrystals of **1a** were prepared as described[46] by Nakanishi and coworkers. These specimens were polymerized with UV light and thermally, and their sizes were studied by transmission electron microscopy and dynamic light scattering. The optical properties, including vibronic structure, of these materials were in agreement with previous reports.

Bromination of **2a** nanocrystals was carried out in multilayers with PDADMAC and **2a** obtained by uv polymerization. The absorbance at 647 nm of these assemblies increased in a linear fashion with the number of bilayers. Ten bilayers of this assembly were exposed to bromine vapor. After 22 hours exposure to bromine vapor, the spectrum of the assembly was not significantly changed. Bulk crystals of **2a** are relatively unreactive[41,42] to bromine vapor. When these assemblies were exposed to liquid bromine for 10 seconds, the visible spectrum of **2a** was replaced by a new spectrum with maxima at 540 and 497 nm. The separation between these two peaks is 1605 cm^{-1}. The Raman spectrum[49] of crystals of PDA-$DCHBr_6$ reveals a shifted line at 1518 cm^{-1}. This shift is characteristic of an extended polyene chain. The broad agreement between the splitting in the optical spectrum of the assembly and the Raman spectrum of bulk crystals suggests that the assembly bromination converted it to a mixed polyacetylene structure. In addition, the carbazole absorption in the polymer shifted to the longer wavelength of 362 nm. This spectral information is displayed in Figure 3 along with spectra of the PDA-DCH assembly exposed to bromine vapor.

Polydiacetylenes: Thermochromic Phase Transitions

The PDA of the *bis*-alkylurethanes of 5,7-dodecadiyn-1,12-diol exhibit thermochromism. At 25°C, single crystals of these materials exhibit peaks in their reflection spectrum between 635-650 nm with light polarized along the polymer chain. When heated at temperatures in the range 120-150°C, the reflection spectrum shifts to higher energy. The issue of changes in the macromolecular structure[8] that occur in the course of this color change has concerned researchers since 1976.

In 1976, it was suggested[50] that the spectral shift was associated with a change in the number of delocalization lengths. Another proposal[51] suggested a change from the acetylenic structure at room temperature to a butatrienic structure above the temperature of the phase transition. Crystallographic data and quantum chemistry calculations ruled out this suggestion. Moreover, a ^{13}C CPMAS NMR study[52] of the PDA of the *bis*-phenyl urethane of 5,7-dodecadiyn-1,12-diol (TCDU, **1d**) showed its backbone to be acetylenic and not butatrienic. An additional proposal[53] involved the formation of an orbital-flip defect or backbone planar-nonplanar transition as temperature is increased. Such a process demands a change in the backbone repeat distance and was excluded[54] by variable temperature X-ray powder diffraction studies of PDA-ETCD, **2e.** Local dielectric effects were proposed[55] to be involved in the transition. While this proposal was not quantitated and the possibility of mechanical strains was not excluded, it has been criticized.[56] The possibility of mechanical

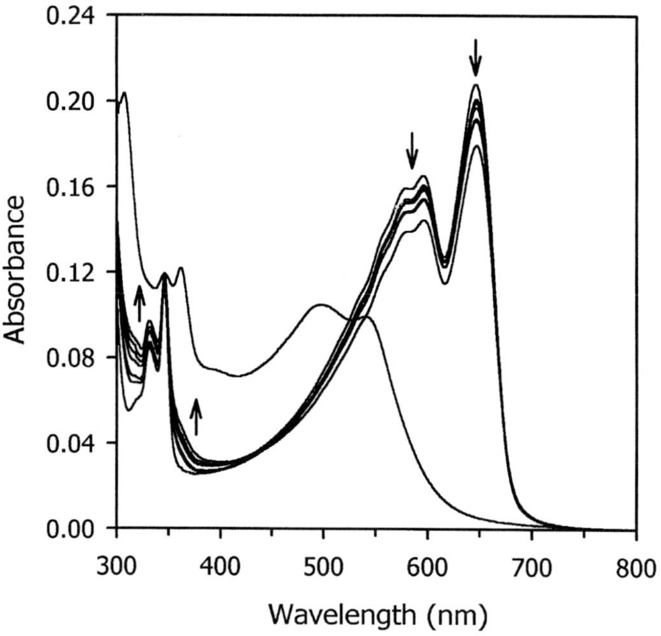

Figure 3. Electronic spectra of a nanocrystal assembly of **2a** exposed to bromine vapor and bromine liquid. The latter gives the maxima at 540 and 497 nm.

strains[57] involving side group and backbone during polymerization to maintain intramolecular hydrogen bonding was advanced. To maintain these hydrogen bonds, the methylene groups would distort resulting in a mechanical strain on the backbone. From theoretical calculations[58] it was deduced that such side group strain could induce a change in PDA electronic structure. Possibly, thermal stimulus during the transition gives the side group sufficient energy to release the mechanical strain by rotation around the single bond between methylene units although such movements of the side group seem somewhat restricted to maintain the necessary crystal unit cell parameters for reversibility. Side group movement decreases the strain on the backbone at the same time. The crystallographic unit cell increases in volume[54] during the transition, making the latter first order, and this results in a release of strain. The increase in unit cell volume[54] in the course of the transition is largely associated with distances involving the side chain and with interlayer features maintaining the backbone repeat distance. Infrared[59] and ^{13}C solid state NMR spectroscopy[60] revealed that the four methylene units between the backbone and urethane function undergo a change in the conformer population in the course of the transition. Change in methylene conformer population was also noted[61] in thermochromic cross polymerized diacetylene-functionalized polyamides. The thermochromic phase transition often occurs at a temperature close to the melting point of the corresponding monomer. The potential role of residual monomer in the transition was ruled out by a study of PDA-IPUDO, **2f**, which has a monomer to polymer conversion greater than 99%.

The availability[62] of a non-thermochromic form of **2e** allows a comparison of its properties to that of the thermochromic materials. The non-thermochromic form of **2e** is obtained by treating the thermochromic material with refluxing chlorobenzene. The chlorobenzene extracts monomer and oligomer, and the exposure to boiling solvent relieves potential mechanical strains. The resulting material retains crystallinity and maintains vibronic structure and anisotropy in its reflection spectrum.

We have recently[63] carried out a study of **2f** and the thermochromic and non-thermochromic forms of **2e** by a combination of thermal techniques and variable temperature ^{13}C CPMAS nmr spectroscopy. In this study, the chemical shift of the acetylenic carbons and its change with temperature and release of mechanical strain is particularly instructive. In general, the ^{13}C chemical shift of acetylenes is sensitive to the length of a conjugated chain[64] and the mechanical strain[65] on the acetylene group. In these cases, increased deshielding signifies a longer conjugation length or greater mechanical strain, as judged by distortion of a bond angle from 180°. Table 1 summarizes the ^{13}C acetylene shifts of these PDA. The increased shielding of the non-thermochromic form of **2e** indicates that mechanical strain has been decreased relative to the thermochromic forms. It follows then that the upfield shifts in going from the blue phase at ambient temperature to the red phase above the thermochromic transition is associated with a decrease in mechanical strain on the backbone. The ^{13}C shift of the acetylene in the non-thermochromic form does not change[63] significantly as temperature is increased. Figure 4 displays the ^{13}C CPMAS spectra of **2f** in a cooling cycle. In the spectrum recorded at 95°C, note the presence of acetylenic resonances corresponding to both the high and low temperature phases.

Table 1. ^{13}C Acetylene Chemical Shifts of Urethane Substituted PDA.

PDA	**^{13}C Acetylene Chemical Shifts, ppm**
2e(blue)	106.1
2e(red)	102.4
2e(non-thermochromic)	100.2
2f(blue)	105.6
2f(red)	101.7

Conclusion

Our studies of **3** and **4** again serve to emphasize the importance of an appropriate crystal structure for the reactant in a solid state polymerization. Our conclusion that a tautomeric form of **3** is involved in the initiation of the solid state reaction is not a common topic in the literature of organic solid state reactivity. The observation of charge transfer absorption in **2b** was not precedented and such spectral features may well be found in related systems. It is possible that such features may lead to new optical and electronic processes in PDA. The bromination of nanocrystals of **2a** led to the observation of spectral features that were not detected in earlier[42,44]studies of brominated bulk crystals of **2a**. Our variable temperature ^{13}C CPMAS nmr studies of the thermochromic PDA allowed study of these materials in both blue and red phases. The shifts of the acetylenic carbons were discussed in terms of contributions of conjugation lengths and mechanical strains. The availability of the nonthermochromic form of **2e** allows us to conclude that mechanical strains are important in the observation of thermochromism. Indeed, in the absence of mechanical strains, it would be expected that these PDA would have absorption maxima in the range 520-550 nm at ambient temperatures.

Acknowledgements

We thank the Petroleum Research Fund (Grants 29604-AC7, 33424-AC7, 37138-AC1), the Office of Naval Research (Multidisciplinary University Research Initiative), and the National Science Foundation (ALC) for partial support of this research. The instrument for solid state NMR was purchased with funds partly provided by the National Science Foundation under

Grant DMR 9802866. M.J. Downey furnished X-ray powder patterns throughout the course of this work. Figure 2 was furnished by Professor C.J. Eckhardt.

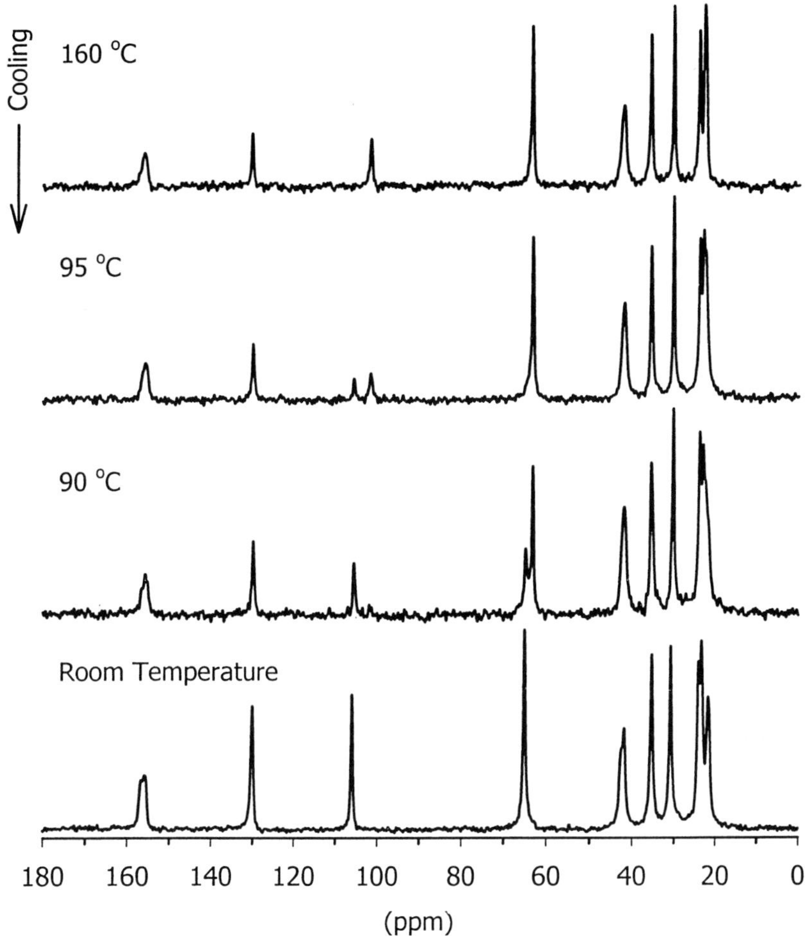

Figure 4. ^{13}C CPMAS spectra of **2e** in a cooling cycle.

[1] H. Shirakawa, *Angew. Chem. Int. Ed. Eng.* **2001,** *40*, 2574.
[2] A.G. MacDiarmid, *Angew. Chem. Int. Ed. Eng.* **2001,** *40*, 2581.
[3] A.J. Heeger, *Angew. Chem. Int. Ed. Eng.* **2001,** *40*, 2591.
[4] G. Wegner, *Z. Naturforsch.,* **1969,** *24B,* 824-832.
[5] H.S. Nalwa, *Adv. Mater.* **1993,** *5*, 341-358.
[6] K. Yang, J. Kumar, D.-C. Lee, D.J. Sandman, S. Tripathy, *Optics Letters* **2000,** *25*, 1186-1188.
[7] D.J. Sandman, *Trends Polym. Sci.* **1994,** *2*,44-55.
[8] N.S. Sariciftci, Ed. *Primary Photoexcitations in Conjugated Polymers: Molecular Exciton versus*

Semiconductor Band Model World Scientific, London, 1997.
[9] D. Moses, J. Wang, A.J. Heeger, N. Kirova, S. Brazovski *Synthetic Metals* **2002**, *125*, 93.
[10] N.S. Sariciftci, B. Kraabel, C.H. Lee, K. Pakbaz, A.J. Heeger, D.J. Sandman, *Phys. Rev. B* **1994**, *50*, 12044-12051.
[11] R.J.O.M. Hoofman, G. P. van der Laan, L.D.A. Sibberles, M.P. de Haas, D. Bloor, D.J. Sandman, *Macromolecules,* **2001**, *34*, 474.
[12] J.H. Schön, Ch. Kloc, B. Batlogg, *Science* 2000, *288*, 2338.
[13] J. H. Schön, Ch. Kloc, A. Dodabalapur, B. Batlogg, *Science* 2000, *289*, 599.
[14] J. H. Schön, Ch. Kloc, & B. Batlogg, *Nature* 2000, *406*, 702.
[15] J. H. Schön, Ch. Kloc, J. Wildeman, G. Hadziioannou, *Adv. Mater.* 2001, *13*, 1273.
[16] J. H. Schön, A. Dodabalapur, Z. Bao, Ch. Kloc, O. Schenker, B. Batlogg, *Nature*, 2001, *410*, 189.
[17] H.J. Cantow, Ed. *Polydiacetylenes* Advances in Polymer Science, Springer Verlag, Vol. 63, 1984.
[18] D. Bloor, R.R. Chance, Eds., *Polydiacetylenes* Martinus Nijhoff, NATO ASI Series, Dordrecht, Boston, 1985.
[19] D.J. Sandman, Ed., *Crystallographically Ordered Polymers* American Chemical Society Symposium Series Vol. 337, 1987.
[20] R.R. Chance, in *Encyclopedia of Polymer Science and Engineering;* J.I. Kroschwitz, Ed., Wiley-Interscience, 2nd ed., 1986, vol. 4, pp. 767-779.
[21] Pope, M.; Swenberg, C.E. *Electronic Processes in Organic Crystals;* Oxford University Press, 2nd ed., 1999, pp. 673-699.
[22] D.J. Sandman, in: *Polymeric Materials Encyclopedia;* J.C. Salamone, Ed., CRC Press, 1996, Vol. 2. pp. 1468-1480.
[23] W.D. Huntsman, in:*The Chemistry of Functional Groups, Supplement C;* S. Patai, Z. Rappoport, Eds., 1983, Wiley.
[24] H. Nakanishi, in: *Polymeric Materials Encyclopedia;* J.C. Salamone, Ed., CRC Press, 1996, vol.10, pp. 8393-8398.
[25] J. Xiao, M. Yang, J.W. Lauher, F.W. Fowler, *Angew. Chem. Int. Ed.*, **2000**, *39*, 2132.
[26] J.S.Brodkin, B.M. Foxman, W. Clegg, T. Cressey, D.R. Harsbron, P.A. Hunt, and B.P. Straughan, *Chem. Mater.* **1996,** *8*, 242-247..
[27] D.J. Sandman, C.S. Velazquez, G.P. Hamill, B.M. Foxman, J.M. Preses, R.E. Weston, Jr., *Mol. Cryst. Liq. Cryst.,* **1988**, *156*, 103-108.
[28] V. Shivshankar, H. Moon, D.J. Sandman, *Mol. Cryst. Liq. Cryst.,* **1998**, *313*, 367-371.
[29] U. Drechsler, D.J. Sandman, B.M. Foxman, *J. Chem. Soc., Perkin Trans. 2*, **2001**, 581-584.
[30] I-B. Kim, M.A. Rixman, Z-H. Tsai, D. Wu, J.M. Njus, J.C. Stark, J. Hankins, D.J. Sandman, *Macromolecules*, **2001**, *34*, 7576-7578.
[31] Z-H. Tsai, D.J. Sandman, *J. Porphrins Phthalocyanines*, **2001**, *5*, 564-568.
[32] D.J. Sandman, M.A. Rixman, Z.-H. Tsai, D. Wu, I-B. Kim, *Polymer Preprints*, **2000**, *41* (2), 1215-1216.
[33] R.W. Begland, D.R. Hartter, F.N. Jones, D.J. Sam, W.A. Sheppard, O.W. Webster, F.J. Weigert, *J. Org. Chem.*, **1974**, *39*, 2341-2350.
[34] J.T. Mague and E.E. Eduok, *J. Chem. Crystallogr.*, **2000**, *30*, 311-320.
[35] I-B. Kim, B.M. Foxman, J. Njus, D. J. Sandman, *Polymer Preprints*, **2001**, *42* (2), 459-460.
[36] D.J. Sandman, G.M. Carter, Y.J. Chen, S. Tripathy, and L. Samuelson, in: *"Molecular Electronic Devices II"*, F.L. Carter, Ed., Marcel Dekker, 1987, pp. 507-521.
[37] J.L. Foley, L. Li, and D.J. Sandman, *Chem. Mater.*, **1998**, *10*, 3984-3990.
[38] J.L. Foley, L. Li, D.J. Sandman, M.J. Vela, B.M. Foxman, R. Albro, and C.J. Eckhardt, *J. Am. Chem. Soc.,* **1999**, *121*, 7262-7263.
[39] M. Morrow, K.M. White, C.J. Eckhardt, and D.J. Sandman, *Chem. Phys. Lett.*, **1987**, *140*, 263-269,
[40] J.M. Pigos, Z. Zhu, and J. Musfeldt, *Chem. Mater.*, **1999**, *11*, 3275-3278.
[41] D.J. Sandman, B.S. Elman, G.P. Hamill, J. Hefter, and C.S. Velazquez, in: *Crystallographically Ordered Polymers*, D.J. Sandman, Ed., American Chemical Society Symposium Series, Volume 337, 1987, pp. 118-127.
[42] H. Eckert, J.P. Yesinowski, D.J. Sandman, and C.S. Velazquez, *J. Am. Chem. Soc.,* **1987**, *109*, 761-768.
[43] D.J. Sandman, Y.J. Chen, B.S. Elman, and C.S. Velazquez, *Macromolecules*, **1988**, *21,* 3112-3113.
[44] D.J. Sandman, B.S. Elman, G.P. Hamill, C.S. Velazquez, and L.A. Samuelson, *Mol. Cryst. Liq. Cryst.*, **1986**, *134*, 89-107.
[45] H.S. Nalwa, H. Kasai, S. Okada, H. Oikawa, H. Matsuda, A. Kakuta, A. Mukoh, and H. Nakanishi, *Adv. Mater.* **1993**, *5*, 758-760.
[46] H. Kasai, H.S. Nalwa, S. Okada, H. Oikawa, and H. Nakanishi, in: *Handbook of Nanostructured Materials and Nanotechnology*, H.S. Nalwa, Ed., Academic Press, **2000**, pp. 433-473.
[47] J.-M. Lehn, *Supramolecular Chemistry,* VCH Weinheim, 1995, pp. 139-144.
[48] J.-A. He, K. Yang, J. Kumar, S.K. Tripathy, L.A. Samuelson, O. Oshikiri, K. Katagi, H. Kasai, S. Okada, H.

Oikawa, and H. Nakanishi, *J. Phys. Chem. B* **1999**, *103*, 11050-11056.
[49] S.H. Hankin and D.J. Sandman in: *Structure-Property Relations in Polymers*, M.W. Urban and C.D. Craver, Eds., American Chemical Society Advances in Chemistry Series, vol. 236, 1993, pp.243-262.
[50] G.J. Exarhos, WM. Risen, Jr., and R.H. Baughman, *J. Am. Chem. Soc.* **1976**, *98*, 481.
[51] R.R. Chance, R.H. Baughman, H. Müller, and C.J. Eckhardt, *J. Chem. Phys.* **1977** *67*, 3616-3618.
[52] D.J. Sandman, B.S. Elman, S.K. Tripathy, and L.A. Samuelson, *Synthetic Metals*, **1986** *15*, 229-235.
[53] R.H. Baughman and R.R. Chance, *J. Appl. Phys.*, **1976**, *47*, 4295.
[54] M.J. Downey, G.P. Hamill, M. Rubner, D.J. Sandman, and C.S. Velazquez, *Die Makromol. Chem.* **1988**, *188*, 1199-1205.
[55].D.J. Sandman and Y.J. Chen, *Polymer*, **1989** *30*, 1027-1031.
[56] A.J. Campbell and C.K.L. Davies, *Polymer* **1995**, *36*, 975.
[57] R.R. Chance, *Macromolecules*, **1980**, *13*, 396.
[58] H. Eckhardt, D.S. Boudreaux, and R.R. Chance, *J. Chem. Phys.*, **1986**, *85*, 4116.
[59] M.F. Rubner, D.J. Sandman, and C.S. Velazquez, *Macromolecules*, **1987**, *20*, 1296-1300.
[60] H.Tanaka, M. Gomez, A.E. Tonelli, and M.K. Thakur, *Macromolecules*, **1989**, *22*, 1208-1215.
[61] H.W. Beckham and M.F. Rubner, *Macromolecules*, **1993**, *26*, 5192-5197; 5198-5201.
[62] C.A. Sandstedt, C.J. Eckhardt, M.J. Downey, and D.J. Sandman, *Chem. Mater.*, **1994**, *6*, 1346-1350.
[63] D.-C. Lee, S.K. Sahoo, A.L. Cholli, and D.J. Sandman, *Macromolecules*, **2002**, accepted.
[64] G. Wenz, M.A. Müller, M. Schmidt, G. Wegner, *Macromolecules*, **1984**, *17*, 837-850.
[65] R. Gleiter and R. Merger in: *Modern Acetylene Chemistry,* P.J. Stang and F. Diederich, Eds., VCH Weinheim, 1995, pp. 285-319.

New Poly(silanes-siloxanes) via Hydrosilation in Supercritical CO_2 and Subsequent Crosslinking

*Hui Zhou, Sreenu R. Venumbaka, John W. Fitch III, Patrick E. Cassidy**

Shell Center for Polymer Science and Technology. Department of Chemistry and Biochemistry Southwest Texas State University, San Marcos, TX 78666, USA

Summary: In order to obtain crosslinkable, low dielectric, low Tg materials, the hydrosilation polymerization was carried out in supercritical carbon dioxide (SC CO_2) using Karstedt's catalyst, and a series of new fluorine-containing poly(carbosilane/siloxane) were synthesized and characterized. The molecular weights of polymers obtained in SC CO_2 were notably greater than those obtained in benzene. Hydrolysis and thermal curing of these polymers leads to materials considerably higher molecular weight which retain their low Tg values.

Introduction

Supercritical carbon dioxide has recently been the subject of extensive investigation for application as an extraction solvent and as a reaction medium. The driving force in these research efforts is the replacement of conventional organic solvents, many of which are toxic, with an inexpensive environmentally benign solvent. Although SC CO_2 has some inherent disadvantage as a potential reaction medium (poor solvent for polar molecules and most high polymers), the potential advantages to using it as a solvent and reaction medium are enormous. These advantages include its low cost, ready availability, easily accessible critical point, tunable density and solvent properties (through temperature and pressure variation), and ease of separation from reaction products. In addition it is a non-flammable, non-toxic naturally occurring component of the atmosphere, which is readily recycled. The use of SC CO_2 as a polymerization solvent, especially for free-radical polymerization, has received extensive study, has been extensively reviewed and remains an area of active investigation.[1]

One potentially important polymerization reaction, hydrosilation, has not been extensively studied in SC CO_2, even though this reaction has received considerable recent attention as a tool for synthesizing novel polymeric architectures. Examples include linear[2-4], comb[5], dendritic[6], hyperbranced[7-8] and crosslinkable[9-10] polymers. The first high molecular weight polymer has been synthesized by using Karstedt's catalyst, platinum-divinyltetramethyl-disiloxane (Pt-DVTMD), in conventional organic solvents[11].

As part of an ongoing research interest in crosslinkable, low dielectric, low Tg materials we

 CCC 1022-1360/00/$ 17,50+.50/0

have utilized Karstedt's catalyst system to conduct hydrosilation polymerization reactions in SC CO_2[12]. In this paper we report more fully on that earlier communication and also report the synthesis and characterization of crosslinkable poly(carbosilane/siloxane)s by hydrosilation in toluene and SC CO_2.

Results and Discussion

Fluorine-Containing Poly(carbosilane/siloxane)s

1,3-Bis(hexafluoro-2-hydroxy-2-propyl)benzene was selected as the source of fluorine in this series of polymers. It was first converted either to its diallyl ether derivative, **1,** or its bis(9-decenyl)ether derivative, **2** (**Scheme 1**). Model compound reactions with phenyldimethylsilane were used to confirm that 1 and 2 undergo

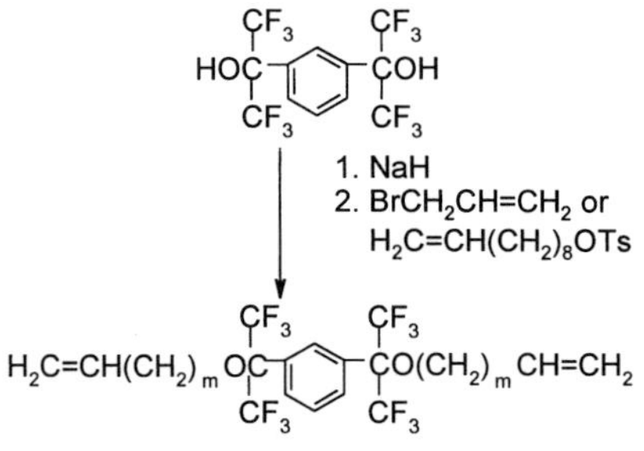

1: m=1 ; **2**: m=8

Scheme 1

hydrosilation in SC CO_2 using either chloroplatinic acid or Karstedt's catalyst, (Pt-DVTMD), which was selected as the catalyst system for subsequent polymerization studies because of its good solubility in SC CO_2.

Monomers **1** and **2** have been reacted with selected disilane monomers, H-Y-H, to give two series of fluorine-containing polymers (**Scheme 2**). In all examples in both series of polymers, the samples prepared in SC CO_2 have greater molecular weights and, therefore, greater degrees of polymerization then analogues prepared in benzene (Table 1). This result indicates that SC CO_2 is a viable substitute solvent for this important reaction. It is also noted that polymers in the series in which x=3 (three methylene groups) are generally of lower molecular weight than analogues in which x=10 (ten methylene groups), and this molecular weight difference appears to be greater in SC CO_2 than in benzene. The origin of this

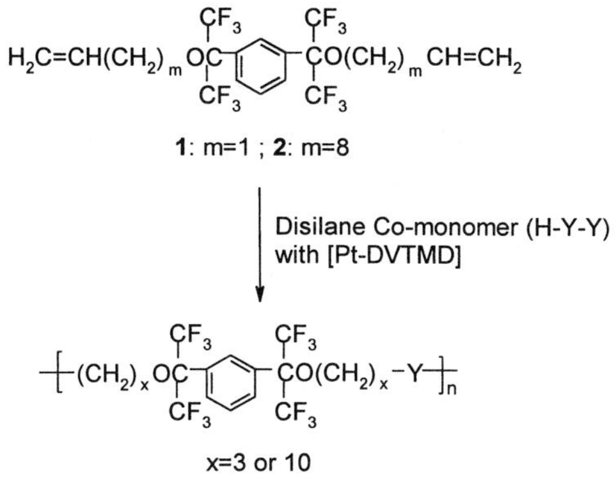

Scheme 2

Table 1. Polymers Molecular Weights – Benzene vs SC CO_2

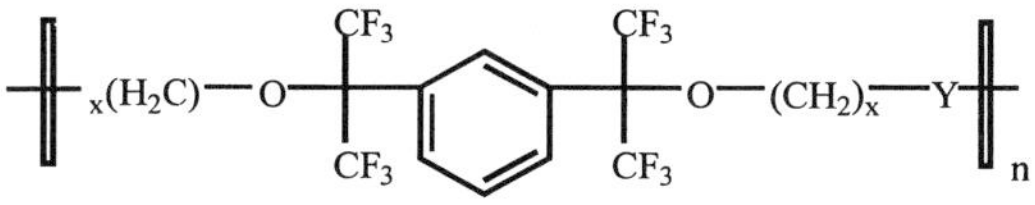

Y	x	Mn*		Mw*		Polydispersity	
		Benzene	SC CO_2	Benzene	SC CO_2	Benzene	SC CO_2
$-Si(CH_3)_2-(CH_2)_2-Si(CH_3)_2-$	3	5200	8000	6600	10900	1.29	1.35
	10	7600	10700	13000	13300	1.70	1.25
$-Si(CH_3)_2-C_6H_4-Si(CH_3)_2-$	3	3500	5300	5100	8600	1.45	1.60
	10	6400	24600	10000	41700	1.53	1.69
$-Si(CH_3)_2-O-Si(CH_3)_2-$	3	2700	3600	3500	4800	1.34	1.28
	10	5900	10000	12400	18700	2.10	2.05
$-(Si(CH_3)_2-O)_3-Si(CH_3)_2-$	3	4900	10000	6300	15000	1.28	1.50
	10	4800	5100	9700	16700	2.01	1.60
$-(Si(CH_3)_2-O-Si(CH_3)_2-(CH_2)_2)_2-$	3	2400	4000	3000	5600	1.24	1.38
	10	9400	10600	13300	24900	1.40	2.34

* by GPC

observation is not clear at this time but could be related to the difference in the steric requirements of monomers **1** and **2**. Monomer **1**, in which the double bond is separated from the bulky hexafluoroisopropoxy groups by a single methylene group, would be expected to have greater steric requirements which could affect its ability to coordinate to the platinum catalyst. Polydispersities are somewhat scattered for both series of polymers but trend higher in the polymers wherein x=10 than in the x=3 analogues. The thermal stability of both series is only moderate as measured by TGA (Table 2). All of the polymers in both series have Tg values below 0°C and exist as colorless viscous liquids (Table 2). The lowest Tg values are found in the x=10 series of polymers containing disiloxane units in the backbone as expected[13].

Therefore, these polymers were selected for further synthetic elaboration into a crosslinkable polymer which retains the low Tg value.

Table 2. Thermal Properties of Polymers

Y	x	Tg (°C)	TGA (10% wt loss)		% Char Yields (600°C)	
			Air	Argon	Air	Argon
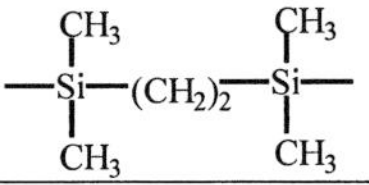	3	-26	287	291	2	0
	10	-38	330	397	2	1
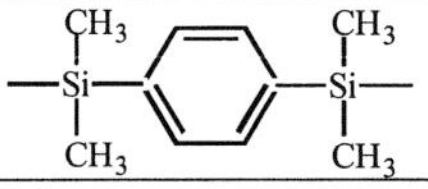	3	-41	321	307	4	0
	10	-32	289	306	2	0
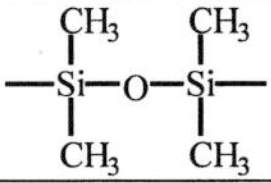	3	-53	249	265	1	1
	10	-67	314	376	4	1
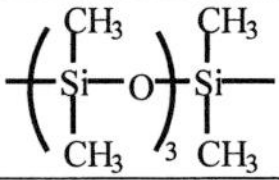	3	-48	317	350	4	0
	10	-59	307	354	6	0
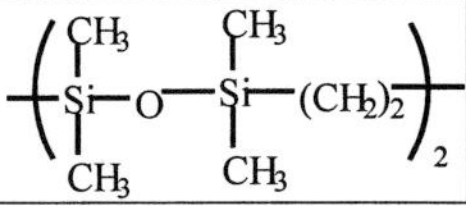	3	-50	268	283	4	1
	10	-59	336	368	3	5

Crosslinkable Ethoxysilane-Modified, Fluorine-Containing Poly(carbosilane /siloxane)s

Crosslinkable ethoxysilane-modified_polymers were prepared as before from monomer **2** except that 1,1,3,3-tetraethoxy-1,3-divinyldisiloxane was added in place of 5% of the stoichiometric amount of the selected disilane, H-Y-H. This mixture was then reacted with disiloxane-containing disilanes in either toluene or SC CO_2 (**Scheme 3**).

The properties of these new polymers are summarized in Table 3. Somewhat higher molecular weight polymers were obtained in toluene solution although the reaction also proceeds in SC CO_2. No attempt was made to optimize reaction parameters in either solvent.

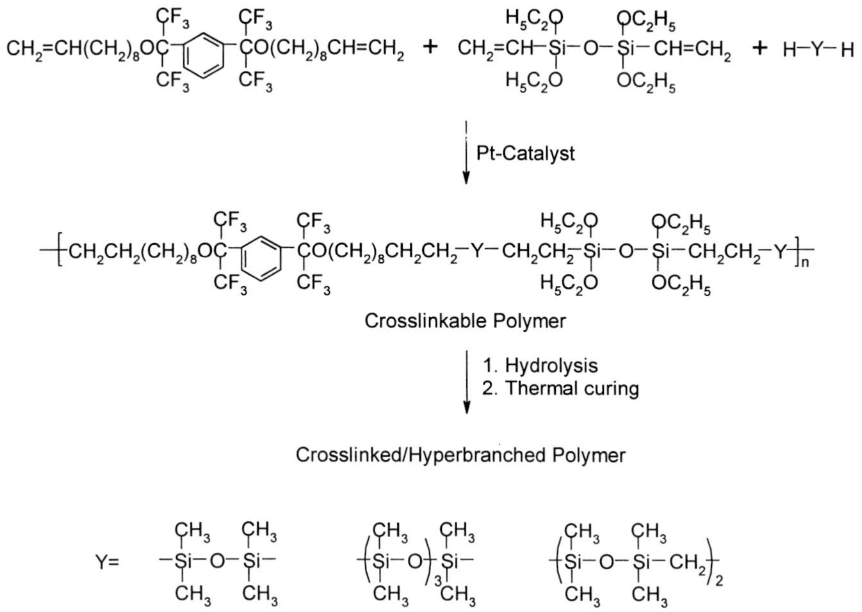

Scheme 3

Table 3. Polymer Properties Prior to Crosslinking (Chain Extension)

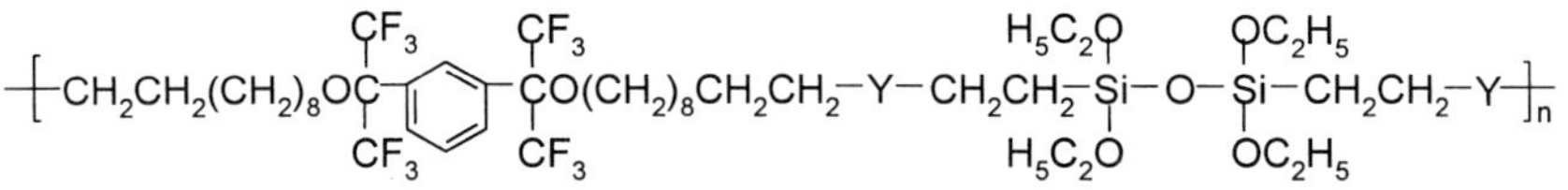

Y	Mw[1] x10^{-3}	PD	Tg [2] (°C)	TGA (°C 10% wt loss) Air	Argon	% Char Yield (@ 600°C) Air	Argon
a	10 SC CO_2	1.3	-54	278	304	7	3
	14 Toluene	1.8	-49	343	398	3	1
b	24 SC CO_2	2.4	-61	309	357	7	2.0
	41 Toluene	3.0	-63	360	416	4	1
c	38 Toluene	2.5	-56	346	405	4	1

[1.] Measured by GPC; [2.] Measured by DSC

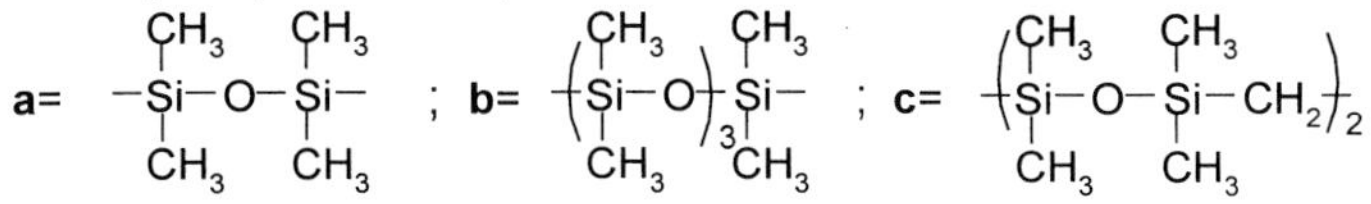

Polydispersities which ranged from 1.3 to 3.0 were more narrow in the two samples prepared in SC CO_2 than in analogous samples prepared in toluene. Thermal stability in this series of polymers is moderate and comparable to the noncrosslinkable analogues (Table 2). Importantly, glass transition temperatures remain low (-49° to -63°) in this series as expected for polymers with very flexible backbones.

The ethoxysilane-modified polymers are readily hydrolyzed by heating in wet diglyme (2% H_20) at 80°C for 24h. The uncured hydrolyzed polymers exhibited a broad IR band at 3200-3700 cm^{-1} which is characteristic of OH stretching in the silanol functionalities formed during hydrolysis. The polymers were then cured *in vacuo* at 135°C for 48h, after which time the intensity of the OH band was greatly diminished, but not totally absent. This thermal cure results in a considerable increase in molecular weight as condensation of silanol groups occurs.

Table 4. Polymer Properties[1] – After Crosslinking (Chain Extension)

Y	$Mw^2 x10^{-3}$	PD	Tg^3 (°C)	TGA (°C 10% wt loss)		% Char Yield (@ 600°C)	
				Air	Argon	Air	Argon
a	16 SC CO_2	2.3	-50	280	347	8	2
	56 Toluene	5.4	-46	320	397	5	0
b	113 SC CO_2	9.0	-60	309	380	8	2
	--[4] Toluene	--[4]	-59	346	417	4	1
c	156 Toluene	8.2	-55	296	403	8	1

1. Cured at 135 °C for 48 h in vaccum; 2. measured by GPC; 3. measured by DSC
4. Dielectric constant = 2.63 @10 GHz

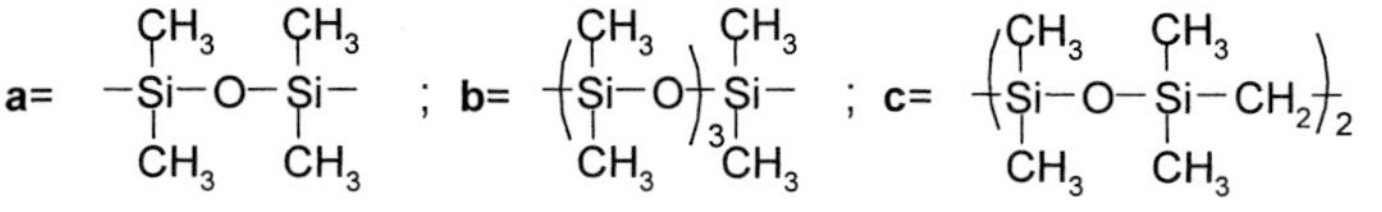

The properties of the thermally cured hydrolyzed polymers are summarized in Table 4. It is noted that the increases in molecular weight of these polymers is accompanied by a large increase in polydispersity as the silanol groups are condensed to disiloxane linkages. However, with the exception of one example, the polymers remain soluble suggesting that the establishment of the disiloxane bonds leads primarily to chain extension rather than crosslinking. It is also important to note that the glass transition temperature of the polymers is essentially unchanged. Even in the case where the polymer became insoluble after hydrolysis and thermal cure, the value of Tg only changed by 4°C., from -63 to -59°C. The dielectric constant of this rubbery material was determined to be 2.63 at 10GHz.

Conclusions

It can be concluded that hydrosilation reactions in SC CO_2 proceed readily without complications. Using hydrosilation in SC CO_2 and in toluene, several new low Tg fluorinated poly(carbosilane/-siloxane)s with crosslinkable (or chain extendable) units have been synthesized. Hydrolysis and thermal curing of these polymers leads to materials of considerably higher molecular weight which retain their low Tg values. The dielectric constant of one of the cured polymers was determined to be 2.63 at 10GHz.

Acknowledgements

The authors are grateful to the Environmental Protection Agency and the National Aeronautics and Space Administration, Langley Research Center for the financial support of this research.

[1] Cooper, A.I. J. Mater. Chem. 2000, 10, 207 and references cited therein.
[2] Budde, H; Hoering, S., Macromol. Chem. Phys. 1998, 199, 2541.
[3] Sargent, J.; Weber, W.P., Macromolecules 1999, 32, 2826.
[4] Nakagawa, Y.; Miller, P.J.; Matyjaszewski, K., Polymer 1998, 39, 5163.
[5] Chiang, W.Y.; Hong, L.D. J. Polym. Sci., Part A: Polym. Chem. 2000, 38, 1609.
[6] Jaffres, P.A.; Morris, R.E., J. Chem. Soc., Dalton Trans. 1998, 16, 2726.
[7] Shim, J.S.; Asthenia, S.: Omar, N.; Kennedy, J.P., J. Polym. Sci., Part A: Polym. Chem. 1998, 36, 2997.
[8] Miravet, J.F.; Frechet, J.M.J., Macromolecules 1998, 31, 3461.
[9] Mera, A.E.; Wynne, K.J., Polymer Preprint 1998, 39, 806.
[10] Furukawa, N.; Yuasa, M.; Kimura, Y., Polymer 1999, 40, 1853.
[11] Dvornic, P.R.; Gerov, V.V.; Govedarica, M.N., Macromolecules 1994, 27, 1068.
[12] Welsch, R., et. Al., Polymer Preprints 2000, 42, 159.
[13] Owen, M.J. Siloxane surface activity. In Zeigler, J.M. and Fearon, F.W.G., Eds; Silicon-Based Polymer Science; ACS Advances in Chemistry series 224; American Chemical Society: Washington D.C. 1990, Chapt. 40, pp. 705-739.

Novel Diacetylene-Containing Polymers for Second Order NLO Applications: Effect of Main Chains and Structure Property Relationships

Susana Hernández[a], *Hirohiko Miura*[a], *Miriam F. Beristain*[a], *Takeshi Ogawa*[a*], *Toshiyuki Watanabe*[b], *Seizo Miyata*[b]

[a]Instituto de Investigaciones en Materiales, Universidad Nacional Autónoma de México, Apartado Postal 70-360, Ciudad Universitaria, México DF 04510, MEXICO E-mail: ogawa@servidor.unam.mx
[b]Faculty of Technology, Tokyo University of Agriculture and Technology, Koganei-shi, Tokyo, JAPAN

Summary: In order to obtain polymers having high second order nonlinear optical (NLO) response, various novel polymers that contain polar dye molecules in side chains and diacetylenic groups in the main chains were synthesized, and their second order NLO properties were studied. Some of these polymers consisting of *para*-benzoate and *para*-cinnamate main chains, showed extremely high NLO coefficients, d_{33} of 200 - 350 pm/V, and the corresponding *meta* polymers showed much inferior NLO coefficients. The polymers having aliphatic main chains had very little NLO response probably because of their low glass transition temperatures and flexibility of main chains. The high SHG responses of these polymers are attributed to their facile orientation of both chromophores and main chains. In the case of *para*-polymers, the main chain orientation of alkoxybenzoate and cinnamate along the chromophore is thought to be a reason for their high NLO coefficients.

Introduction

There are many studies on the polymers, which possess second order nonlinear optical (2-NLO) susceptibility in the last decades, and their majority is concerned with polymers having polar chromophore groups in the side chains.[1-3] Many of them are vinyl polymers such as polymethacrylates containing dye molecules such as Disperse Red 1. However, the monomers having large dye molecules have poor polymerizability, and copolymerization with ordinary monomers is often required, and the concentration of the dye-containing monomers becomes low, and the distribution of the dye monomers in the polymer is not homogeneous due to difference in reactivity of two monomers. In order to obtain polymers with high chromophore concentrations and homogeneous structures it is necessary to obtain polymers consisting of only chromophore-containing monomers. The dye molecules in the polymers are then poled

 CCC 1022-1360/00/$ 17,50+.50/0

for orientation of the chromophores, but the relaxation is inevitable and the NLO effect fades away with time. To avoid the relaxation of poled chromophores, cross-linking is thought to be a solution.[4, 5] However, the introduction of monomer units containing cross-linking groups in copolymerization is not recommendable because it decreases the concentration of chromophore-containing monomers, and the molecular motion between the cross-linked sites cannot be avoided unless very high cross-linking density is attained. Furthermore, irregular cross-linking results in relatively large optical error. Use of high Tg polymers such as polyamides and polyimides has been also studied.[6,7] However, the orientation of chromophores by poling for such high Tg polymers is not easy, and results in poorly orientated polymers having low 2-NLO coefficients.

In order to obtain polymers which might satisfy the above requisitions; high chromophore concentration and cross-linking, Dalton and coworkers have reported four different types of polymers containing chromophore of Disperse Red 19 (Aldrich #21,576-7). A polymer consisting of *p*-phenylenediacrylate main chain showed a d_{33} value of 250 pm/V.[8] They attributed the high NLO coefficients to the high chromophore concentration. However, irradiation of their films with a UV light drastically decreased the absorption height due to the trans-cis transformation of the azo dye.

Diacetylenes are known to polymerize in crystalline state by irradiation or heat to give corresponding crystalline polydiacetylenes. They also undergo polymerization when heated in the amorphous state although the products may not be completely 1,4-polydiacetylenes. Therefore the diacetylene groups are considered to be excellent cross-linking sites. The present authors have reported previously the synthesis of several polymers which contain 2,4-hexadiynylene groups in the main chains and polar chromophores in their side chains,[9,10] and also a few results of the 2-NLO measurements of some of the polymers.[11] In this work, as continuation of the previous work, more polymers which contain diacetylene groups in main chains and polar chromophores in side chains, were synthesized and characterized, and the effects of main chain structures on high 2-NLO coefficients of some of these polymers are discussed in this article.

Experimental

Synthesis

The synthesis of the polymers consisting of the benzoates main chains, were reported in the previous papers.[9, 10] Their structures are shown with Scheme 1.

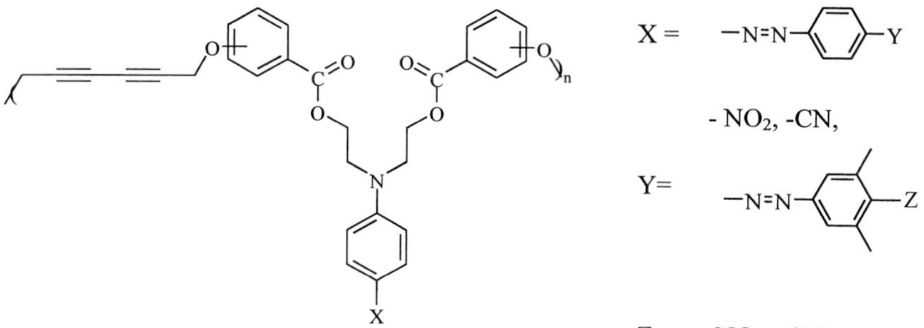

Z= $-NO_2$, -CN

Scheme 1

The polymers having aliphatic main chains have the following structures (Scheme 2):

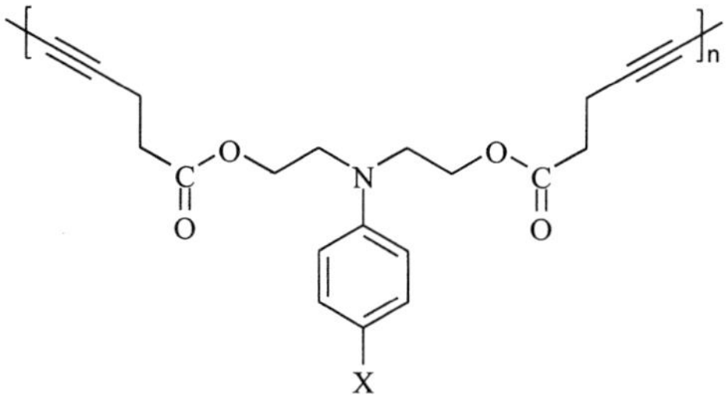

Scheme 2

They were synthesized by the reaction of pentynoyl chloride with the *N,N*-diethanol amino dyes, and the bisacetylenic monomers were polymerized by the oxidative coupling reaction similarly to the cases of the benzoates.

The poly(cinnamates) containing the dye Disperse Red 19 shown with Scheme 3, were synthesized by the reaction of *para*- and *meta*-cinnamic acids or their methylesters with propargyl bromide, followed by the chlorination with thionyl chloride.[12] The propargyloxy cinnamoyl chlorides were then reacted with Disperse Red 19 to obtain the corresponding bisacetylenic monomers, which were polymerized at room temperature in N-methylpyrrolidinone (NMP) for 10 hours by the oxidative coupling reaction using copper (I) chloride and *N,N,N'N'*-tetramethylethylenediamine as catalysts, gently bubbling oxygen into the system. The polymers were precipitated in methanol and purified by reprecipitation from DMF/methanol system. The polymers had inherent viscosity of 0.3-0.7 at 25 °C in NMP.

Films with excellent optical quality were obtained by spin coating from DMF solution at 40 - 50 °C. Langmuir-Blodgett membranes could be prepared from chloroform solutions. The *para*-polymers crystallizes when kept in DMF solution for several hours, and the crystalline polymer is no longer soluble in DMF, but soluble in NMP.

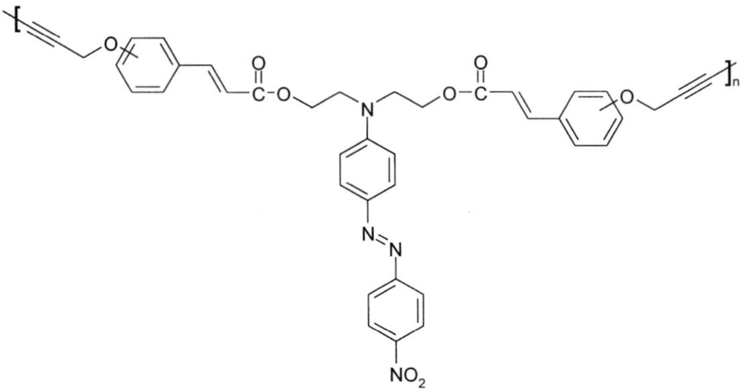

Scheme 3

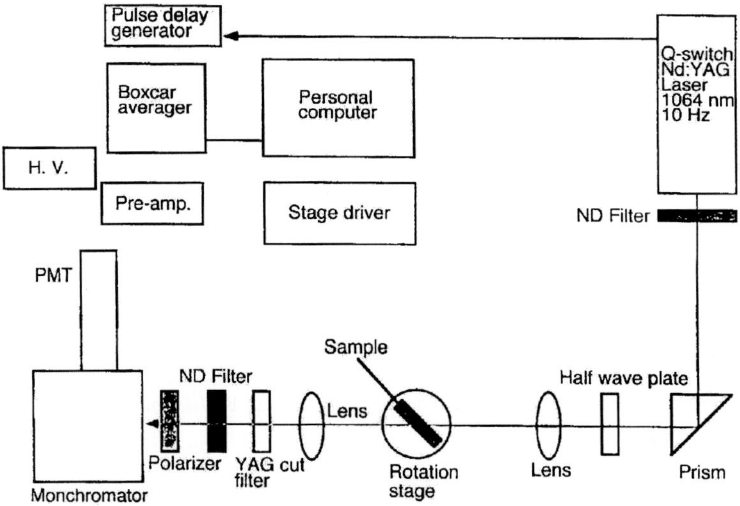

Fig. 1. Schematic diagram of SHG measurement set up.

The polymers were characterized by DSC, NMR, IR. The irradiation of films was carried out using a 400W medium pressure Hg lamp of Ace Glass Co.

The refractive indexes and film thickness were determined by prism coupling and elipsometry.

The 2-NLO coefficients were measured by the set-up shown in Fig.1. The chromophore density was calculated by the ratio of molecular weight of chromophore over polymer molecular weight.

Results and Discussion

The poly(benzoates) and poly(pentynoates) are shown in Table 1. The hyperpolarizability of the chromophores used in this work, are shown in Table 2, and the 2-NLO coefficients are shown in Table 3.

Table 1. Poly(benzoates) and poly(pentynoates) of which NLO properties were determined.

Polymer	Main Chain	Chromophore	Tg (DSC)
1	*p*-benzoate	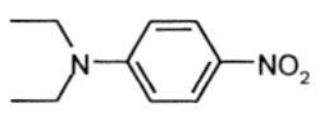	100
2	Pentynoate		30
3	*p*-benzoate	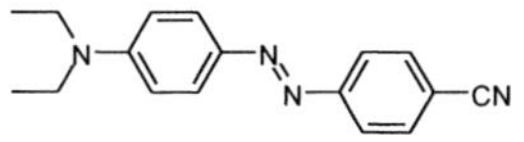	121
4	*m*-benzoate		50
5	*p*-benzoate	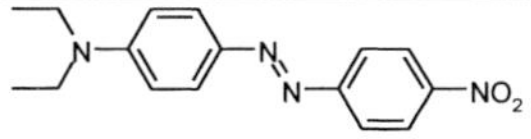	100
6	*p*-benzoate	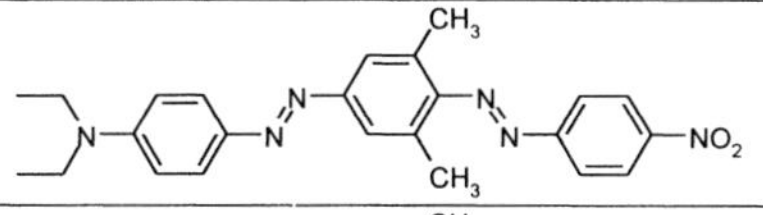	115
7	*p*-benzoate	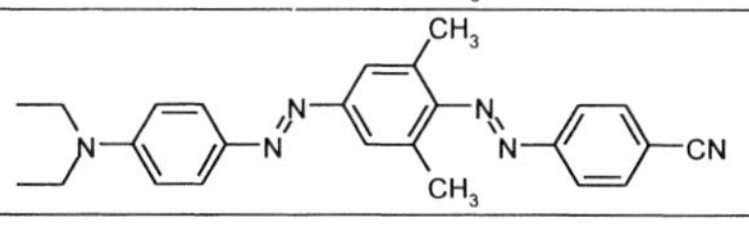	120
8	Pentynoate		63

It can be seen clearly that the aliphatic main chain polymers have poor NLO susceptibility compared with those having benzoate main chains, although same chromophores are attached (Polymers 1 and 2). This is probably because the flexible aliphatic main chain polymers have low T_g and the orientation of the chromophore cannot be maintained, and rapid relaxation may be taking place. A significant difference in d_{33} values is observed between the Polymers 3 and 4. The T_g of Polymer 3 (*para*-benzoate) is 121 °C and that of Polymer 4 (*meta*-benzoate) is 50°C, and the relaxation of chromophore will take place more rapidly in the case of Polymer 4. Another possible explanation is that the Polymer 3 contains *para*-alkoxy benzoate group, which will respond to the electric field, and it is also orientated in parallel with the chromophores, and the whole polymer becomes an orientated material. The Fig. 2 shows the lowest energy conformers of model molecules optimized at HF/6-31G level for the *para*- and *meta*-poly(benzoates). It can be seen that the *para*-polymer can be more easily orientated. Imase, et al have reported that strong SHG activities are observed for polar rod-like polyesters, such as poly(4-hydroxybenzoic acid).[13] The fairly large hyperpolarizability of the aromatic polyester is ascribed to the large electronic transition moment. Therefore, in the case of poly(*para*-benzoates), such effect can be taken into account for their high 2-NLO coefficients.

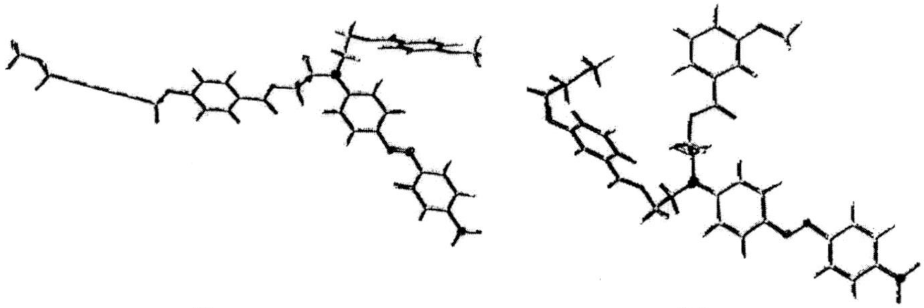

Para　　　　Meta

Fig. 2. Lowest energy conformers of model molecules optimized at HF/6-31G level.

Table 2. Dipole moment and molecular hyperpolarizability of the chromophores calculated by MOPAC-PM3.

CHROMOPHORES	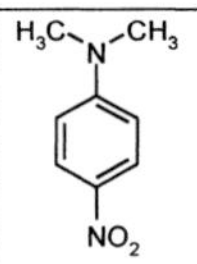	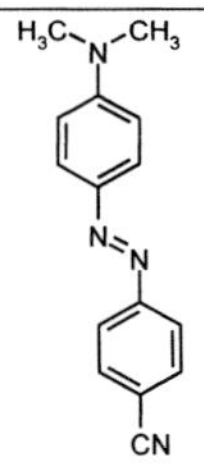	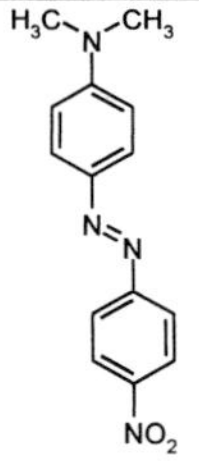	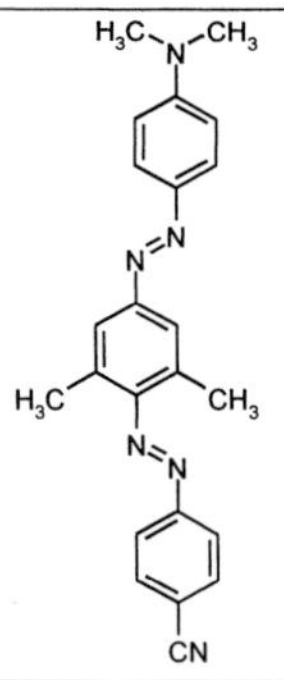	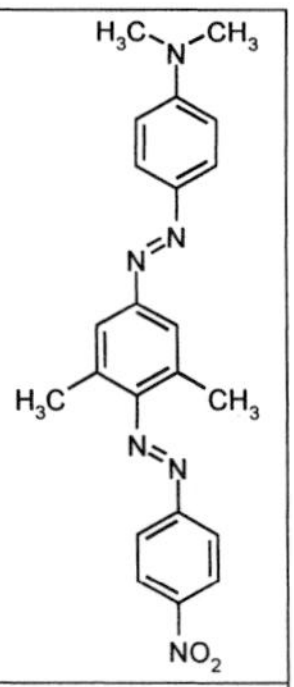
A	7.01	5.65	8.11	4.88	7.75
B	8.32	22.8	26.9	14.4	41.10
C	27.0	87.0	98.3	77.2	184

A : Dipole moment/debye. **B:** Hyperpolarizability (ω=0)/10^{-30} esu.
C: Hyperpolarizability (1064 nm)/10^{-30} esu

A tremendous difference is observed between Polymers 6 and 7. The former showed the highest ever 2-NLO coefficient, d_{33}, of 350 pm/V. The both polymers have *para*-benzoate main chain, and the only difference is the electron acceptor groups nitro and cyano. The high value of Polymer 6 is due to the high molecular hyperpolarizability of the chromophore, together with the effect of the *para*-benzoate main chain. In the case of Polymer 7, the low d_{33} value may be attributed to the lower dipole moment and hyperpolarizability of the cyano-chromophore. Further studies are being made to clarify this problem. In the case of Polymer 8, the low T_g of the pentynoate main chain cannot maintain the orientation of the chromophore.

Table 3. NLO coefficients of novel NLO polymers

Sample	λ_{cutoff} (nm)	n	NLO coefficient (pm/V) d_{31}	d_{33}	d_{33}/d_{31}	S	Chromophore Density %
Polymer 1	490	1.6865 (532 nm)	7	18	2.6	-	41.5
Polymer 2	606	1.6594 (532 nm)	-	-	-	-	81.7
Polymer 3	610	1.7382 (633 nm)	21	100	4.8	0.59	50.5
Polymer 4	605	1.7328 (633 nm)	11	39	3.5	0.46	50.5
Polymer 5	635	1.8058 (633 nm)	32	224	7.0	0.61	50.9
Polymer 6	675	1.8837 (780 nm)	118	350	3.0	-	58.5
Polymer 7	650	1.7224 (780 nm)	2	12	6.0	0.54	58.2
Polymer 8	668	1.7391 (780 nm)	0.7	2	2.9	-	89.8

Some of the results on the 2-NLO property of poly(cinnamates) are shown in Fig. 3. In this case also, the *para*-polymers showed 4 - 5 times greater d_{33} value than the *meta*-polymer. A copolymer of *para*/*meta* 60/40 mole% showed intermediate value. The results agree with those of the benzoates, the *para*-mains chain being the key for the high NLO property. The *para*-polymer maintained its initial coefficient by nearly 80% after 2000 hrs, while the *meta*-polymer lost by 80% after 2000 hrs. The reason why the *para*-polymers possess high 2-NLO properties is not yet clear, but one possible explanation is that the *para*-main chains can orientate by poling in the same direction to the chromophore orientation, as shown in Fig. 4.

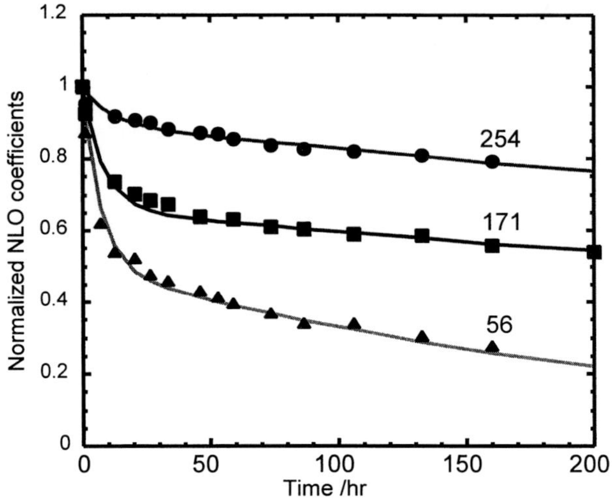

● : 100% *para* — d_{33}=0.10*exp(-t/100) + 0.90*exp(-t/12000)
■ : 60% *para* and 40 % *meta* (mole%) — d_{33} = 0.34*exp(-t/80) + 0.66*exp(-t/10000)
▲: 100% *meta.* — d_{33} =0.5*exp(-t/70) + 0.5*exp(-t/2400)

Fig. 3. NLO coefficients as a function of elapsed time after poling for Poly(hexa-2,4-diynylene-1,6-dioxydicinnamates) containing Disperse Red 19 dye. The d_{33} values in the figure are the initial value after poling.

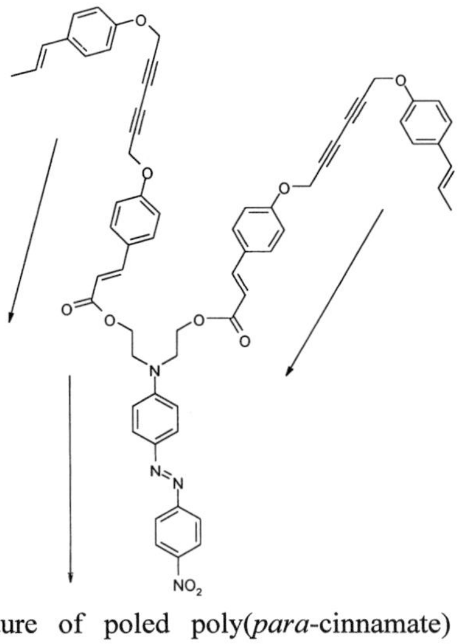

Fig. 4. A schematic picture of poled poly(*para*-cinnamate) indicating orientation of chromophore and main chain component.

Another possibility is that poly(hexa-2,4-diynylene-1,6-dioxydicinnamates) have crystallization tendency. The poly(*para*-cinnamate) having *para*-nitro-N,N-diethanol aniline is highly crystalline and insoluble in organic solvents. Therefore, the rigid structure inhibits relaxation of orientated chromophores.

The discharge dynamic study[14] showed that the *para*-benzoate polymers have a polymer structure very favorable for chromophore orientation by poling, as very little discharge is observed. On the other hand, the *para*-cinnamate polymer showed vigorous discharge below the T_g, and required 4.5 kV or higher voltage for poling the chromophore while the *meta*-cinnamate could be poled with 4 kV. The *para*-cinnamate polymer with Disperse Red 19 crystallizes from its DMF solution. When the solution is left to stand at room temperature for more than 24 hours, its crystalline polymer precipitates, which is soluble only in NMP. Amorphous polymer is recovered by reprecipitation with methanol. Langmuir-Blodgett membranes of the poly(*para*-cinnamate) and poly(*para*-benzoate) are well orientated as observed by the atomic force microscopy of their 2-layer LB membranes.[15] The *para*-benzoate polymer self-assembled on the water surface. It seems that they apparently have a self-assembly character. The d_{33} of the poly(*para*-cinnamate) is somewhat higher than that of poly(*para*-benzoate), 254 and 224 pm/V, respectively. The atomic force microscopy of Langmuir-Blodgett membranes showed that the cinnamate polymer is better orientated than the benzoate analogue. It can be said that the geometrical structures of these 4 polymers (*para*- and *meta*-benzoates and cinnamates) may be quite different, and further studies are being made to clarify the reason for high SHG coefficients.

With respect to the chromophore density, Dalton[8] attributed the high d_{33} values, 108, 250, 19, and 223 pm/V) of their 4 polymers to the high chromophore density (48, 68, 47, and 52%, respectively). However, high chromophore density alone cannot explain high NLO coefficients. In the case of this work also the chromophore densities are high, being around 50%, but d_{33} values depend more on the type of main chains polymers. The type of chromophore is important as can be seen in the comparison among Polymers 1, 3, 5, 6 and 7. It is obvious that when the main chains are the same each other, the greater the hyperpolarizability of chromophores, the greater the 2-NLO effect.

These polymers can be cross-linked thermally through the diacetylene groups in the main chains. However, as can be seen in Fig. 5, the temperature at which the thermal cross-linking takes place is much higher than T_g, and therefore, the orientation cannot be maintained due to vigorous thermal motion. Therefore, simultaneous thermal cross-linking and poling cannot be performed. Dalton and coworkers[8] have irradiated their films with UV light and cross-linked their polymers. In their work, the absorption peak due to the Disperse Red 19 at 468 nm drastically decreased it intensity due to the trans to cis transformation of the azo group by the irradiation. In the case of this work, the trans-to-cis transformation was not appreciable when irradiated with a UV lamp at 100-120°C (poling temperature), as reported in the previous papers for poly(benzoate)[10] and for poly(cinnamate).[12] This is probably due to the thermal cis-to-trans transformation which compete the photochemical trans-to-cis transformation.

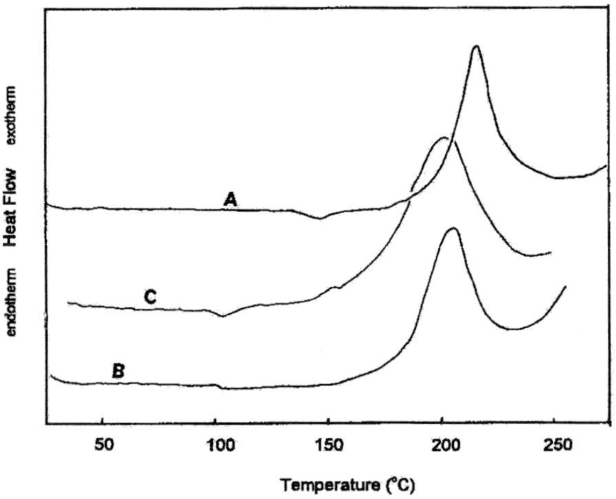

Fig. 5. DSC Curves of Polymer 6 (**A**), Polymer 5 (**B**), and poly(*m*-cinnamate) (**C**) .

After the irradiation the films were completely cross-linked, and converted to thermoset resin films which are totally insoluble in DMF, and did not swell nor pealed off from the substrate when immersed in DMF. The irradiation at room temperature showed consistent decrease in absorbance with irradiation time. This observation suggests that the simultaneous poling and cross-linking can be done satisfactorily.

Conclusion

It was shown that the main chain structure is very important for designing the SHG polymers. It seems that the poly(*para*-cinnamates) and poly(*para*-benzoates) are so far the best candidates for obtaining polymers with high 2-NLO coefficients. The aliphatic main chain polymers such as polypentynoates may not give high NLO coefficients even if they have proper T_g of around 100-120 °C, although their chromophore density is very high. The nitro group was found much better acceptor group than cyano group, as expected from the difference in their hyperpolarizability and dipole moment.

The NLO stability of these polymers is being studied, and the results will be reported in future. It looks that little relaxation takes place for *para*-benzoates and *para*-cinnamates even without cross-linking.

Considering from extremely fine optical quality of the films, very high NLO coefficients, and the NLO durability of these polymers, the *para*-polymers can be satisfactorily used for constructing various photonic devices.

Acknowledgement

This work has been supported financially by the CONACYT (National Council of Science and Technology of Mexico) under contract numbers 25259-A and 33001-U, and by the DGAPA (UNAM Research Fund) under the contract number IN-102401. The simulation of Fig.1 was made by Dr. Sergei Fomine of Instituto de Investigaciones en Materiales, UNAM. Thanks are also due to Carmen Vázquez for her DSC analysis.

[1] H.S.Nalwa and S.Miyata. Edts. "Nonlinear Optics of Organic Molecules and Polymers." CRC Press, Inc. Boca Raton, FL, **1977.**

[2] S.Miyata and H.Sasabe, edts. "Poled Polymers and Their Applications to SHG and EO Devices," Gordon and Breach Sci. Publishers, Amsterdam, **1977**

[3] G.A.Linsay and K.D.Singer, edts. "Polymers for Second-Order Nonlinear Optics," ACS Symposium Series 601, ACS, Wasington, DC **1995**

[4] R.Reck, M.Eich, D.Jungbauer, R.J.Twig, C.G.Wilson, D.Y.Yoon, and G.C.Bjorklund, *SPIE Proc.,* **1989** *1147*, 74.

[5] D.Jungbauer, I.Teraoka, D.Y.Yoon, B.Reck, J.D.Swalen, and R.Twieg, *J.Appl.Phys.,* **1991**, *69,* 8011.

[6] R.D.Miller, D.M.Burland, M.Jurich, V.Y.Lee, P.M.Lundquist, C.R.Moylan, R.J.Twieg, J.I.Thackara, T.Verbiest, Z.Sekkat, J.Wood, E.F.Aust, and W.Knoll, in *Photonic and Optoelectronic Polymers,* pp100-122, ACS Symp. Ser. 672, American Chemical Society, Washington, DC, **1995**.

[7] D.Y.Wenjie, L.A.Gharavi, and L.Yu, , in *Photonic and Optoelectro-nic Polymers,* pp123-132, ACS Symp. Ser.672, AmericanChemical Society, Washington, DC. **1995**

[8] M.Chen, L.Yu, L.R.Dalton, Y.Shi, and W.H.Steiner, *Macromolecules,* **1991**, *24*, 5421.

[9] S.Hernández and T.Ogawa, *Polym. Bull.,* **2001**, *47*, 127.

[10] S.Hernández, M.F.Beristain, and T.Ogawa, *Designed Monomers and Polymers,* **2002**, *5,* No.1, 125.

[11] S.Hernández, T.Ogawa, T.Watanabe, and S.Miyata, *Mol.Cryst.Liq.Cryst.,* **2002**, *374*, 503.

[12] H.Miura and T.Ogawa, *Polym. Bull.,* submitted for publication (2002).

[13] T.Imase, S.Kawauchi, and J.Watanabe, *Mol.Cryst.Liq.Cryst,* **2000**, *346*, 107.

[14] S.Castillo-Rojas and T.Ogawa, unpublished results.

[15] M.P.Carreon and T.Ogawa, unpublished results.

Novel Silicon-Based Alternating Copolymers: Synthesis, Photophysical Properties, and Tunable EL Colors

Hwan Kyu Kim, Kyung Lim Paik, Nam Seob Baek, Youngil Lee*[†], *Katsumi Yoshino*[‡]

National Creative Initiative Center for Smart Light-Harvesting Materials and Department of Polymer Science & Engineering, Hannam University, Daejon 306-791, Korea; Email: hwankkim@mail.hannam.ac.kr
[†]Dongbu Research Council, Daejon 103-2, Korea
[‡] Department of Electronic Engineering, Osaka University, Osaka 565-0871, Japan

Summary: We synthesized novel silicon-based alternating copolymers for tunable electroluminescent (EL) colors by Heck synthetic method. Their thermal, photophysical and electroluminescent properties were studied. Most of them exhibited a blue-green EL color at the operating voltage of lower than 12 V. Unusually, we observed the white EL color from a EL device based on SiPhPVK. From photophysical studies and the time-resolved PL spectroscopies, it might be attributed to the formation of stabilized excited state in SiPhPVK. Furthermore, in order to reduce the operating voltage of their LED with increasing the electron affinity of the main chain in silicon-based alternating copolymers, we synthesized the silicon-based copolymers containing electron transporting oxadiazole units in main chain. We also studied their photophysical and electroluminescent properties.

Introduction

Since electroluminescence from π-conjugated polymer was discovered,[1] the polymeric light-emitting diodes (PLEDs) have attracted much attention because of academic interests and potential utility of this technology in a wide variety of applications such as flat panal displays and light-emitting diodes.[2,3] EL polymeric materials offer a number of advantages, such as low operating voltages, three primary R/G/B colors with the control of π-π* energy gap through the manipulation of the molecular structure, fast response time, high quality of display, and ease of device processability with semiconductor technologies, compared to inorganic EL materials and organic dye molecules.[4,5] However, the realization of potential uses is limited by their low quantum efficiency and EL efficiency. The EL efficiency of

 CCC 1022-1360/00/$ 17,50+.50/0

devices can be affected by the homogeneity of film quality, the injection efficiency of carriers from electrodes to organic materials, and the material's EL efficiency. The two former can be solved by thin film fabricating processes, but the latter can be only achieved by developing materials with high EL efficiency.

Recently, we have studied the development of novel silicon-based copolymers exhibiting blue-white emissive colors. The synthesis of silicon-based alternating copolymers was carried out using well-known Heck reaction.[6-8] Instead of the Wittig method used in the early stage of this research [6, 9], the Heck synthetic routes to the preparation of silicon-based copolymers was used mainly for the following reasons. (1) The Heck route could overcome the problem of low quantum efficiency due to the formation of the triplet state arising from the unreacted or remained aldehyde functional groups in the copolymers obtained from the Wittig reaction. (2) To obtain quantitatively trans-double bond from cis-double bond, the Wittig reaction requires a further post-reaction of isomerization step, achieved by heating the crude polymers with a trace of iodine in toluene. However, the Heck reaction directly produces the desired polymers with trans configuration, which is important to optimize the luminescence efficiency and the emissive wavelength. According to our results, the introduction of organosilicon units into π-conjugated systems also improves their processability and limits the π-conjugation length, resulting blue light emission. And, silicon-based copolymers exhibit the blue LEDs operated at the low voltages, due to the lowering of the LUMO levels and d-orbital participation of silicon atoms.[10] In this paper, we describe the Pd-catalyzed direct synthesis, photophysical, and EL properties of the silicon-based copolymers.

Results and Discussion

We have synthesized the silicon-based alternating copolymers by the Heck reaction of alkyl/aryl-substituted distyrylsilanes with various aromatic or heteroaromatic dibromides, as shown in Scheme 1 and 2. The polymerization results, thermal and photophysical properties of silicon-based copolymers are summarized in Table 1. All the copolymers have the glass transition temperature (T_g) in the range of 94 to 126 °C. They did not show any definite melting points, implying that the silicon-based copolymers could be amorphous. All of the polymers showed good thermal stability up to 300 °C, since a small amount (or less than 5 %) of weight loss began to occur at 300 °C in a nitrogen atmosphere. The maximum absorption wavelengths (λ_{max}) of SiHMPPV and SiPhPPV exhibit at 322 and 365 nm. The absorption spectra of SiHMPVK and SiPhPVK show two peaks around 322 and 356 nm, which are attributed to π-π* transition of the carbazole segments and the π-conjugated segment (see Figure 1). Also the absorption spectra of SiHMFPV and SiPhFPV show the UV absorption bands around 380 nm. The maximum absorption wavelength (λ_{max}) of SiHMThV, SiPhThV and SiPhThThV appears at the longer wavelength of 400, 407, and 416 nm, due to the strong delocalization of the π-conjugated thiophene units, respectively.

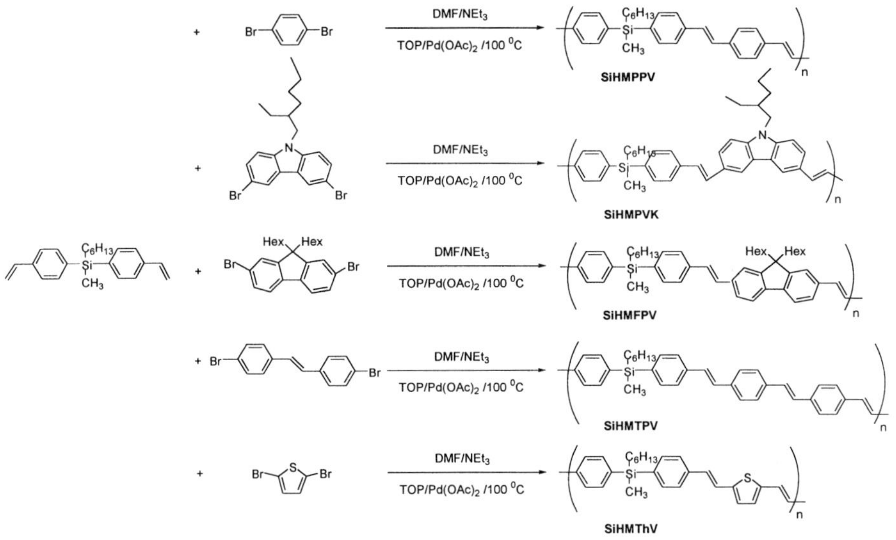

Scheme 1. Synthesis of silicon-based copolymers using alkyl substituted distyrylsilane monomer.

Their photoluminescence (PL) spectra similarly shifts to those observed in the absorption spectra. With an excitation wavelength of 365 nm, the SiHMPPV, SiPhPPV, SiHMFPV and SiPhFPV spectra show a peak around 470 nm, indicating a blue emission. And, the PL spectra of both SiHMPVK and SiPhPVK give a peak in the blue emission region at 440 nm. With an excitation wavelength of 410 nm, the PL spectra of SiHMThV, SiPhThV and SiPhThThV exhibit a peak at 520, 526, and 512 nm in the green region, respectively.

The single-layered light-emitting diode of an Al/SiPhPPV or SiHMPPV/ITO glass was fabricated. The threshold voltage is in the range of 10 to 12 V from the I-V curve. The EL spectrum showed a peak at 450 nm, when the operating voltage of 12 V was applied. Furthermore, the multi-layered light-emitting diodes of an Al/LiAl/SiPhPVK or SiHMPVK/

PANI/ITO glass were fabricated. The threshold voltage is in the range of 6 ~ 12 V. The EL spectrum of SiHMPVK exhibits a broad band around 460 nm in the blue region at the operating voltage of higher than 12 V. However, the EL property of SiPhPVK depends strongly on the applied voltage. Unusually, we observed the white EL color from a EL device based on SiPhPVK at the applied voltage of higher than 11V (see Figure 2). The EL spectrum gives a peak at 467 nm, when the operating voltage of 7 V was applied. When the operating voltage of 11 V was applied, however, the EL spectrum gives a peak at 467 nm and an additional strong EL emissive band in the red region. Therefore, two EL peaks may combine to produce a white EL color at the operating voltage of higher than 11 V. Also, we observed that the intensity of two EL peaks increases with the applied voltage.

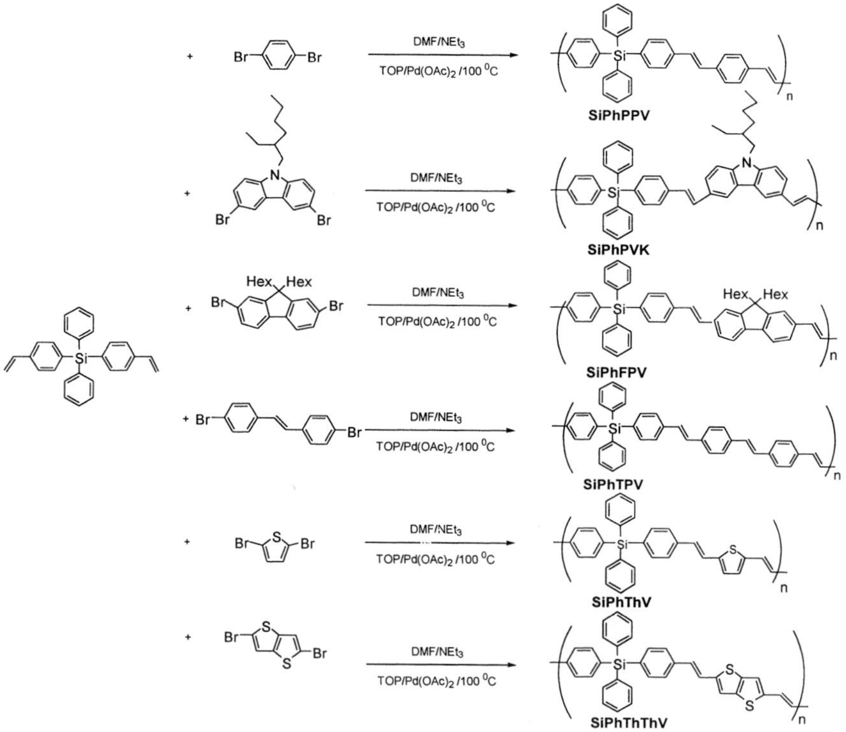

Scheme 2. Synthesis of silicon-based copolymers using aryl substituted distyrylsilane monomer.

In order to understand the origin of an additionally new red EL band, we measured the emission and the excitation spectroscopies of SiPhPVK. [7, 9] Figure 3 (a) shows the PL spectra of SiPhPVK with photoexcitation at two different wavelengths. These two excitation wavelengths at 300 and 400 nm were chosen for selective photoexcitation of carbazole and

Table 1. Summary of polymerization results, thermal and photophysical properties of silicon-based alternating copolymers.

Polymers	Yield (%)	$M_w\times10^{-3}$	T_{ID} (°C)	T_g (°C)	UV(λ_{max}) (nm)	PL(λ_{max}) (nm)	PL efficiency	EL(λ_{max}) (nm)
SiHMPPV[a)]	52	8.5	180	94	356&365	470	-	450
SiHMPVK[a)]	51	2.3	175	106	322&356	440	-	460
SiHMFPV	70	9.8	185	96	378	470	12.7	-
SiHMTPV	66	2.7	200	112	386	470	-	-
SiHMThV	50	10.3	150	94	400	520	-	-
SiPhPPV[a)]	48	6.3	250	127	355&365	470	26.4	450
SiPhPVK[b)]	43	3.7	253	102	325&356	440	-	467& 620
SiPhFPV	77	12.7	250	109	380	476	11.6	-
SiPhTPV	45	4.5	253	110	374&388	463	-	-
SiPhThV	55	13.3	225	112	407	526	5.3	-
SiPhThThV	50	18.6	200	116	416	512	-	-

a) EL data from single-layered LED of Al/SiPHPPV or SiHMPPV/ITO
b) EL data from multi-layered LED of Al/LiAl/SiPHPVK/PANI/ITO

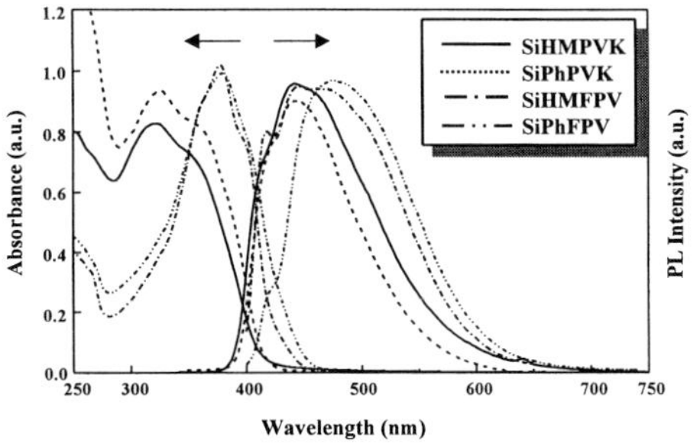

Figure 1. UV-visible and PL spectra of SiHMPVK, SiPhPVK, SiHMFPV and SiPhFPV in $CHCl_3$.

SiPh segments in SiPhPVK, respectively. The PL spectra exhibit the emission maximum band around 440 nm, in addition to the emission band at 520 nm. Upon photoexcitation of carbazole unit at 300 nm, the intensity ratio between these two bands is different from that obtained with photoexcitation of SiPh segment at 400 nm. From this result, we can suggest that the origin of these two emission bands is different. Thus, we measured the excitation spectra in the wavelength range from 280 to 420 nm at the emission wavelengths of 440 and 520 nm. We could observe two excitation spectra were obviously different (see Figure 3 (b)). Considering the above results, it is reasonable to suppose that the shorter wavelength emission band results from the carbazole segments. Meanwhile, it is not straightforward to assign the origin of the emission band at low energy side to the conjugated SiPh segments in SiPhPVK, because no such emission band is found in either PL or EL spectra from a SiPhPPV thin film. Therefore, it may be considered that the longer wavelength emission band in SiPhPVK could result from the formation of either the exciplex between excited state carbazole and PPV segment or the excimer between a ground state and an excited state of carbazole segments. In fact, it is well known that a carbazole unit with small substituent group in the polymer could easily form an excimer between ground and excited states of carbazole units. In our system, however, the longer wavelength emission is mainly due to the optical transition of SiPh main chain because the carbazole unit in our system has a branched and bulky alkyl chain which should hinder the excimer formation between carbazole units. The other probable explanation can be drawn from the stabilized excited state arising from the intramolecular interaction. Since the carbazole unit in SiPhPVK has a high hole affinity, the SiPh unit may possess the electron affinity with respect to the carbazole unit. Thus, we can suggest that the long wavelength emission in SiPhPVK originates from other stabilized excited state, e.g. CT state.

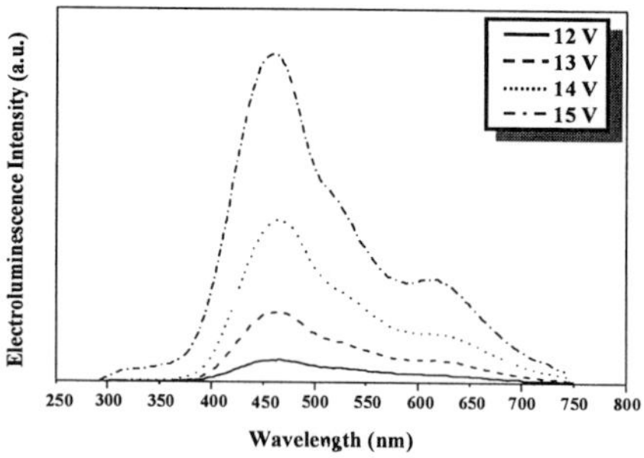

Figure 2. Electroluminescence spectra of SiPhPVK as a function of applied voltages.

In order to gain further insight into the nature of the long wavelength emission, we measured the PL decay profiles of SiPhPVK with photoexcitations at 300 and 400 nm. Figure 4 shows the PL decay profiles of SiPhPVK monitored at 430, 480 and 550 nm. The long wave-

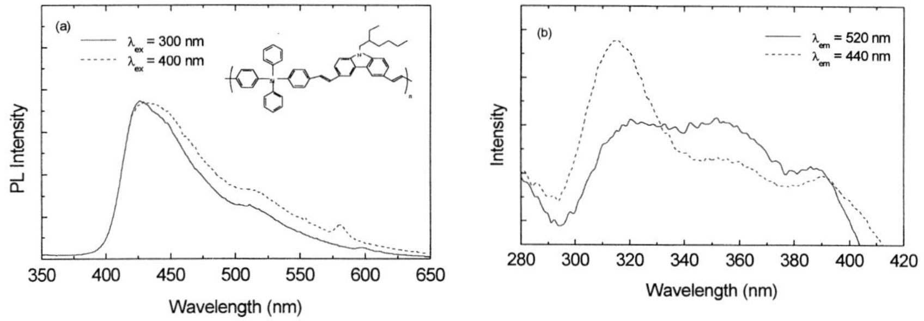

Figure 3. (a) The PL spectra of SiPhPVK with two photoexcitation wavelengths at 300 and 400 nm and (b) the PL excitation spectra for SiPhPVK at two probes of 440 and 520 nm.

length PL lifetime of SiPhPVK, which is believed to be due to the charge transfer process, is approximately 1.5 ns. However, the PL lifetime at 550 nm seems to be too long to be explained in terms of the energy migration process. The long emission lifetime is characteristic feature of the charge transfer formation along with a long tail emission. Thus, we can suggest that the charge transfer process between carbazole and SiPh units in the same chain induces the long wavelength emission of SiPhPVK. It is generally accepted that upon photoexcitation with high energy the PL decay in polymers becomes fast as compared to the photoexcitation with lower energy because of defect sites. However, upon photoexcitation with high energy (300 nm), the PL decay time of SiPhPVK is longer than that obtained by photoexcitation at lower energy (400 nm). Therefore, we can suggest that the charge transfer process is facile upon photoexcitation of a carbazole unit in SiPhPVK as compared with the photoexcitation of SiPh unit.

From time-resolved PL measurement and photophysical studies, we proposed that the additional new PL band, unlikey SiHMPVK, might be attributed to the formation of a certain charge transfer complex, as shown in Scheme 3. [11] The stable resonance structure in the excited state was formed through stabilizing it with phenyl side groups in the SiPh unit of SiPhPVK. Phenyl side groups in the SiPh unit of SiPhPVK behave like electron-withdrawing group.

Similarly, to increase the electron-withdrawing property of the main chain in silicon-based alternating copolymers, we synthesized the silicon-based copolymers containing oxadiazole units (SiHMOXD/Cz 10) to balance the hole-electron charges injected and to see the structure-property relations (see Scheme 4).[12] In the UV-visible spectrum, the absorption maximum wavelength (λ_{max}) of SiHMOXD/Cz 10 appears at 355 and 357 nm in chloroform solution and thin film, respectively.

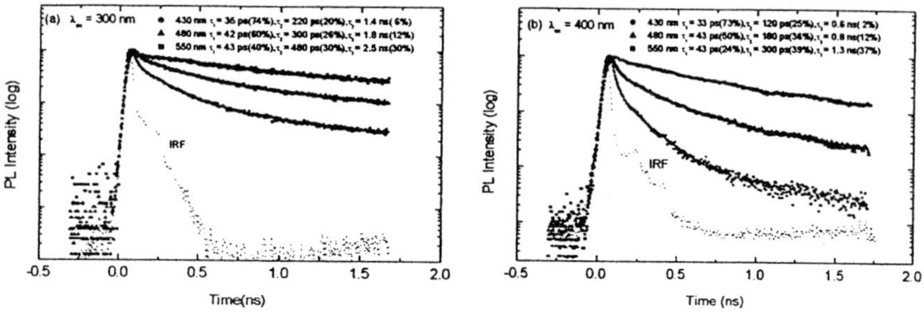

Figure 4. Transient PL decay profiles for SiPhPVK monitored at 430, 480 and 550 nm with two photoexcitation wavelengths of (a) 300 nm and (b) 400 nm.

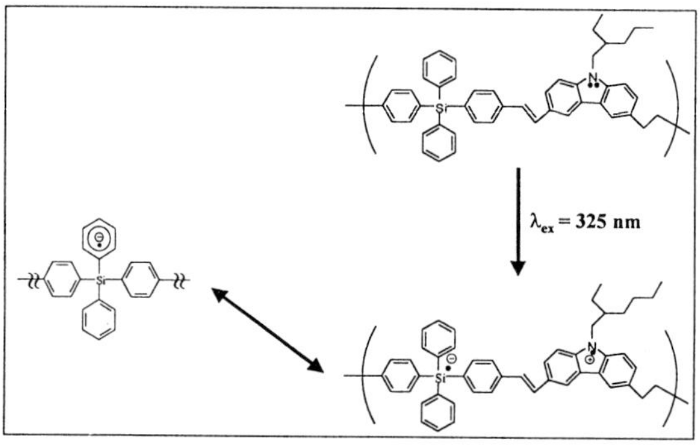

Scheme 3. Proposed scheme for the formation of a charge complex in SiPhPVK.

And, the PL spectra show a strong band around 435 and 451 nm in chloroform and thin film, respectively. The multi-layered light emitting diode of Al (200nm)/Ca (50 nm)/ SiHMOXD/Cz 10 (80nm)/PEDOT (50nm)/ITO glass was fabricated. The I-V curve shows the turn-on voltage of 7 V (see Figure 5). This value is a little lower than any other silicon-based copolymers developed by our research group, due to the more electron affinity of the oxadiazole units. Its EL spectrum, as shown in Figure 6, exhibits a broad band around 460 nm in the blue region and a very broad, weak band in the red region at the operating voltage of 12.5 V. With the applied voltage, the emissive EL bands were red-shifted from blue region to red region. When the higher voltage was applied, however, the broad band in the blue region decreases and a new broad band in red region increases. The new red EL peak is generated only by electric field, since this new red band is exhibited only in EL, but not in PL spectra (see Figure 7). In other word, a certain charge complex, more like electroplex, is formed under a strong electric field inside the device and cannot be produced by photoexcitation. A similar phenomenon has been reported in an oxadiazole-containing conjugated polymer.[13] Therefore, two broad emissive bands combine to produce the white emissive color above 13 V. And, the maximum luminance of the white emissive color was 3.71 cd/m^2 at the applied voltage of 12 V.

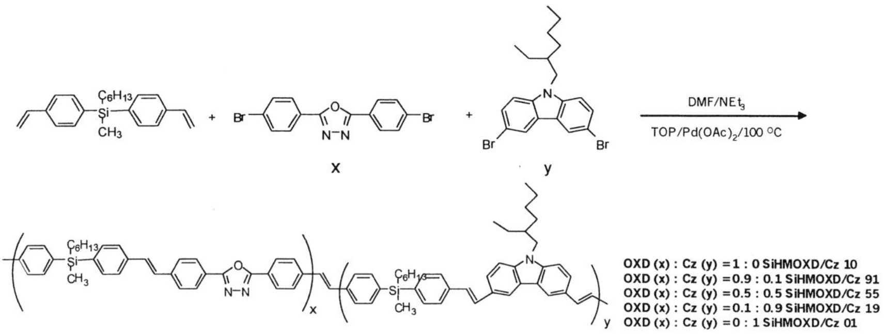

Scheme 4. Synthetic routes of silicon-based copolymers containing oxadiazole units.

And, to overcome the poor solubility of the resulting polymer containing only an oxadiazole unit and to improve the quantum efficiency and luminescent properties, we introduced the carbazole units into the SiHMOXD/Cz 10 with various molar ratios, yielding novel silicon-based copolymers containing both electron-transport oxadiazole and hole-transport carbazole moieties in the main chain (SiHMOXD/Cz xy) (see Scheme 4). [14] The maximum absorption wavelength (λ_{max}) of SiHMOXD/Cz 91, 55, and 19 appears at the wavelength of 357, 360, and 349 nm in film state, respectively. The PL spectra of these polymers exhibit a strong band around 455, 483, and 475 nm in the blue region, respectively. The multi-layered light-emitting diodes with the same device structure of SiHMOXD/Cz xy were fabricated. Their EL properties depend strongly on both the applied voltage and the loading amount of hole-transport carbazole moieties in the present copolymers. The turn-on voltage is in the range of 6 ~ 7 V (see Figure 5). The EL spectra from the device based on the copolymers containing only electron-transport oxadiazole moieties in the main chain (SiHMOXD/Cz 10) exhibit a strong band around 460 nm in the blue region and a very broad, weak band in the red region, at the operating voltage of lower than 12.5 V. With the applied voltage, these emissive EL bands were red-shifted from blue region to red region, exhibiting a very broad, weak band in the blue region and a very strong band in the red region.

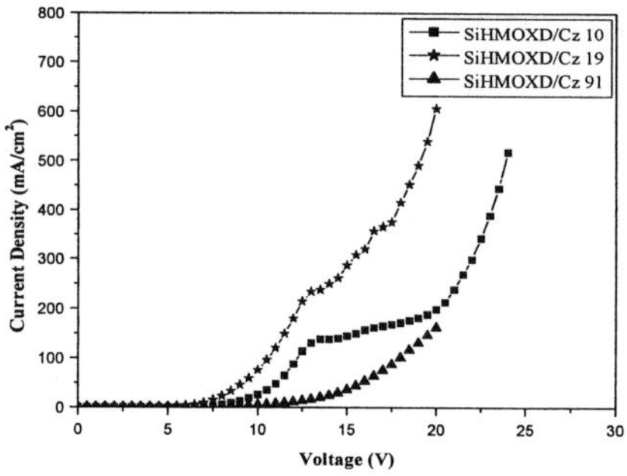

Figure 5. I-V curves of SiHMOXD/Cz 10, SiHMOXD/Cz 91, and SiHMOXD/Cz 19.

Also, the EL spectra from the devices based on SiHMOXD/Cz 91 – 19 have a similar EL spectral feature with the applied voltage. The intensity of a blue EL band at the relatively high operating voltages increases with the loading amount of carbazole units. Eventually, the LED device from the copolymer containing the mole ratio of electron-transport oxadiazole moiety to hole-transport carbazole moiety = 1/9 exhibits the almost same intensity of two bands. Surprisingly, the intensity of two emissive EL bands increases concomitantly with the applied voltage, showing two strong emissive bands, like two crests. They combine together and emit a strong white color. And, the maximum luminance of the white emissive color was 6.04 cd/m^2 at the applied voltage of 17 V. Two obvious crests in the blue region and the new red peak at about 650 nm can be assigned to emissions from the individual lumophore and a certain charge complex like excimer, exciplex, or electroplex, respectively.

Also, in order to gain further insight into the nature of the certain charge complex, we fabricated two LED devices from the blend systems of SiHMOXD/Cz 19 with polystyrene (PS). Even with the device from the blend system of SiHMOXD/Cz 19 with the large amount of PS (80%), in which intermolecular interaction between the polymer chains can be avoid, we still obtained two crests similar to that of SiHMOXD/Cz 19. It indicates that these unusual EL properties, originating from the formation of a certain charge complex, do not exclude intra-chain interaction.

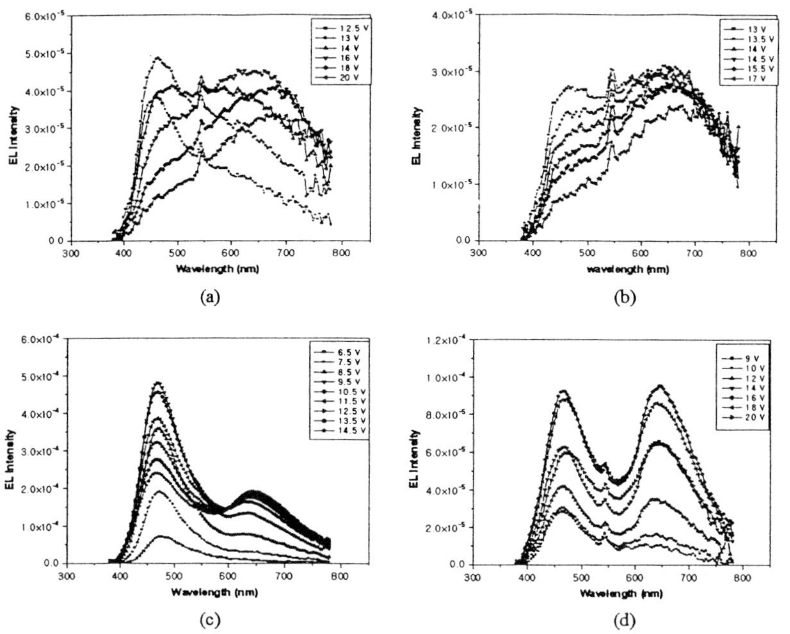

Figure 6. EL spectra of SiHMOXD/Cz 10 (a), SiHMOXD/Cz 91 (b), SiHMOXD/Cz 55 (c), and SiHMOXD/Cz 19 (d).

Then, we measured the emission and the excitation spectroscopies for SiHMOXD/Cz 01, SiHMOXD/Cz 10, SiHMOXD/Cz 55, and SiHMOXD 19 at the various wavelengths (see Figure 7). These two excitation wavelengths at 300 and 350 nm were chosen for the selective photoexcitation of the intrinsic carbazole unit (Cz units: D_{Cz}) or the intrinsic oxadiazole unit (OXD unit: A_{OXD}) and the π-conjugated segments with the carbazole unit (D_{π}) or the oxadiazole unit (A_{π}). The PL spectra of the SiHMOXD/Cz 01 and the SiHMOXD/Cz 10 show the same emission maximum band around 430 nm and 455 nm, which are attributed to photoexcitation of the carbazole unit and the oxadiazole unit, respectively. However, these PL spectral features behave quite differently from SiPhPVK (see Figure 3 (a) and Scheme 3), while the stable resonance structure in the excited state was formed through stabilizing it with phenyl side groups in the SiPh unit of SiPhPVK.

From these different results, we suggest the electron withdrawing power of oxadiazole units in the SiHMOXD is not enough to stabilize the excited state of SiHMOXD for the formation of the stable resonance structure in the excited state, since the oxadiazole units are connected

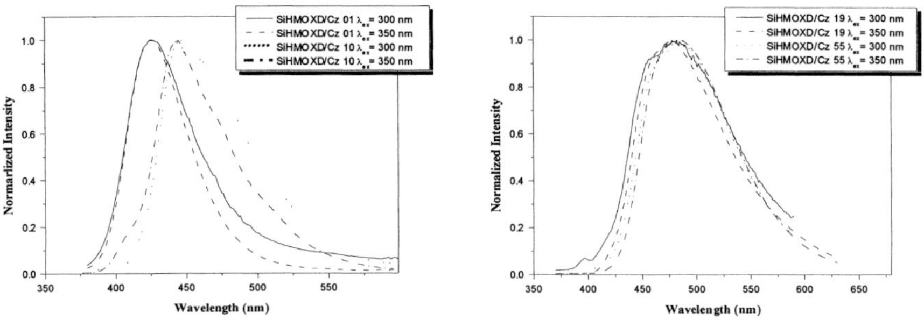

Figure 7. Emission spectra of SiHMOXD/Cz 10 and SiHMOXD/Cz 01 (a), SiHMOXD/Cz 55 and SiHMOXD/Cz 19 with two photoexcitation wavelengths at 300 & 350 nm.

indirectly to silicon atoms via styryl groups, unlike SiPhPVK (see Scheme 3). Their PL spectra of SiHMOXD/Cz 55 and SiHMOXD/Cz 19 show the same emission maximum band around 465 nm, which is attributed to photoexcitation of the intrinsic oxadiazole unit (OXD unit: A_{OXD}) as well as the π -conjugated segments with the oxadiazole unit (A_{π}). Even though the carbazole units (D_{Cz}) as well as the π -conjugated segments with the carbazole unit (D_{π}) were photoexcited at 300 nm, the emission maximum band around 430 nm, corresponding to carbazole moieties, was not observed. It might due to the energy transfer of the excited state of carbazole moieties to the ground state of the oxadiazole units. From these emission studies, we could not observe the red PL spectra, ascribed to the formation of a certain intramolecular charge complex. Therefore, we could propose the formation of a certain intramolecular charge complex for the silicon-based copolymers containing both electron-transport oxadiazole and hole-transport carbazole moieties, corresponding to a red color, as shown in Scheme 5. The proposed scheme can be deduced as follows: In the LED device of SiHMOXD/Cz 10, this new red band is exhibited only in EL, but not in PL spectra. And, in the LED device of SiHMOXD/Cz 01, the blue EL band is exhibited in both EL and PL spectra. So, the blue EL color and the new red EL color come from the silicon-based copolymer with only the carbazole moiety and the silicon-based copolymer with only the oxadiazole moiety, respectively. Thus, a certain charge complex, more like excimer, exciplex, or electroplex, is formed under a strong electric field inside the device and cannot be produced by photoexcitation. From the blending study, the charge complex can be formed intramolecularly between the oxadiazole units and the carbazole units. When the voltage was applied, the electrons can inject into the oxadiazole units (A_{OXD}) and π-conjugated segments with the oxadiazole unit (A_{π}), yielding a certain charge complex of the negative polarons in the oxadiazole units (A_{OXD}) and π-conjugated segments with the oxadiazole unit (A_{π}).

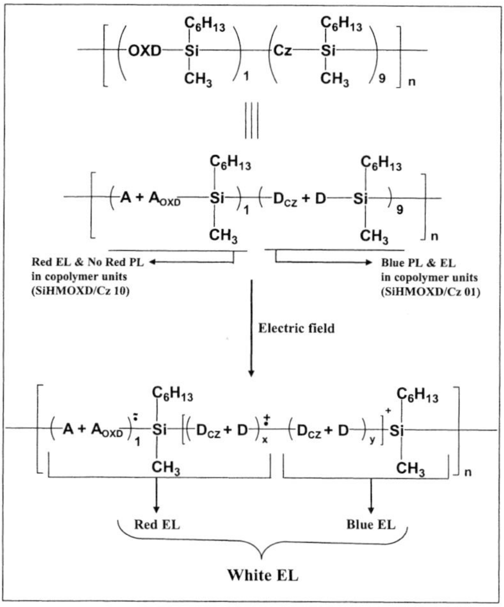

Scheme 5. Proposed scheme for the formation of charge complex in SiHMOXD/Cz 19.

Alternatively, the holes can inject into the carbazole units (D_{Cz}) as well as the π-conjugated segments with the carbazole unit (D_{π}), forming a certain charge complex of the positive polaron in the carbazole units (D_{Cz}) as well as the π-conjugated segments with the carbazole unit (D_{π}). As seen from Scheme 5, a certain charge complex emits the longer emissive bands, which corresponds to the red EL color in EL spectra. In 1996, this kind of the charge complexes from the photoinduced charge separation occurring donor-acceptor substituted silanes (A_{π}-$SiMe_2$-D_{π}) was reported.[15] Also, some of a certain charge complex of the positive polaron in the the carbazole units (D_{Cz}) as well as the π-conjugated segments with the carbazole unit (D_{π}) emits a blue EL color. Therefore, two broad EL bands combine to produce the white emissive color.

Conclusions

We successfully synthesized novel silicon-based alternating copolymers for tunable electroluminescent (EL) colors using Heck synthetic method. Their thermal, photophysical and electroluminescent properties were reported. Most of them exhibited a blue-green EL color at the operating voltage of lower than 12 V. Unusually, we observed the white EL color from a EL device based on SiPhPVK. From photophysical studies and the time-resolved PL spectroscopies, it might be attributed to the formation of stabilized excited state in SiPhPVK. Also, the silicon-based copolymers containing electron transporting oxadiazole and hole-transport carbazole units in main chain were successfully synthesized using Heck synthetic method. We studied their photophysical and electroluminescent properties. The resulting polymers exhibit a strong UV-visible absorption band in the range of 345 ~ 356 nm in chloroform solution and in film state. Upon a photoexcitation around the absorption maximum wavelength, their PL spectra show a maximum band around 435 ~ 485 nm in the blue region. The light-emitting diodes of Al (200nm)/Ca (50nm)/EL polymer (80nm)/PEDOT (50 nm)/ITO were successfully fabricated. And, I-V curves show the turn on voltage of 6 ~ 7V. Their EL properties depend strongly on both the applied voltage and the loading amount of hole-transport carbazole moieties in the present copolymers. The EL spectra from a device based on a copolymer containing hole-transport carbazole moieties (SiHMOXD/Cz 01) exhibit a broad band around 460 nm in the blue region at the operating voltage of higher than 13 V. The EL spectra from the device based on the copolymers containing electron-transport oxadiazole moieties in the main chain (SiHMOXD/Cz 10) exhibit a strong band around 460 nm in the blue region and a very broad, weak band in the red region, at the operating voltage of lower than 12.5 V. With the applied voltage, these emissive EL bands were red-shifted from blue region to red region, exhibiting a very broad, weak band in the blue region and a very strong band in the red region. Two EL bands combine to exhibit a moderate white color. The EL spectra from the devices based on SiHMOXD/Cz 91 – 19 have a similar EL spectral feature with the applied voltage. Also, the intensity of a blue EL band at the relatively high operating voltages increases with the loading amount of carbazole units. Eventually, the LED device from the copolymer containing the mole ratio of electron-transport oxadiazole moiety to hole-transport carbazole moiety = 1/9 exhibits the almost same intensity of two bands, like two crests, giving a strong white color. And, its maximum luminance of the white emissive color was 6.04 cd/m^2 at the applied voltage of 17 V. Two obvious crests in the blue region and the new red peak at about 650 nm can be assigned to emissions from the individual lumophore and a certain charge complex like excimer, exciplex, or electroplex, respectively: The blue EL color comes from the carbazole segments in these silicon-based copolymers. The latter red EL color comes from a certain charge complex arising from oxadiazole (and carbazole) segments. The new red band is exhibited only in EL, but not in PL spectra. From the photophysical studies, a certain intramolecular charge complex was proposed.

Acknowledgements.

The authors acknowledge that this research was financially supported from Hannam University through KyoBi Research Program (2001).

[1] (a) J. H. Burroughes, D. D. C.Bradley, A. R. Brown, R. N. Marks, K. Mackay, R. H. Friend, P. L. Burns, A. B. Holmes, *Nature* **1990**, 347, 539 (b) Y. Ohmori, M. Uchida, K. Muro, K. Yoshino, *Jpn. J. Appl. Phys.* **1991**, 30, L1941.(c) G. Gustafasson, Y. Cao, G. M. Treacy, F. Klavetter, N. Colaneri, A. J. Heeger, *Nature* **1992**, 357, 477. (d) G. Grem, G. Leditzky, B. Ullrich, G. Leising, *Adv. Mater.* **1992**, 4, 36.

[2] (a) M. Berggrenm, O.Inganas, G. Gustafsson, J. Rasmusson, M. R. Andersson, T. Hjertberg, O. Wennerstrom, *Nature* **1994**, 372, 444. (b) A. Kraft, A. C. Grimsdale, A. B. Holmes, *Angew. Chem., Int. Ed. Engl.* **1998**, 37, 402. (c) D. Y. Kim, H. N. Cho, C. Y. Kim, *Prog. Polym. Sci.* **2000**, 25, 1089.

[3] See http://www.kodak.com/go/oled

[4] (a) C. Zhang, H. Von. Seggern, K. Parbaz, B. Kraabel, H. W. Schmidt, A. J.Heeger; *Synth. Met.* **1994**, 62, 35. (b) T. A. Fisher, D. G. Lidzey, M. A. Pate, M. S.,Weaver, D. M. Whittaker, M. S. Skolnick, D. D. C.Bradley, *Appl. Phys. Lett.* **1995**, 67, 1355. (c) X. Zhang, S. A. Jenekhe, *Macromolecules* **2000**, 33, 2069.

[5] (a) S. C. Chang, J. Bharathan, Y. Yang, R. Helgeson, F. Wudl, M. B. Ramey, J. R. Reynolds, *Appl. Phys. Lett.* **1998**, 73, 2561. (b) J. M. Lupton, I. D. W. Samuel, R. Beavington, P. L. Burn, H. Bassler, *Adv. Mater.* **2001**, 258.

[6] (a) H. K. Kim, M.-K. Ryu, S.-M. Lee, *Macromolecules* **1997**, 30, 1236.; (b) H. K. Kim, M. K. Ryu, K. D. Kim, S. M. Lee, S. W. Cho, J. W. Park, *Macromolecules* **1998**, 31, 1114. (c) K. D. Kim, J. S. Park, H. K Kim, T.B. Lee, K.T. No, *Macromolecules* **1998**, 31, 7267.

[7] (a). S. H. Jung, H. K. Kim, S. H. Kim, S. C. Jeoung, Y. H. Kim, D. Kim,; *Macromolecules* **2000**, 33, 9277. (b) N. S. Baek, S. H. Jung, D. J. Oh, H. K. Kim, G. T. Hwang, B. H. Kim, *Synth. Met.* **2001**, 121, 1743.

[8] Y. Yoshida, Y.Nichihara, R. Ootake, A. Fujii, M. Ozaki, K. Yoshuno, H. K. Kim, N. S. Baek, S. K.Choi, *J. Appl. Phys.* **2001**, 90, 6061.

[9] H. K. Kim, M. K. Ryu, K. D. Kim, J. H. Lee, *Synth. Met.* **1997**, 91/1-3, 297.

[10] H. K. Kim, K. D. Kim, D. S. Park, H. H. Song, *Nonlinear Optics* **1999**, 20, 297.

[11] Lee, S. N.; Baek, N. S.; Kim, H. K.; Shim, S.D.; Joo, T. Mol. *Cryst. & Liq. Cryst.*, in press (2002).

[12] K. L. Paik, N. S. Baek, H. K. Kim, J.-H. Lee, *ACS Polymer Preprints* **2002**, 43/1, 77.

[13] Y.–Z., Lee, X. Chen, M.-C. Chen, S.-A. Chen, *Appl. Phys. Lett.* **2001**, 79, 308.

[14] K. L. Paik, N. S. Baek, H. K. Kim, J.-H. Lee, Y. Lee, *Macromolecules*, accepted (2002/5).

[15] C. A. V. Walree; M. R. Roest, W. Schuddeboom, L. W. Jenneskens; J. W. Verhoeven, J. M. Warman, H. Kooijman, A. L. Spek, *J. Am. Chem. Soc.* **1996**, 118, 8395.

Suppression of Secondary PL Emission by Indirect Photoexcitation

H.Y. Byun,[1] *I.J. Chung,*[1] *Y.-S. Suh,*[1] *H.K. Shim,*[1] *D.Y. Kim,*[2] *C.Y. Kim**[3]

[1]KAIST, Daejeon 305-701, Korea
[2]KIST, Seoul 130-650, Korea
[3]Kyungsung University, Pusan 605-736, Korea

Summary: A blend of a newly synthesized polyfluorene(PDHBF) and polyvinylcarbazole(PVK) exhibits a photoluminescence(PL) emission spectrum of PDHBF without an increase in the PL intensity on photoexcitation at 340 nm, the UV-visible absorption maximum of PVK, despite of a substantial spectrum overlap. However, the indirect photoexcitation of the blend suppresses the secondary emission of the PL with the maximum at 520 nm. The chromophores generating the secondary emission are formed when the chromophores are photoexcited above the critical energy level of an excited state. The chromophores formed by the energy transfer have energy lower than the critical energy and fail to form the excimers. A low temperature PL study of the blend in a cryogenic chamber proves that the energy transfer in the system takes place mainly between the excimers of PVK generated by the partially eclipsed dimeric states of two carbazole units and the fluorophores of PDHBF.

Introduction

A progress is made to introduce ink jet printing technology in polymer light emitting diodes(PLEDs)[1-2] and a great deal of efforts is also put into applying organic transistors imbedded on a flexible substrate to the devices.[3] It seems apparent that PLEDs have advantages on fabrication and mass production over organic light emitting diodes(OLEDs). It has been announced that a mobile phone equipped with a PLED will be in market this year [4] and larger-screen PLEDs drived by an active matrix system are introduced in 2005.[5] However, device fabricators of PLEDs have a difficulty to receive reliable light emitting polymers and suppliers of blue light emitting polymers are hard to be found, yet.

The latest trend in development of light emitting polymers has shifted to optimizing molecular structures of fluorene-based polymers.[6] Emitting colors of the polymers are tuned by modifying the fluorene unit or by copolymerizing with monomers of specific functions.[7] There are reports of syntheses of fluorene-based light emitting polymers covering the three principle colors.[8] There are also reports of LEDs fabricated with blue light emitting fluorene-

 CCC 1022-1360/00/$ 17,50+.50/0

based polymers with lifetime of 10,000 h.[9] The polymers with fluorene units show high efficiencies of power and luminescence. However, the materials are still short in lifetime and its extension is the main huddle to overcome for PLEDs at the moment.

Photoluminescence(PL) emission spectra of polymers generally show a broad secondary emission in addition to the main emission maximum with a vibronic feature. The broad emission is known as either excimeric [10] or intrinsic excitonic emission. [11] Color purity of emitting light, which is important in fabrication of full color LEDs, is reduced by the secondary emission. Most of fluorene-based light emitting polymers shows the secondary emission of a longer wavelength. The present work has tried to understand the mechanism of the secondary emission generated from a polyfluorene with least molecular modification but keeping good solubility and high light emitting efficiency.

Experimental

Synthesis

1-bromo-4-cyclohexylbutane was prepared following the literature procedure.[12]

Preparation of 2,7-dibromo-9,9'-di(cyclohexylbutyl)fluorene

To a 250 mL two-neck round-bottomed flask containing 2,7-dibromofluorene (3.466 g, 10.7 mmol), 1-bromo-4-cycloethylbutane (5.67 g, 32.0 mmol), 50 mL of toluene, and 0.16g of tetrabutylammonium bromide as a phase-transfer catalyst was added 50 mL of 50-wt% NaOH aqueous solution. This two-phase mixture was refluxed at 100°C with vigorous stirring and nitrogen gas bubbling. TLC monitored the progress of reaction. After 2,7-dibromofluorene, the starting compound, had completely disappeared, the organic layer was washed with aqueous $NaHCO_3$ solution two times to remove excess NaOH and then with water. The organic layer was separated, dried over $MgSO_4$, and concentrated by solvent evaporation. The crude product was washed with warm ethanol several times(yield of 87 %, white crystal). ^1H NMR ($CDCl_3$, 400MHz); 7.5 (q, 2H), 7.4 (q, 4H), 1.9-1.8 (q, 4H), 1.7-1.5 (m, 10H), 1.2-0.9 (m, 16H), 0.8-0.7 (m, 4H), 0.6-0.5 (m, 4H), ^{13}C NMR ($CDCl_3$, 100MHz): 152.6, 139.1, 130.1, 126.2, 121.5, 121.1, 55.7, 40.1, 37.6, 36.9, 33.3, 27.0, 26.7, 26.4, 23.9

Polymerization of poly-9,9'-dicyclohexylbutyl-2,7-fluorene (PDHBF)

The homopolymer was synthesized by nickel-mediated polymerization.[13-14] A solution containing 5 mL of anhydrous DMF, bis(1,5-cyclooctadienyl)nickel(0), 2,2'-dipyridyl, and 1,5-cyclooctadiene (the latter three in a molar ratio of 1:1:1 by 2.672 mmol each) was heated at 80 °C for 30 minutes under argon. The monomer (1.35 mmol) dissolved in 5 ml of toluene was added dropwise to the solution. After 3 days of heating at 80 °C, 9-bromoanthracene (0.1 g, 0.39 mmol), the end capper in an excess amount, dissolved in 3 ml of anhydrous toluene was added to the flask. The reaction mixture was further heated at 80 °C for 24 hours. The polymer was precipitated on addition of an equivolume mixture of concentrated HCl, acetone, and methanol. The filtered polymer was extracted with chloroform; K_2CO_3 solution was used instead of distilled water until the aqueous layer became neutralized. The organic layer was separated, dried over $MgSO_4$, concentrated by solvent evaporation, and reprecipitated in methanol. The polymer was further purified by a Soxhlet extraction in methanol/acetone for 2 days. Repeating precipitation of the polymer with methanol did final purification. Filtration and drying under vacuum gave the yield of 0.314g(40 % yield); ^{1}H NMR ($CDCl_3$, 400MHz): 7.6-7.4 (m, 6H+aromatic protons of end-caps), 1.9-1.8 (d, 4H), 1.7-1.5 (m, 10H), 1.2-0.9 (m, 16H), 0.9-0.7 (m, 8H), ^{13}C NMR ($CDCI_3$, 100MHz): 151.8, 140.5, 140.0, 126.2, 121.5, 119.9, 55.4, 40.3, 37.6, 37.1, 33.4, 27.2, 26.7, 26.4, 24.2.

Instrumentation

^{1}H and ^{13}CNMR spectra were recorded on Brucker Avance 300 and 400 spectrometers, respectively. Molecular weights of the polymers were determined with a Waters GPC-150C calibrated with polystyrene as the standard and THF as the eluent. A Dupont 9900 was used for thermal characterization of the polymer. UV-visible absorption spectra were made on a Jasco V-530. The PL spectra of films spin-coated on quartz plates from the polymer solutions were performed on a Spex Fluorog-3 luminescence spectrometer. PL efficiency values were calculated by comparing to quinine sulfate (ca. $1x10^3$ M) in 0.10 M H_2SO_4 as standard. A time-correlated single-photon counting (TCSPC) technique was employed for the time-resolved PL measurements. A detailed description of an instrumental setup for TCSPC is found elsewhere.[15]

Results and Discussion

The number and weight average molecular weights are 45,900 and 118,000 g/mol, respectively, and the polymer dispersion index is 2.6. The glass transition temperature(Tg) and decomposition temperature(Td) are 103 and 423 ^{0}C, respectively. The fluorescence quantum yield of the polymer in chloroform is 0.93.

PDHBF has a simple electronic structure, indicating that there is a uniform chromophore absorbing the UV-visible light with the absorption maximum at 390 nm as shown in Figure 1-a. The fine resolution of the PL spectrum of a PDHBF film with a shoulder peak at 450 nm as shown in Figure 1-b indicates that the polymer chains are rigid in the solid state.[16] The PL spectrum, however, still shows a rather strong emission at 520 nm on photoexcitation at 390 nm, the UV-visible absorption maximum, despite of the two bulky side chains on position 9 of the fluorene unit where stacking of molecules might be inhibited by the bulky side chains. There is little shift of the PL emission maximum but a strong vibronic feature in the PL spectrum of the solution in chloroform as shown in Figure 1-c. However, the PL emission with the maximum at around 520 nm is shrunk in the polymer solution indicating a reduction of the secondary emission. The small Stokes' shift and the distinctive vibrational structure of the PL spectrum of the solution imply that the polymer chains are reasonably stiff in the solution.

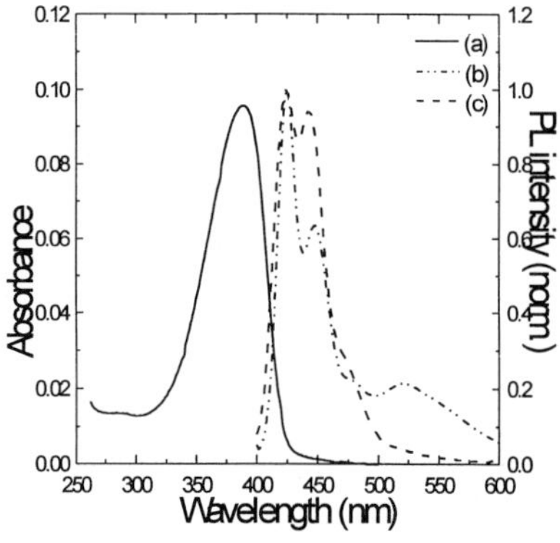

Figure 1. Spectra of UV-Vis absorption(a) and PL emission of film(b) and solution(c) of PDHBF on photoexcitation at 390 nm.

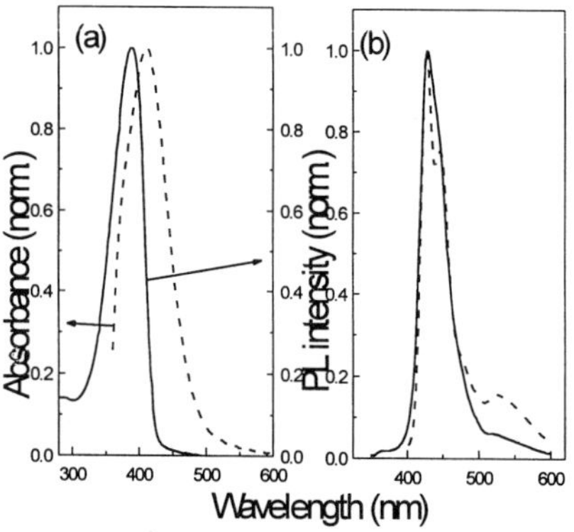

Figure 2-a. UV-Vis absorption spectrum of PDHBF(solid line) and PL emission spectrum of PVK(dashed line) on photoexcitation at 340 nm.
Figure 2-b. PL spectra of a PVK/PDHBF(9/1) blend film on photoexcitation at 340 nm(solid line) and at 390 nm(dashed line), respectively.

Suppression of the PL emission with the emission maximum at 520 nm is tried by blending PVK and the polymer with a blending ratio of PVK/PDHBF(9/1). The spectrum overlap between the PVK emission and PDHBF absorption is substantial for energy transfer between them as shown in Figurer 2-a. The PL spectrum of a blend film on photoexcitation at 340 nm, the UV-visible absorption maximum of PVK, displays a high intensity of the overall PL emission with the emission maximum at 425 nm as shown in Figure 2-b. However, direct photoexcitation of PDHBF in the blend at 390 nm also generates an intensive PL spectrum. This implies a rather poor result of the indirect photoexcitation of PDHBF in the blend system. It is, nevertheless, important to note that the emission with the peak at 520 nm is substantially reduced. It is also noticed that the PL spectrum of the blend lost the vibronic feature, which suggests that the molecular interaction between the two polymers derives a new environment at the interface.

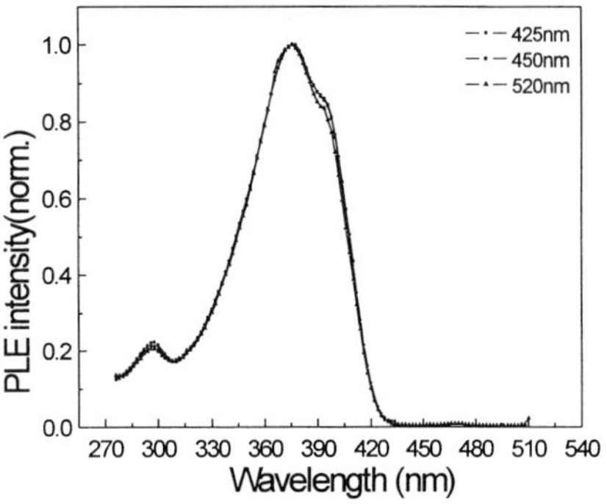

Figure 3. Excitation spectra of PDHBF at three wavelengths of 425, 450 and 520 nm, respectively

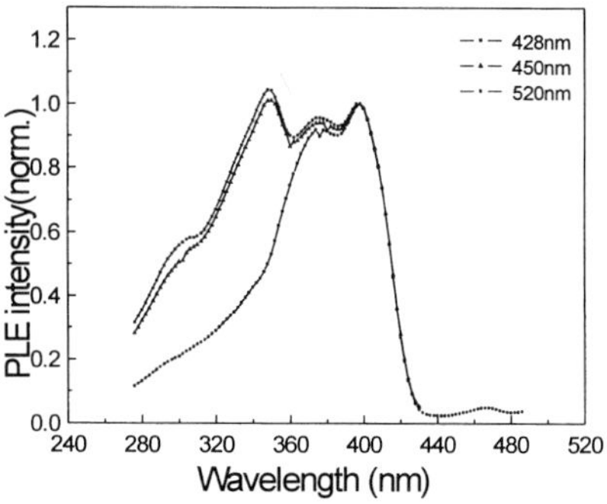

Figure 4. Excitation spectra of a PVK/PDHBF(9/1) blend film at three wavelengths of 425, 450 and 520 nm, respectively.

The excitation spectra of PDHBF fixed at 425 and 520 nm reveal that the photoexcitation of the polymer at around 390 nm generates the highest PL intensities at both 425 and 520 nm as shown in Figure 3. The secondary emission with the maximum at 520 nm might be assigned either as an excimeric radiative decay[10] or as a singlet excitonic emission by an intrinsic structure with a longer conjugation length in the polymer chain.[11] However, the two chromophores for the emission at 425 and 520 nm seems to have a correlation with each other since the normalized PL spectra are fully superposed with each other. It is expected that the population of the overexcited chromophores are present as long as the original chromophores in the polymer are directly photoexcited. The excited chromophores in the S_1 state may derive a change in molecular conformation from the vibrating molecular segments and two sequenced chromophores, one excited and the other at the ground state, may straighten out to form a new chromophore with a longer conjugation. The chromophore decays to the ground state and gives an emission with the PL maximum at 520 nm. The excitation spectra from fixing the two PL emission maxima at 425 and 520 nm display the shallow peak at 295 nm. The peak may indicate the direct excitation of the chromophores to the higher state than S_1. The exciton spectrum at 450 nm, the peak of vibronic feature, fully superposes with those at 425 and 525 nm , which indicates that the three excitation spectra have the same origin.

Figure 4 shows the two different excitation spectra of the polymer blend of PVK and PDHBF with a blending ratio of 9 to 1 on fixing the excitation energy at either 425 and 450 or 520 nm. PDHBF in the blend still generates a strong PL emission with the peak at 520 nm on the direct photoexcitation of PDHBF at 390 nm, the UV-visible absorption maximum of PDHBF and shows no dilution effect on blending. The PL intensity is increased but slightly on the indirect photoexcitation at 340 nm, the UV-visible absorption maximum of PVK. The indirect photoexcitation of PDHBF in the blend reveals that the energy transfer from PVK to PDHBF is not so efficient as observed previously in the different systems.[18] It is, however, demonstrated that the secondary emission with the emission maximum at 520 nm is greatly reduced on the indirect photoexcitation of PDHBF in the blend at 340 nm while the direct photoexcitation of PDHBF in the blend at 390 nm still generates the secondary emission.

It is incorrect to assign the PL emission with the peak at 520 nm as an excimeric one since the direct photoexcitation of PDHBF in the blend at 390 nm generates the emission at 520 nm while the indirect photoexcitation at 340 nm shows no emission at the same wavelength. The chromophores with the same molecular conformation hardly generate different PL emission spectra on photoexcitation at the different wavelength. It is also incorrect, based on the same argument as above, to assign the PL emission at 520 nm as a singlet radiative decay by a chromophore with a longer conjugation. The chromophores photoexcited directly, it is suggested, have energy enough to extend their conjugation length by coupling of the sequenced chromophore in the polymer chain prior to reaching the lowest state of S_1 by loosing the absorbed energy on vibration. The new excitons decay radiatively for the PL emission at 520 nm. However, the transferred energy from PVK to PDHBF is not high enough to form extended chromophores, which results in showing no PL emission at 520 nm. It is significant to notice that the indirect photoexcitation of a chromophore in a polymer blend suppresses a secondary PL emission at a longer wavelength.

Figure 5-a suggests that there are two fluorophores in PVK. The emission maxima of the PL spectra show a blue shift from 410 to 380 nm on lowering temperature from 300 to 53 K. The PL spectrum with the emission maximum at 380 nm is assigned as the emission of excimers with partially eclipsed carbazole units while that at 410 nm with fully eclipsed ones.[19] The intensity of the PL spectrum of the cold film with the emission maximum at 380 nm is much higher than that with the maximum at 410 nm at room temperature. It shows that the photoexcitation of PVK generates excimers of the partially eclipsed structure more than those of the fully eclipsed one. The excimers in the higher energy state lose little energy by chain vibration at low temperature and decay radiatively to the ground state in their lifetime. However, the excimers in the high energy state either form excimers of the fully exclipsed structure on its way to a lower energy state via vibration at room temperature or decay to the ground state unradiatively.

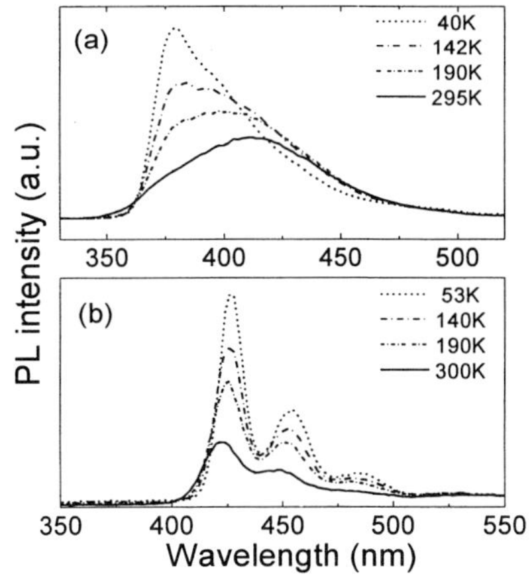

Figure 5. PL emission spectra of PVK(a) and PDHBF(b) photoexcited at 340 and 390 nm, respectively, at various temperature.

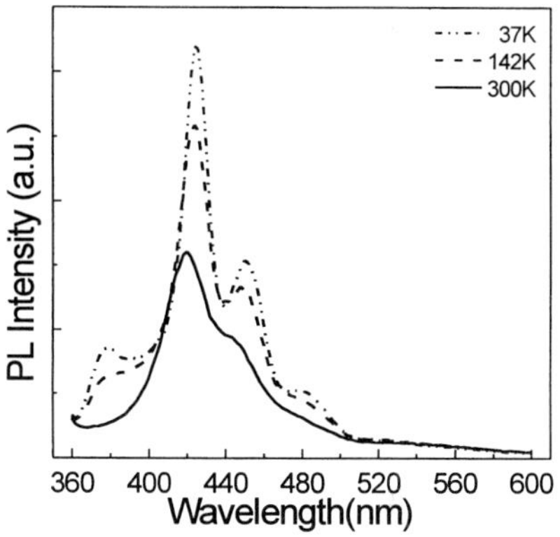

Figure 6. PL spectra of a PVK/PDHBF(9/1) blend film photoexcited at 340 nm at various temperature.

The intensity of PL emission of homopolymer PDHBF is substantially.enhanced at lower temperature as shown in Figure 5-b. It is apparent that vibration of molecular segments in the polymer at the excited state is detrimental to the PL emission. There are a redshift of the emission maximum of PL and a strong vibronic feature on lowering temperature. It implies that the polymer molecules become stiffer at lower temperature and the conjugation length of the chromophres becomes extended. The new peak at 485 nm grows on lowering temperature and is assigned as the second vibronic structure.

The energy transfer from the PVK excimers to the fluorophores of PDHBF is carried out but inefficiently at room temperature and also inefficiently at lower temperature as shown in Figure 6. The shrunk emission peak at 380 nm suggests that the excimers with the partially eclipsed structure failed to complete the energy transfer to PDHBF at lower temperature, which implies that the excimers of the partially eclipsed structure with a long lifetime survive long enough to decay radiatively. It might be expected that a strong emission with the peak maximum at 380 nm would be observed at room temperature if the lifetime of the excimers with the partially eclipsed structure would be short compared to that of the fully eclipsed one. The time correlated single photon counting(TCSPC) technique is utilized to pursuit the mechanism of charge transfer between two chromophores with different band gaps. The emission intensity both at 410 nm, the PL emission maximum of PVK, and 430 nm, that of PDHBF decrease exponentially as shown in Figure 7. The exponential decay curve of PDHBF at 430 nm shows a single component with a lifetime of 190 ps as shown in Table 1. The exponential decay curve of PVK at the emission maximum of 410 nm has two components of 2.64 and 21.31 ns. The PVK/PDHBF blend with a ratio of 9/1 shows a faster decay of the PL intensity at 410 nm with three components of the decay lifetime; 650 ps, 3.42 and 18.06 ns. When a blend with a ratio of 1/1 is photoexcited at 340 nm, the PL intensity at 410 nm decreases much faster with two decay components of 100 ps and 5.57 ns than that of the blend of 9/1.

PDHBF has a short lifetime compared to that of PVK, which is a necessary condition for a good acceptor in the blend system. The decay curves of the two blends of PVK/PDHBF with blend ratios of 1/1 and 9/1 are controlled almost single-handedly by PDHBF. However, the blend with a higher concentration of the acceptor with the blend ratio of 1/1 receives the energy faster from the excited chromophores of PVK(lifetime of about 20 ns). There is no component of the lifetime of 20 ns in the exponential decay curve of the 1/1 blend while the exponential curve is still controlled partially by the PVK component of 5.57 ns. All the energy of the PVK chromophore with the lifetime of 20 ns is transferred completely to the acceptor before contributing to the exponential decay of the intensity. When the acceptor concentration in the blend is reduced to 10 %, the energy transfer is incomplete and there is still a different amount of contribution to the exponential decay curve between the two decay components in PVK. It is observed that the component with the longer lifetime loses energy faster to the acceptor than the one with a shorter lifetime.

It suggests that the excimers with the longer lifetime of 21.31 ns are assigned as the partially eclipsed ones and the shorter one of 2.64 ns as the fully eclipsed ones. If the assignment might be made the other way around, there should be two excimer emission maxima at room temperature. Then, it would be correct to predict that excimers of the fully eclipsed structure with a longer lifetime would be saturated on photoexcitation and the excimers of the partially eclipsed ones with shorter lifetime might decay to the ground state before transferring the excess energy to the less populated carbazole units at the circumstance leading to having two emission maxima in the PL spectrum. In reality, the PL spectrum of PVK is rather broad but has a single emission maximum although there are two excimers with different energy. Since the lifetime of the excimers with the fully eclipsed structure is shorter, the excimer formation cycle is short and the density of the excimers with higher energy decreases fast before exhibiting radiative decay at the higher energy.

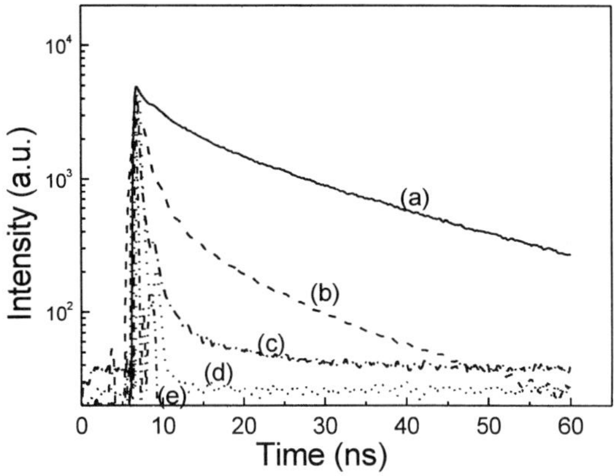

Figure 7. PL intensity profiles of polymer films of PVK(a) and blends of (9/1)(b) and (1/1)(c) at 410 nm on photoexcitation at 280 nm, and of PDHBF(d) at 430 nm on photoexcitation at 280 nm.

Table 1. Fluorescence lifetimes of PVK and PDHBF blend films.
*Lifetimes determined with excitation at 280-283nm on quartz.

System	τ_1 (ns)	% composition	τ_2 (ns)	% composition	τ_3 (ns)	% composition	λ_{max} /nm
PVK	-	-	2.64	52.1	21.31	47.9	410
PVK(9)PDHBF(1)	0.65	83.5	3.42	12.3	18.06	4.2	410
PVK(5)PDHBF(5)	0.10 0.58	88.2 0.6	5.57	11.1	-	-	410
PDHBF	0.19	100.0	-		-	-	430

Conclusion

Two bulky side chains of 4-cyclohexyl-n-butylene attached to position 9 of fluorene units in PDHBF render good solubility in organic solvents but fail to suppress the secondary PL emission with the maximum at around 520 nm on photoexcitation at 390 nm, the UV-visible absorption maximum of the polymer. The emission intensity at 520 nm is greatly reduced only on indirect photoexcitation of PDHBF blended with PVK at 340 nm. It is suggested that the excitons generating the PL emission spectrum with the maximum at 520 nm are formed by coupling of two sequenced fluorophores in the higher state of S_1 by direct photoexcitation at 390 nm while those formed in PDHBF after accepting energy from excited PVK have energy not high enough to extend the polymer chains and result in a radiative decay without the tail at 520 nm in the PL spectrum. However, it is observed that the PVK/PDHBF blend system has a poor energy transfer between the two polymers despite of a reasonable emission/absorption overlap.

[1] T. Hebner, C. Wu, D. Marcy, M. Lu, J. Sturm, **Appl.Phys.Lett.** 1998, *72,*519.
[2] S. Chang, J. Liu, J. Bharathan, Y. Yang, J. Onohara, J. Kido, **Adv.Mater.** 1999, *11*, 734.
[3] G. Gustafsson, Y. Cao, G. M. Treacy, F. Klavetter, N. Colaneri, A. J. Heeger, **Nature** 1992, *357*, 477.
[4] http://www.optics.org/articles/news/8/6/5/1.
[5] http://www.optics.org/articles/news/8/3/30/1.
[6] D. Y. Kim. H. N. Cho, C. Y. Kim, **Prog. Polym. Sci.** 2000, *25*, 1089.
[7] J. I. Lee, G. Klaerner, M. H. Davey, R. D. Miller, **Syn.Met**. 1999, *102*, 1087.
[8] I. S. Millard, **Syn.Met.** 2000, *111-112*, 119.
[9] R. H. Friend, R. W. Gymer, A. B. Holmes, J. H. Burroughes, R. N. Marks, C. Taliani, D. D. C. Bradley, D. A. Dos Santos, J. L. Brédas, M. Lögdlund, W. R. Salaneck, **Nature** 1999, *397*, 121.
[10] N. T. Harrison, D. R. Baigent, I. D. W. Samuel, R. H. Friend, A. C. Grimsdale, S. C. Moratti, A. B. Holmes, **Phys. Rev. B** 1996, *53,* 15815.
[11] C. J. Collison, V. Treemaneekarn, W. J. Oldham, J. H. Hsu, J. L. Rothberg, **Synth. Met.** 2001, *119,* 515.
[12] C. Xia, R. C. Advincula, **Macromolecules** 2001, *34,* 5854.
[13] B. S. Furniss, A. J. Hannaford, P. W. G. Smith, A. R. Tatchell, **Vogel's Textbook of Practical Organic Chemistry, 5th ed.,** Prentice Hall, London, 1989.
[14] T. Yamamoto, A. Morita, Y. Muyuzaki, T. Maruyama, H. Wakayama, Z.-H. Shou, Y. Nakamura, T. Kanba, S. Sasaki, K. Kubota, **Macromolecules** 1992, *25*, 1214.
[15] M. Y. Lee, D. Kim, **J. Opt. Sci., Korea** 1990, *1*, 52.
[16] M. Belletête, J.-F. Morin, S. Beaupré, M. Ranger, M. Leclerc, G. Durocher, **Macromolecules** 2001, *34*, 2288.
[17] C. Wang, M. Kilitziraki, J. A. H. MacBride, M. R. Bryce, L. E. Horsburgh, A. K. Sheridan, A. P. Monkman, I. D. W. Samuel, **Adv. Mater**. 2000, *12*, 217.
[18] K.Kim, J. W.Yu, H. N. Cho, D. Y. Kim, C.Y. Kim, **Mol.Cryst.Liq.Cryst.** 1999, *327*, 165.
[19] A. Itaya, K-I. Okamoto, S. Kusabayashi, **Bull.Chem.Soc.Jpn.** 1979, *52(12),* 3737; A. J. Roberts, C. G. Cureton, D. Philips, **Chem.Phys.Lett.** 1980, *72(3),* 554.

Conducting Polymers Electromechanical Actuators and Strain Sensors

Geoffrey M. Spinks, Gordon G. Wallace, Lu Liu, Dezhi Zhou*

Intelligent Polymer Research Institute, University of Wollongong, NSW 2522, Australia
E-mail: gspinks@uow.edu.au

Summary. Electromechanical actuators are being investigated for a wide range of applications in medical, electronics and industrial areas. One attractive application is to incorporate conducting polymer fibre actuators into fabrics for use in prosthetic applications. In this paper, the design of polypyrrole fibre actuators for use in a glove to open and close the human hand (for assisting those with paralysis or hand injuries) is described. The key requirements for this application are the simultaneous generation of 16 mm of contractile movement and 2.9 N of force. Although not critical in the first prototypes, eventually it will also be necessary to produce a rate of movement of around 10 mm sec^{-1}. The effect of the geometry of polypyrrole actuators is examined in this paper and it is shown that a tubular geometry is superior to conventional flat films. Another aspect of the practical use of actuator materials is their control. Fabric strain gauges with polymer actuators is a convenient means for providing feedback control to the actuating element. The fabric strain gauges ideally articulate with fibre actuators to give both the actuating and sensing function in the same fabric structure.

Introduction

In this paper we describe two essential features needed for practical movement systems using polymer actuators: the actuator materials, based on polypyrrole (PPy); and a displacement sensor based on a conducting stretch-fabric material. The two elements work together in as much as the actuators produce motion and the fabric provides a feedback system to control the motion. We focus here on an actuating glove that can be used to open and close the human hand. This kind of device would be greatly beneficial to persons with paralysis, hand injuries or when recovering from hand surgery. A simple biomechanical analysis of the muscle displacement and force needed to lift the human fingers through their full motion (up to 100°) shows that the "artificial muscle" needs to simultaneously produce a contraction of 16mm and 2.9 N force, Table 1 [1]. Ideally, this motion should be completed within ~1-2 seconds (giving a displacement rate of ~10mm sec^{-1}) although in first prototypes a much slower rate of

 CCC 1022-1360/00/$ 17,50+.50/0

movement is acceptable.

Table 1 Muscle force and tendon excursion required to cause bending of the fingers on an adult human hand.

Finger	Joint	Muscle Contraction Required for full flexion (mm)	Force Required for full flexion (N)
Digit 2 (Index)	Proximal Phalange	15.7	2.1
	Middle Phalange	13.1	1.4
	Distal Phalange	7.8	0.5
Digit 3 (Middle)	Proximal Phalange	15.7	2.9
	Middle Phalange	13.1	1.4
	Distal Phalange	7.8	0.6
Digit 4 (Ring)	Proximal Phalange	15.7	2.8
	Middle Phalange	13.1	1.2
	Distal Phalange	7.8	0.6
Digit 5 (Little)	Proximal Phalange	15.7	1.6
	Middle Phalange	13.1	0.8
	Distal Phalange	7.8	0.4

Conducting polymers have been extensively studied as actuators and been shown to generate useful strains (around 3%) and stresses (up to 10 MPa) [2]. The strain rate of conducting polymers has been reported to be as high as 3.2% sec^{-1} [3]. Using these performance characteristics as design parameters, it is possible to design actuators for the glove application: length = 533mm; cross-sectional area = 0.29 mm^2 (corresponding to a film 5 mm wide and 58 μm thick). The problem with this design is that the available length (from the knuckle to the elbow) is only ~ 350mm in an adult male. This limitation means that the strain needed from the artificial muscle should be at least 4.5%.

Analysis of the data from the literature reveals that the reported stress and strains from polypyrrole materials are not for simultaneous conditions. That is, the strains are usually reported for near-zero loads and the stresses are for isometric (zero strain) conditions. We have recently shown [4] that the strain generated by PPy actuators decreases when the applied isotonic stress increases. Although it is possible to generate 5% strains, this only occurs at stress levels < 0.1 MPa. To provide full movement for the human fingers, a parallel stack of 100 films (5mm x 350mm x 58 μm) would be required. Obviously, it is a considerable technical challenge to build such a device.

In this paper we report improvements to PPy actuator performance by adopting a hollow fibre (tube) geometry. The method of preparation of the tubes is described as is the actuator performance in terms of strain under load and strain rate.

Finally, we are also interested in control systems for mechanical actuators. Fabric strain gauges [5] are interesting devices that may be useful in providing direct feedback control to actuators. The operation of fabric strain gauges is also briefly described in this paper.

Experimental

Polypyrrole films and fibres were prepared by electropolymerisation using platinum working electrodes in an electrolyte containing 0.06M pyrrole, 0.05M tetrabutylammonium hexafluorophosphate (TBA.PF_6) in propylene carbonate. The polymer was electrodeposited on the electrode using galvanostatic conditions (current density of 0.15mA/cm^2) and at –28°C. The hollow fibres were prepared by electrodeposition on Pt wire (125-250 μm in diameter) to give a coating thickness of ~50 μm. The Pt wire was removed after the electrodeposition to give a free-standing hollow tube of PPy. Flat films prepared in the same fashion were thicker, in the range 120-140 μm. Mechanical and electrical properties of the film and tube samples are given in Table 2.

Table 2 Basic electrical and mechanical properties of tube and film samples of PPy/PF_6.

	Tube(1)	**Flat Film**(2)
Conductivity (Scm^{-1})	170 Scm^{-1}	85
Tensile strength	19.6 ~ 27.3	6.0
Elongation to break	15.4 ~ 19.0	8.5
Electrolytic efficiency	10%	5.0%

Actuator testing was conducted by attaching samples to one side of a balance beam and fixing at the bottom of a container for electrolyte. The tension applied to the sample could be adjusted by changing the weights on either side of the balance beam. The test apparatus is illustrated in Figure 1. Various testing conditions were used, but in most cases the electrolyte was 0.25M TBA.PF_6 in propylene carbonate. For practicality it is beneficial to operate actuators in a two-electrode arrangement using a simple DC power supply.

Coated fabrics for strain gauge applications were prepared by first soaking the fabric (eg nylon/polyurethane) in a solution containg ferric chloride oxidant and then exposing the soaked fabric to pyrrole vapour. Fabrics were tested by stretching in an Instron testing machine with the resistance change in the fabric measured using a Wheatstone Bridge circuit.

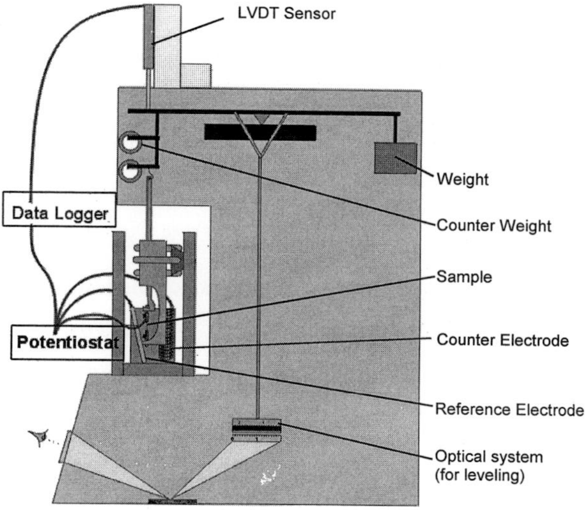

Figure 1 Balance beam apparatus for isotonic actuator measurements: the actuator displacement is determined by the linear variable distance transducer (sensor) for fixed isotonic loads. The load can be varied by removing counter-weights.

Results

Actuators

Figure 2 shows how the strain (at constant stress) changes with increasing applied potential for polypyrrole films and tubes. The actuation strain increases with increasing voltage, but the maximum voltage is limited by the stability of the polymer. For the tube geometry the maximum voltage range was +/- 3.5 V, while for the film a wider range of +/-5V could be applied. The higher conductivity of the tube accounts for its lower voltage stability range, since the electrochemical potential achieved at the polymer working electrode (E) depends upon the polymer resistance:

$$E = Eapp - iR_T$$
$$R_T = R_s + R_p$$

[R_T = total resistance, R_s = solution resistance, R_p = polymer resistance]

where Eapp is the potential applied by the external power source and *i* is the current. Since the tube has a lower resistance (R_p) than the film, a smaller external voltage (E_{app}) is needed to ensure that the polymer reaches the breakdown potential (E).

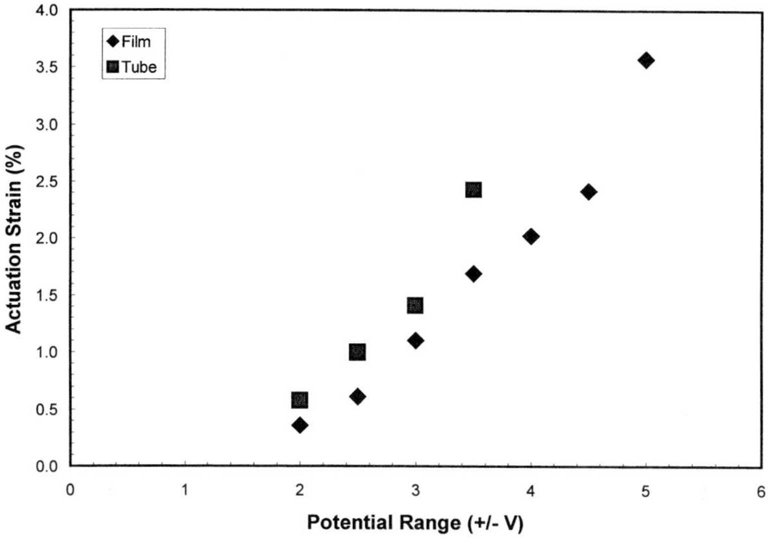

Figure 2 Isotonic actuator strain measured for polypyrrole film and tube geometries for different applied potential limits using a voltage scan rate of 50 mV/s (stress = 0.05 MPa for the film and 0.72 MPa for the tube).

In Figure 3 the inherent slowness of the actuation process in polypyrrole (dominated by ion-diffusion kinetics) is illustrated by the rapid decrease in actuation strain as the voltage scan rate is increased. The kinetics appear to be similar for the two geometries, with the tube slightly less affected by increasing scan rates compared with the film. Again, this observation can be explained by the higher conductivity of the tube compared with the film.

The effect of increasing applied load on the actuation strain for the two different geometries is shown in Figure 4. At all loads the tube produces substantially higher strains due to its higher electrical conductivity, so that the polymer potential (E) for the tube is higher than for the film. The tube can also be tested to higher isotonic stress levels than the film. Mechanical failure limits the stress that can be applied and the higher tensile strength of the tubes compared with the films means that actuation can be measured up to 3.5 MPa for the tubes compared to <1.5 MPa for the films.

Finally, the strain rate achieved by the tubes is significantly higher than that produced by the films under the same stimulus conditions. As shown in Figure 4, the tubes produce a strain rate in the order of 0.05 % sec^{-1} compared to the films with 0.01% sec^{-1}.

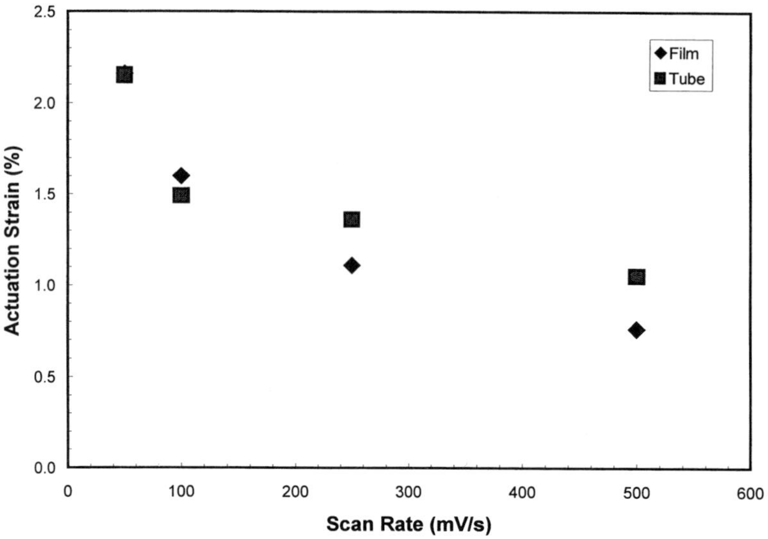

Figure 3 Isotonic actuator strain for $PPyPF_6$ film and tube samples tested at different voltage scan rates between the limits of +/- 3.5V (stress = 0.10 MPa for the film and 1.0 MPa for the tube).

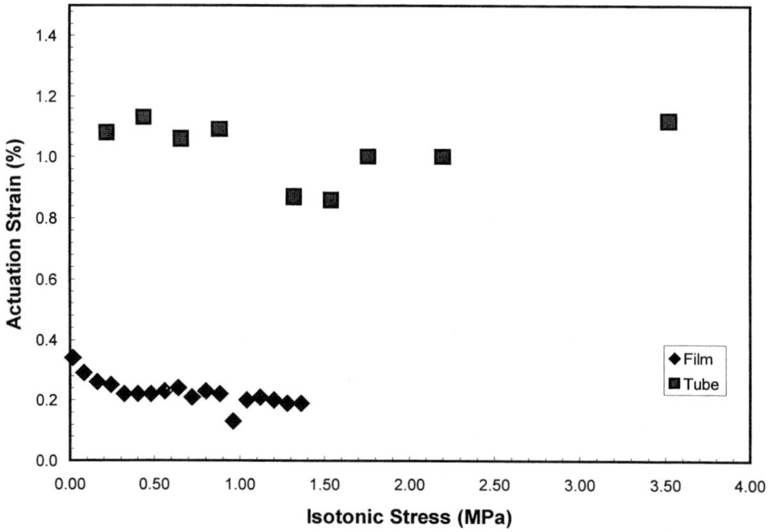

Figure 4 Actuator strain obtained isotonically at different applied stress levels for $PPyPF_6$ films and fibres. A cyclic voltage signal of +/-2.5V at 50 mV/s was used in both cases.

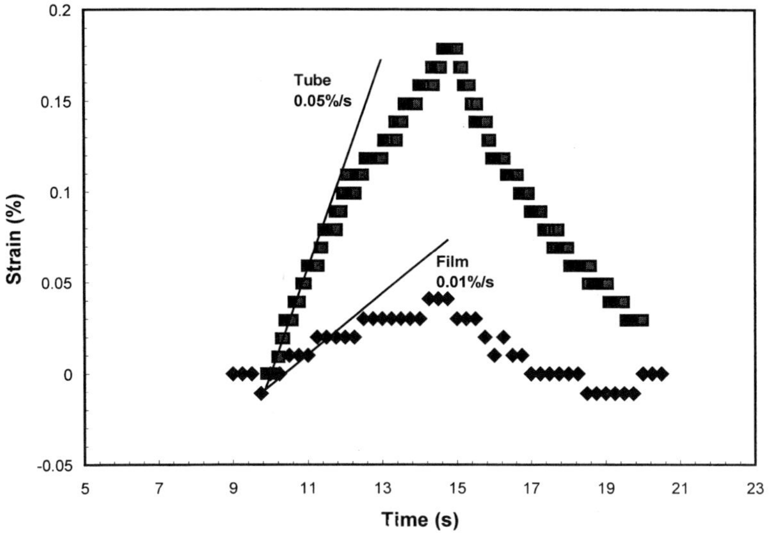

Figure 5 isotonic strains induced by a square wave voltage cycle at +/-1.5V at 0.1 Hz.

Fabric Strain Gauges

Polypyrrole coated stretch fabrics have been shown to act as useful strain gauges having a wide dynamic range [5]. The resistivity of the fabric changes dramatically when the fabric is stretched (Figure 6). The direction of stretch produces different strain gauge response; with a more linear change in resistance occurring when stretching is conducted in the course direction. The gauge factor of these fabrics is approximately 2, which is similar to conventional strain gauges. However, the dynamic range is up to 100%, which is an order of magnitude greater than conventional strain gauges. Examination of the deformation process (Figure 7) for stretch fabrics shows that the weave structure opens up as stress is applied. The opening of the weave changes the number of interfibre contacts and this is most probably the reason for the change in resistance with stretch.

The combination of strain gauge functionality with wearable fabrics opens up the possibility of manufacturing garments for monitoring human movement. In Figure 8 a prototype sensor glove is shown. Coated fabric strips are sewn across the proximal interphalangeal joint in this example and bending of the finger causes a measurable change in fabric resistance.

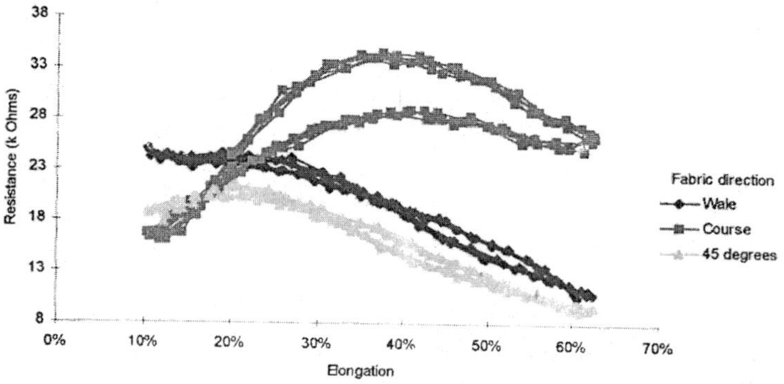

Figure 6 Change in resistance of polypyrrole coated fabric as a function of elongation when stretched in different directions.

Figure 7 Scanning electron micrograph of coated fabric with the right side under tension in the course direction.

Figure 8 Prototype sensor glove with stretch fabric strain gauges attached to the metacarpo-phalangeal joints of each finger.

Conclusions

This study has focused primarily on the effect of geometry of PPy actuators on the (isotonic) actuator strain produced. The strain increased for hollow tubes compared with flat films, as did the strain rate. The increase in performance is attributed to the higher electrical conductivity of the tubes. Strains were obtained at higher isotonic stresses in the tubes compared with the films and this was attributed to the higher tensile strength of the former. Despite these performance improvements, the tube actuators still do not meet the requirements necessary to produce full movement of human fingers. We are now working on improving the electrical connection to the tube geometry in order to address the current limitations.

Acknowledgements

The authors wish to thank Mr. Philip Whitten, Mr. Mark Eckermann and Ms. Melanie Werner for assistance with experimental work. The authors also gratefully acknowledge the Cooperative Research Centre for Hearing Aid and Cochlear Implant, Royal North Shore Hospital, Sydney and the Australian Research Council for financial assistance and Dr. Tim Scott, and Professor Danilo De Rossi for valuable technical discussions.

[1] Spinks, G.M., et al. Electroactive Polymers and Rapid Prototyping. in Materials Research Society Fall Meeting: Electroactive Polymers and Their Applications as Actuators, Sensors, and Artificial Muscles. 2001. Boston: Materials Research Society.

[2] Bar-Cohen, Y., ed. Electroactive Polymer (EAP) Actuators as Artificial Muscles- Reality, Potential and Challenges. . 2001, SPIE - The International Society for Optical Engineering: Bellingham, Washington, USA.

[3] Madden, J.D., et al., Fast contracting polypyrrole actuators. Synth. Met., 2000. **113**(1-2): p. 185-192.

[4] Spinks, G.M., et al., Strain response from polypyrrole actuators under load. Advanced Functional Materials, 2002. **In Press**.

[5] De Rossi, D., A. Della Santa, and A. Mazzoldi, Dressware: wearable hardware. Mater. Sci. Eng, 1999. **C7**(1): p. 31-35.

Macromol. Symp. **192**, *171–181 (2003)*

Durable Chiral Sensor Based on Quartz Crystal Microbalance Using Self-Assembled Monolayer of Permethylated *β*-Cyclodextrin

Siu-Choon Ng, Tong Sun, Hardy S. O. Chan**

Department of Chemistry, National University of Singapore, Lower Kent Ridge Road, Singapore 117543
Email: chmngsc@leonis.nus.edu.sg; chmcsoh@nus.edu.sg

Summary: A novel chiral sensor based on quartz crystal microbalance (QCM) with a self-assembled monolayer of permethylated β-cyclodextrin as the enantioselective coating has been derived which exhibit improved enantioselectivity and excellent long-term environmental stability when used in gas phase sensing.

Introduction

The detection of individual enantiomer remains a challenging task in analytical chemistry, since enantiomers exhibit their individualities only in a chiral environment. Although chromatographic methods like HPLC, GC, CE, using chiral stationary phases (CSPs), are efficient for enantioseparations, they are time-consuming and not readily adaptable for real time determination of enantiomers.[1] Meanwhile, gravimetric sensors like quartz crystal microbalance (QCM) and surface acoustic wave (SAW) devices have shown promise as rapid, sensitive and portable transducer elements for chemical sensing.[2] Several appeared recently describing the application of QCM for chiral sensing in gas phase. Ide et al.[3] reported on the use of quartz-resonator gas sensor coated with cyclodextrins films to discriminate optical isomers of rose oxide and citronellal, although the results were questioned by others.[4] Thereafter, Bodenhöfer et al.[4,5] carried out detailed studies of gas phase discrimination of enantiomers on QCM spin-coated with Chirasil-Val film. These studies show that the design of chiral sensor based on QCM transducer with an appropriate chiral coating is feasible in the gas phase. However, in these works, the chiral coatings were deposited using physical method, which has the drawback of poor durability that could impair the discriminative efficiency of the chiral sensor. Unlike GC or HPLC, which has thousands of "plates" in the column on which adsorption/desorption processes take place, QCM chiral sensor has only one

 CCC 1022-1360/00/$ 17,50+.50/0

"theoretical plate" available for interactions with guest molecules.[4] Consequently, the surface arrangement of the enantioselective elements on quartz surface can be crucial in achieving premium chiral discrimination performance.[6] Spin-coating and airbrushing methods are unlikely to afford an ordered surface of functional coating, thus the effective chiral host-guest interactions might be impaired. Moreover, these coating methods could also result in the easy loss of functional materials over time, as we have noticed.[7] It was envisaged that a chemically bonded enantioselective layer would overcome this disadvantage. However, to the best of our knowledge, there has been no reported work on the design of QCM chiral sensor coated with self-assembled functionalized *β*-cyclodextrins (*β*CD).

In the present work, we undertook a systematic approach to demonstrate this design of QCM chiral sensor based on self-assembled permethylated *β*CD for gas phase chiral discrimination with enhanced chiral discriminatory performance and improved durability of functional coatings. Permethylated *β*CD molecule was immobilized on Au electrodes of a 10 MHz AT-cut quartz plate through single thiol linkers, $NH_2(CH_2)_nSH$, of different lengths, n=2, 6, 11, respectively. The surface concentration of resulted monolayers was calculated. From the real-time responses of the QCM upon exposure to enantiomeric analytes, the chiral discriminatory ability of the different permethylated *β*CD monolayers was compared.

Experimental

The mercaptyl functionalizaed *β*-cyclodextrins were synthesized in accordance to Scheme 1 using an extended application of the Staudinger reaction.[8]

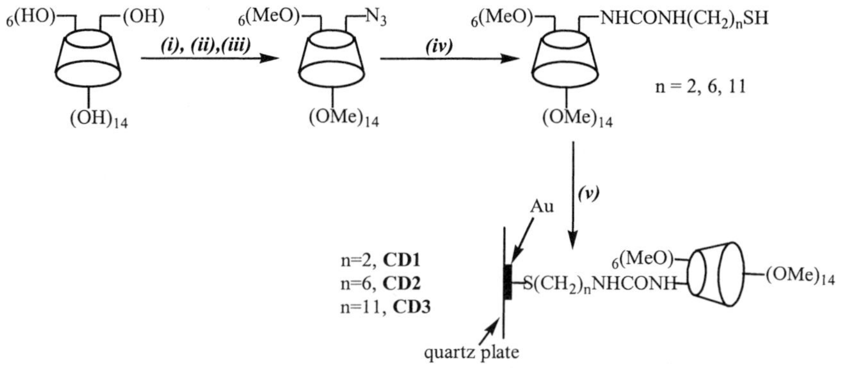

Reagents and Conditions: (i) p-TsCl / Pyridine (dry); (ii) NaN_3 (aq.) / Δ; (iii) MeI / NaH / DMF (dry); (iv) $NH_2(CH_2)_nSH$ / CO_2 / Ph_3P / THF (dry); (v) absolute ethanol / 24 hrs.

In this scheme, $NH_2(CH_2)_nSH$ with n=2, 6, 11, respectively, was attached to the 6^A position of (6^A-azido-6^A-deoxy)heptakis(2,3-di-*O*-methyl)- 6^B, 6^C, 6^D, 6^E, 6^F,6^G-hexa-*O*-methyl-*β*-cyclodextrin, respectively. The products are denoted as **CD1**, **CD2**, and **CD3**, accordingly. The structures and purities of the *β*-CD derivatives were confirmed by NMR spectroscopy (Bruker DPX 300 spectrometer), Electron-spray-ion mass spectroscopy (ESI-MS) (Finnigan TSQ7000 mass spectrometer), FTIR spectroscopy (Perkin Elmer 1600 spectrometer), and Elemental analysis (Perkin Elmer 240C elemental analyzer. C, H, N, and S, within ±0.5% errors). Detailed structural data are given in the last part of this paper. Using established self-assembled monolayer (SAM) procedure compounds **CD1-3** were immobilized onto the surface of Au electrodes of QCM.[9]

Time-of-flight secondary ion mass spectra (TOF-SIMS) of the monolayers were obtained with an ION-TOF GmbH instrument using a pulsed primary beam of Ar ions (10 KeV) under static conditions. Spot size is 10×10 μm^2.

The topographic images of the immobilized *β*-CD monolayers were obtained with a Nanoscope II (Digital Instruments, Santa Barbara, CA) Atomic Force Microscopy (AFM) in non-contact mode, 1 μm×1 μm scanning area.

The QCM employed is a commercially available 10 MHz, AT-cut quartz, purchased from International Crystal Manufacturer, Inc., Oklahoma City, USA. The quartz crystal plate was deposited with Au electrodes (diameter 0.201") on both sides (area 0.20 cm^2). The QCMs were cleaned with "*pirhana*" solution of 30% H_2O_2/70% conc. H_2SO_4 (v/v) at 70°C just prior to monolayer deposition (**Caution:** *pirhanan* solution reacts violently with organic compounds and must be used with extreme caution!). For TOF-SIMS and AFM experiments, commercial Au mirrors (Bio-Rad, Inc.) were used for monolayer deposition. After rinsing with copious Milli-Q water, the QCMs/Au mirrors were immersed into 10^{-3} M ethanol solution of **CD1-3** for 24 h. After the treatment with copious ethanol and Milli-Q water, the QCMs/Au mirrors were dried in N_2 stream until the frequency shift of the quartz plate remained constant, indicating the complete removal of physisorbed small molecules.

The gas phase measuring system used was similar in design to that in ref. 10. It includes a microscale vaporizer to allow for the evaporation of minute amounts of enantiomer samples and a thermostated flow-through measuring cell housing the QCM devices. Predried N_2 was used as carrier gas and controlled by an Alicat mass flow meter. The temperature of the vaporizer was maintained at 35 ± 0.1 °C with the measuring cell at 25±0.1 °C. The volume of the thermostated measuring cell (66.0 cm^3) can ensure the temperature of arrival N_2 gas stream on the surface of QCM to be 25.1±0.1 °C, no condensation of sample analytes was observed during the measurements. A Lab Crystal Oscillator of ICM Co. was used to generate the signal, and a TF830 universal counter to measure the frequency shifts. All data were record by an online PC.

Results and Discussion

CD1-3 Monolayers

The successful self-assembly of derived βCD molecules on Au surface was confirmed using TOF-SIMS. Figure 1 displays a typical spectrum of **CD3**/Au under 10 KeV Ar ionization. The

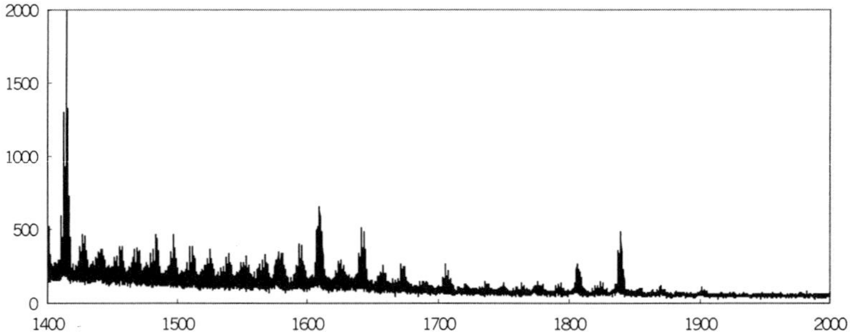

Figure 1. Positive TOF-SIMS spectrum of **CD3** monolayer on Au mirror.

presence of peaks at 1807 $[M+Au-S]^+$ and 1839 $[M+Au]^+$ (where M is the molecular mass of permethylated β-CD with a thiol linker, e.g. M_{CD3} =1643.7 g/mol) corroborates the successful confinement of **CD3** molecules on the Au surface.[11] Further evidence confirming the formation of **CD1-3** monolayer on Au surface was illustrated in Figure 2. As can be seen, there is no presence of particles at the height of 12nm on the surface of commercial bare gold (Figure 2a). After being immobilized with **CD1**, the surface looks rougher (Figure 2b). Typical particles with the diameter of 10-50nm are well observed at the height of 12-15nm.

The surface coverage of the monolayers can be directly calculated from the decrease of the resonant frequency (Δf) of quartz plate using the equation derived by Sauerbrey.[12] For AT-cut 10MHz quartz crystal, the mass-frequency response can be expressed as:

$$\Delta f = (-2.26 \times 10^{-6}) f_0^2 (\Delta m / A) \qquad (1)$$

where f_0 is the fundamental frequency of the quartz resonator (Hz), and A the surface area of the resonator Au electrode (0.20 cm^2). Accordingly the mean surface concentrations of the three permethylated βCD derivatives were calculated to be 6.20 × 10^{-11}, 7.08 × 10^{-11} and 8.98 × 10^{-11} mol cm^{-2}, for **CD**1, **CD**2 and **CD**3, respectively. These data are consistent with a surface concentration approximating to the hexagonal closed packing model,[9b] where the CD torus lies parallel to the surface. Amongst the three monolayers, **CD**3 which has the longest linker appears to be more closely packed.

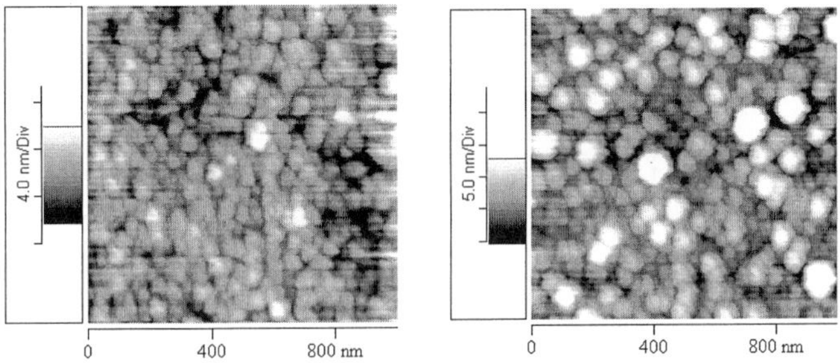

Figure 2. Topographic images of (a) bare Au surface; (b) **CD1** on Au surface.

Chiral Discrimination at Monolayers

Five pairs of pure enantiomers with different functional groups and various molecular size were selected to evaluate the enantiomeric discriminatory ability of the three derived sensors in gas phase. The frequency change of three QCM sensors (Δf) versus time plot are shown in Figure 3 with respect to the chiral analyte gas composition (R- and S-enantiomer). The overlay of the respective sensor signals affords better comparison between two enantiomeric forms. Meanwhile the repeatability of the sensor response is also depicted in Figure 4. As can be seen, the QCM sensor performances towards both forms of the chiral analyte are similar but differ in the extent in frequency reduction indicating the preferential binding presented by the permethylated βCD monolayers between R- and S-enantiomers, which affords the differences in the inclusion complexation amount (Δf) of the respective analytes on the monolayer. As being summarized in Table 1, the R-enantiomer of all five pairs of analytes invariably affords larger Δf than does the S-form indicating the higher binding affinity depicted by the permethylated βCD monolayers towards the R-enantiomer irrespective to the probably varied surface monolayer structure caused by the different length of thiol linkers.[13]
The chiral discriminatory factor, $\alpha_{R/S}$, of the three QCM sensors was calculated from the

equation: $\alpha_{R/S} = \frac{\Delta f_R}{\Delta f_S}$,[4] and listed in Table 1 along with the sensor recovery time, *t*. The chiral discriminatory performances of our three sensors appear to fall into two categories: for

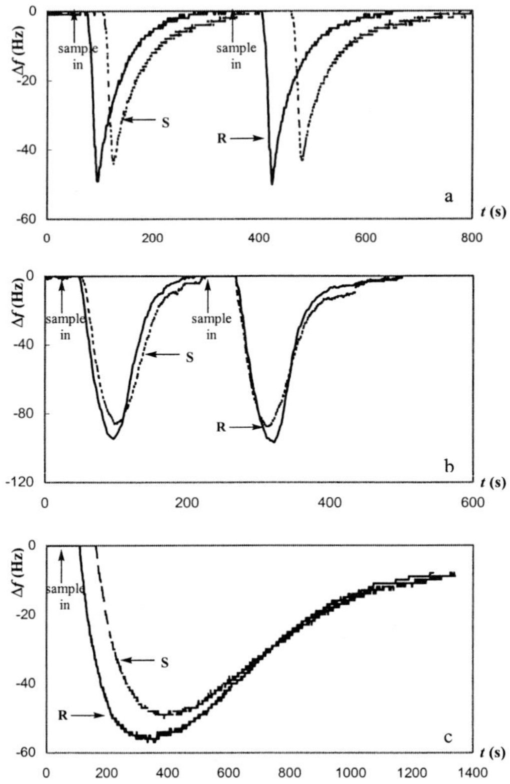

Figure 3. Typical time course of frequency change of three QCM sensors towards different chiral analytes. a) **CD1** vs methyl lactate (2.0 μl); b) **CD2** vs 2-octanol (0.5 μl); c) **CD3** vs carvone (1.0 μl)

analytes such as 2-butanol, methyl lactate, and ethyl lactate, which has relative small molecular size, the $\alpha_{R/S}$ increases from **CD1** to **CD3,** corresponding to the enhanced inclusion complexation of the guest molecules (Δf) on the monolayer from **CD1** to **CD3**. This can be attributed to the increased surface concentration of immobilized βCD molecules on Au. On the other hand, for 2-octanol and carvone, which has relatively large molecular size or rigid structure, the inclusion complexation of both guest molecules decreases from **CD1** to **CD3**, supported by the low $\alpha_{R/S}$ value of **CD3**. This phenomenon could be explained by the formation of a quasi-two-layer structure in **CD3** monolayer,[14] which renders **CD3** monolayer less accessible interaction sites for larger guest molecules in the monolayer structure.

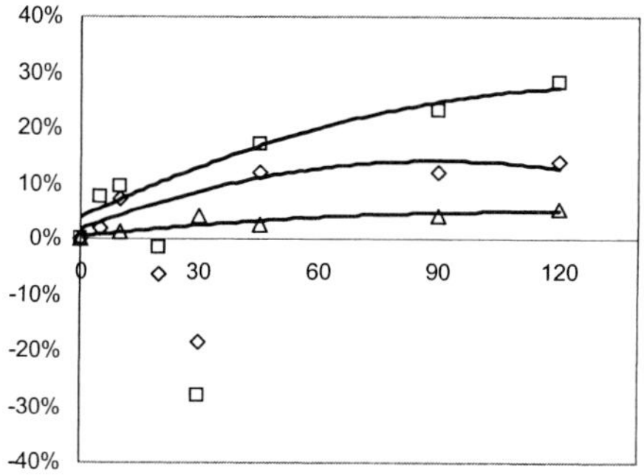

Figure 4. The long-term stability of the chiral coating of three sensors. The frequency deviation percentage $\Delta f_t / \Delta f_0$ over time (day). Δf_t, the amount of coating loss in terms of frequency deviation (Hz); Δf_0, the total coating amount measured in frequency shift (Hz) at Day 0.

Table 1. Chiral discrimination factor ($\overline{\alpha}_{R/S}$) of three QCM chiral sensors towards 5 pairs of enantiomers (Temp., 25.0°C, N_2 flowing rate, 200ml/min).

Devices		Analytes				
		2-butanol	2-octanol	Methyl lactate	Ethyl lactate	Carvone
		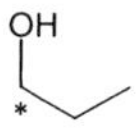	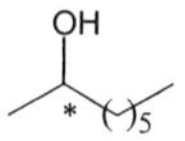	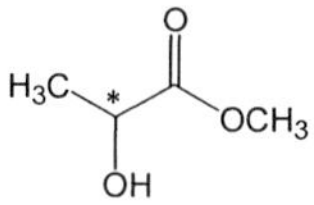	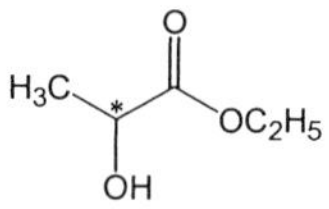	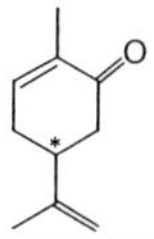
		2.0μl[a]	0.5μl	2.0μl	1.0μl	1.0μl
CD1	Δf (Hz) (R/S)	-58/-54	-115/-103	-49/-44	-67/-60	-61/-56
CD2		-58/-53	-95/-86	-65/-58	-68/-61	-60/-55
CD3		-60/-54	-56/-51	-72/-64	-65/-58	-56/-52
CD1	t(s)	46	152	157	108	1110
CD2		47	168	165	111	1154
CD3		49	140	159	104	1065
GC[b]	$\overline{\alpha}_{R/S}$	1.081[c]	1.020[c]	1.119[e]	1.119[e]	1.02[d]
CD1		1.074	1.116	1.114	1.117	1.089
CD2		1.094	1.105	1.121	1.115	1.090
CD3		1.111	1.098	1.125	1.121	1.076

[a] sample injection volume; [b] reported optimal separation factors by GC.
[c] on 2,3,6-Tri-O-β-CD/OV-1701 column;[15b]
[d] on permethyl-O-(S)-2-hydroocypropyl-derivatized α-CD capillary GC[15a];
[e] on QCM coated with chiral amides by Bodenhöfer et al.[4]

In comparison with the reported GC separation factors towards 2-butanol, 2-octanol and carvone,[15] the three QCM sensors in this study exhibit superior chiral discriminatory ability, in which the improvement of $\alpha_{R/S}$ is up to 9.5% by **CD1**. Although there has been no clear explanation thus far as to the actual mechanism behind the chiral discrimination process in GC, several factors such as operation temperature, substrate structure, mobile phase, etc. are known to affect enantioselectivity of chiral receptors to a certain extent.[16] In this study, the improved chiral discrimination ability could be the result of an ordered surface morphology in the monolayers. The permethylated βCD molecules self-assembled on the Au(111) surface afforded an ordered packing on the surface that would result in the increased number of binding sites accessible by chiral analytes. In particular, the interference from substrates and mobile phases would have been minimized in contrast to the situation of CSPs in GC.

Amenability to working under ambient temperature is another advantage of these chiral QCM sensors, and lower working temperature may also contribute to the improved chiral discriminative performances. While in GC, it is necessary for the columns to operate at elevated temperature to ensure optimal separation that usually result in lower separation factors.

In comparison with the chiral discriminatory results achieved on the chiral amides coated QCM sensors for methyl lactate and ethyl lactate,[4] the permethylated βCD deposited QCM sensors only achieve marginal improvement, about 0.5% in the case of methyl lactate. This could be attributed to the weak interaction presented between these guest molecules (methyl lactate, ethyl lactate, and 2-butanol) and permethylated βCD cavities due to the small size of the analyte molecule. Further studies on the influence of modified βCD cavity size on the complexation of these analytes in gas phase are currently undergoing in our laboratory on other modified βCD monolayers.

Chiral Sensor Recovery

For sensor recovery, since different sample sizes were involved, an easy comparison seems to be difficult here. Generally, all three QCM sensors can be recovered within a few minutes except in the case of carvone where the sensors experienced very slow recovery. The sensor recovery follows the sequence: $t_{\text{2-butanol}} \ll t_{\text{2-octanol}}, t_{\text{methyl lactate}} < t_{\text{ethyl lactate}}$, and t_{carvone} is the largest, where t is the time required for 95% recovery of frequency decrease in terms of the R-enantiomer of the analytes. In the case of carvone, t was measured only at 85% recovery of the sensor's resonant frequency change. Apparently the sensor recovery is strongly affected by the interactions between analyte molecules and βCD host molecules, which is closely related to the molecular size of analytes due to the formation of inclusion complexes by βCD molecules. The bulky ring structure of carvone enhances the interactions with the interior wall of the permethylated βCD cavity, thus producing more stable transient inclusion complexes between carvone and permethylated βCD on the surface.[17]

Chiral Coating Stability

It is envisaged that the formation of Au-S bond between the cyclodextrinthiol derivatives and gold surface would render the enantioselective coating more environmentally durable. In addition, the length of the thiol spacer would exert an influence on the spatial arrangement and/ or surface concentration of the monolayer, which in turn will affect sensor properties. Accordingly, we examined the stability of chiral coatings having differing lengths of thiol spacers. The results are plotted in Figure 4. The deviation in the resonant frequency of **CD 1-3** (Δf_t) measured over time was compared with the resonant frequency of freshly coated QCMs (Δf_0 at Day 0). These coatings showed improved stability over time in N_2 stream, especially **CD3** with a longer spacer, which only displayed an increase of 9 Hz after 120 days corresponding only to a 5.4 % loss of the original surface coating. Data points with negative frequency deviation were recorded immediately after the chiral sensing measurement indicating that some guest molecules have remained inside the cavities of βCDs despite being purged with N_2 for prolonged periods.

Conclusion

In summary, we have demonstrated the applicability of self-assembled permethylated βCD monolayers as the enantioselective material for QCM chiral sensing. The resulting chiral sensors exhibit enhanced enantiomeric discriminatory ability and excellent long-term stability making it a potential candidate for real-time chiral sensing applications. Moreover, this approach provides a simplified model for a facile study of host-guest interactions: one theoretical plate instead of thousands in GC, room temperature instead of at elevated conditions of GC, and no interference from the substrate.

Support Structural Data

(6^A-ω-mercapto-ethylureado-6^A-deoxy)heptakis(2,3-di-*O*-methyl)-6^B, 6^C, 6^D, 6^E, 6^F, 6^G-hexa-*O*-methyl-*β*-cyclodextrin (CD1) IR (cm^{-1}): 3396 (N-H str), 2931 (C-H str), 1668 (C=O str, urea), 1040 (C-O-C str), ^{13}C-NMR ($CDCl_3$, 75 MHz) δ (ppm): 158.72 (-NH-<u>C</u>O-NH-), 98.73 (C-1), 81.66 (C-2), 80.24 (C-3), 80.02 (C-4), 71.16 (C-6b), 70.83 (C-5), 67.73 (C-6a), 61.11 (H_3<u>C</u>O-C3), 58.81 (H_3<u>C</u>O-C6), 58.28 (H_3<u>C</u>O-C2), 40.82 (-NH<u>C</u>H_2-), 25.43 (-<u>C</u>H_2-SH). ESI-MS: Calcd 1517.4, Found 1517.2. Elemental analysis: Calcd C 51.45%, H 7.70%, N 1.84%, S 2.11%; Found C 51.54%, H 7.51%, N 1.93%, S 1.95%.

(6^A-ω-mercapto-hexanureado-6^A-deoxy)heptakis(2,3-di-*O*-methyl)-6^B, 6^C, 6^D, 6^E, 6^F, 6^G-hexa-*O*-methyl-*β*-cyclodextrin (CD2) IR (cm^{-1}): 3399 (N-H str), 2930 (C-H str), 1657 (C=O str, urea), 1037 (C-O-C str). ^{13}C-NMR ($CDCl_3$, 75 MHz) δ (ppm): 159.00 (NH-<u>C</u>O-NH), 98.76 (C-1), 81.66 (C-2), 81.32 (C-3), 80.22 (C-4), 71.40 (C-6b), 70.84 (C-5), 67.54 (C-6a), 61.18 (H_3<u>C</u>O-C3), 58.90 (H_3<u>C</u>O-C6), 58.47 (H_3<u>C</u>O-C2), 41.22 (-NH<u>C</u>H_2-), 38.67, 30.07, 28.99, 28.14 (-(<u>C</u>H_2)$_4$-), 26.44 (-<u>C</u>H_2-SH). ESI-MS *m/z* Calcd 1573.5, Found 1596.5 for $[M+Na]^+$. Elemental analysis: Calcd C 52.67%, H 7.94%, N 1.78%, S 2.03%; Found C 52.71%, H 7.79%, N 1.76%, S 1.92%.

(6^A-ω-mercapto-undecanylureado-6^A-deoxy)heptakis(2,3-di-*O*-methyl)-6^B, 6^C, 6^D, 6^E, 6^F, 6^G-hexa-*O*-methyl-*β*-cyclodextrin (CD3) IR (cm^{-1}): 3396 (N-H str), 2928 (C-H str), 1659 (C=O str, urea), 1039 (C-O-C str). ^{13}C-NMR ($CDCl_3$, 75 MHz) δ (ppm): 159.02 (NH-<u>C</u>O-NH), 98.70 (C-1), 81.64 (C-2), 81.35 (C-3), 80.26 (C-4), 71.85 (C-6b), 70.97 (C-5), 67.68 (C-6a), 61.14 (H_3<u>C</u>O-C3), 58.83 (H_3<u>C</u>O-C6), 58.36 (H_3<u>C</u>O-C2), 41.27 (-NH<u>C</u>H_2-), 40.36-26.80 (-(<u>C</u>H_2)$_9$-), 24.44 (-<u>C</u>H_2-SH). ESI-MS *m/z* Calcd 1643.6, Found 1666.1 for $[M+Na]^+$. Elemental analysis: Calcd C 54.05%, H 8.22%, N 1.70%, S 1.95%; Found C 54.49%, H 7.82%, N 1.29%, S 1.70%.

Acknowledgements

The authors thank the National University of Singapore for financial support through the research grant R-143-000-070-112. Tong Sun is grateful to NUS for the award of a research scholarship.

[1] S. Ahuja, *Chiral Separation by Chromatograohy* Oxford University Press, Washington D.C., **2000**, pp7-14.

[2] M.T. Cygan, G.E. Collins, T.D. Dunbar, D.L. Allara, C.G. Gibbs, and C.D. Gutsche, Calixarene Monolayers as Quartz Crystal Microbalance Sensing Elements in Aqueous Solution, *Anal. Chem.* **1999**, 71, 142.

[3] J. Ide, T. Nakamoto and T. Moriizumi, Discrimination of aromatic optical isomers using quartz-resonator sensors, *Sens Actuators A* **1995**, 49, 73.

[4] K. Bodenhöfer; A, Hieriemann; J. Seemann; G. Gauglitz; B. Koppenhoefer; W. Göpel, Chiral discrimination in gas phase using different transducers: Thickness shear mode resonators and reflectometric interference spectroscopy, *Anal. Chem.* **1997**, 69, 3058.

[5] K. Bodenhöfer; A, Hieriemann; J. Seemann; G. Gauglitz; B. Koppenhoefer; W. Göpel, Chiral discrimination using piezoelectric and optical gas sensors, *Nature*, **1997**, 387, 577.
[6] S. Flink, F. C. J. M. van Veggel, D.N. Reinhoudt, Sensor functionalities in self-assembled monolayers, *Adv. Mater.* **2000**, 12, 1315.
[7] From our own measurements, a spin-coated 10 MHz QCM device with permethylated βCD, which resulted in 600Hz decrease of resonant frequency after drying in N_2, would display 60-100Hz frequency increases during a period of 45 days after the initial measurement.
[8] L. F. Zhang; Y. C. Wong; L. Chen; C. B. Ching, and S. C. Ng, A facile immobilization approach for perfunctionalised cyclodextrin onto silica via the staudinger reaction, *Tetra. Lett.*, **1999**, 40, 1815.
[9] M.T. Rojas, R. Königer, J.F. Stoddart, and A.E. Kaifer, Supported monolayers containing preformed binding sites. Synthesis and interfacial binding properties of a thiolated β-cyclodextrin derivative, *J. Am. Chem. Soc.* **1995**, 117, 336. (b) M. Weisser, G. Nelles, P. Wohlfart, G. Wenz, and S. Mittler-Neher, Immobilization kinetics of cyclodextrins at gold surface, *J. Phys. Chem.* **1996**, 100, 17893. (c) J.-Y. Lee and S.-M. Park, Electrochemistry of guest molecules in thiolated cyclodextrin self-assembled monolayers: Implication of size-selective sensors, *J. Phys. Chem. B* **1998**, 102, 9940.
[10] H.S. Ji, S. McNiven, K.H. Lee, T. Saito, K. Ikebukuro, I. Karube, Increasing the sensitivity of piezoelectric odour sensors based on molecularly imprinted polymers, *Biosens. Bioelec.* **2000**, 15, 403. (b) T. Nakamoto, A. Iguchi, T. Moriizumi, Vapor supply method in odor sensing system and analysis of transient sensor responses, *Sens. Actuators B*, **2000**, 71, 155.
[11] L. Van Vaeck, A. Adriaens, R. Gijbels, Static secondary ion mass spectrometry (S-SIMS) Part I: Methodology and structural interpretation, *Mass Spectrom. Rev.* **1999**, 18, 1.
[12] A.W. Czanderna and C. Lu, *Applications of Piezoelectric Quartz Crystal Microbalances* Elsevier, New York, **1984**, pp 145.
[13] S. Mitter-Neher, J. Spinke, M. Liley, G. Nelles, M. Weisser, R. Back, G. Wenz, W. Knoll, Spectroscopic and surface-analytical characterization of self-assembled layers on Au, *Biosens. Bioelec.* **1995**, 10, 903.
[14] J. Qian, R. Hentschke, W. Knoll, Superstructures of cyclodextrin derivatives on Au(111): A combined random planting-molecular dynamics approach, *Langmuir* **1997**, 13, 7092.
[15] D. W. Armstrong, W. Li, C. D. Chang, Polar-liquid, derivatized cyclodextrin stationary phases for the capillary gas chromatography separation of enantiomers, *Anal. Chem.* **1990**, 62, 914. (b) A. Berthod, W. Li, D. W. Armstrong, Multiple enantioselective retention mechanisms on derivatized cyclodextrin gas chromatography chiral stationary phase, *Anal. Chem.* **1992**, 64, 873.
[16] T. O'Brien, L. Crocker, R. Thompson, K. Thompson, P.H. Toma, D.A. Conlon, B. Feibush, C. Moeder, G. Bicker, and N. Grinberg, Mechanistic aspects of chiral discrimination on modified cellulose, *Anal. Chem.* **1997**, 69, 1999.
[17] K. B. Lipkowitz, G. Pearl, R. Coner, M. A. Peterson, Explanation of where and how enantioselective binding takes place on permethylated β-cycclodextrin, a chiral stationary phase used in gas chromatography, *J. Am. Chem. Soc.* **1997**, 119, 600.

Formation of Intragap States and Achieving Balance in Carrier Mobilities for a PPV Derivative Bearing the Carbazole Pendants

*Kyungkon Kim[1], Yongsup Park[2], Cheol Eui Lee[3], Jae Won Jang[3], Jung-Il Jin[1]**

[1]Division of Chemistry and Molecular Engineering, Center for Photo- and Electro-Responsive Molecules, Korea University, 5-1, Anam-Dong, Seoul 136-701, Korea
[2]Surface Analysis Group, Korea Research Institute of Standards and Science, Yusung 305-600, Taejon, Korea
[3]Department of Physics, Korea University, 5-1, Anam-Dong, Seoul 136-701, Korea

Summary: We have found by NEXAFS spectroscopic studies that a PPV derivative (CzEh-PPV) bearing the carbazole pendants form new intragap states when contacted with calcium metal. Moreover, the polymer exhibited much more balanced mobilities of holes and electrons than PPV and MEH-PPV. Such characteristics are believed to bring about low turn-on electric field and improved efficiency for the LED device fabricated with the polymer.

Introduction

Since the pioneering work by the Cambridge group on the electroluminescence (EL) properties of ligh-emitting diodes (LEDs) based on poly(p-phenylenevinylene) (PPV),[1] there have been many reports on the EL characteristics of a wide variety of PPV derivatives.[2-8]Recently, we reported[9] that the following polymer, CzEh-PPV, exhibited a very high device efficiency, low turn-on electric field, and a very high maximum brightness when the LED device had the configuration of ITO/ POEDOT/ polymer/ Ca.

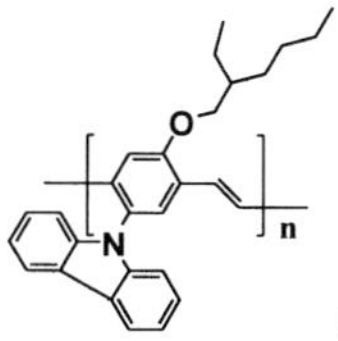

CzEh-PPV

The polymer bears the 2-ethylhexyloxy and carbazole pendant groups, both of which are electron-donating groups. This polymer could be easily prepared by the Gilch polymerization method[10] and are soluble in common organic solvents such as tetrahydrofuran, 1,1,2,2-

 CCC 1022-1360/00/$ 17,50+.50/0

tetrachloroethane, and chloroform. The LED device constructed with this polymer emits green light. The HOMO and LUMO levels of the polymer determined by the UPS method[11] are 5.5 and 3.2 eV, respectively.[9]

In this communication, we would like to report the evolution of the near-edge X-ray absorption fine structure (NEXAFS) spectrum during the deposition of calcium metal on the thin film of CzEh-PPV and carrier mobilities of the polymer measured by the time-of-flight method.

Characteristics of the LED Device Fabricated with CzEh-PPV

Table 1 summarizes the LED device characteristics of CzEh-PPV. The maximum emitted light intensity is observed at 528 nm. For the sake of comparison characteristics of LED devices constructed with PPV and MEh-PPV also are included in the table.

Table 1. LED device characteristics of CzEh-PPV

Device Structure	Φ_{Ext} (%)	Electric Field for 0.1 mA/m^2 (MV/cm)	Maximum Brightness (cd/m^2)
ITO/ PPV/ Al	2.0×10^{-4}	1.1	5
ITO/ CzEh-PPV/ Al	0.010	0.63	20
ITO/ CuPc/ CzEh-PPV/ Li:Al	0.040	0.4	350
ITO/ CzEh-PPV/ Alq_3/ Li:Al	0.38	1.0	1650
ITO/ PEDOT/ MEh-PPV/ Ca/ Al	1.1	0.4	20540
ITO/ PEDOT/ CzEh-PPV/ Ca/ Al	4.4	0.5	30390

A close examination of the experimental data given in Table 1 teaches us several important points: (1) The metal cathodes of lower work function exhibit better device performance. (2) The device of CzEh-PPV reveals a lower turn-on electric field than that of PPV. (3) Utilization of a conducting layer between the anode and the light emitting layer lowers the threshold electric field of the device. (4) The device constructed with CzEh-PPV performs much better than PPV and MEH-PPV. The observations described in (1) and (3) are rather well known facts.[12] Those described in (2) and (4), however, require proper explanations. We believed that NEXAFS spectroscopy and XPS measurement of carrier mobility will be provide us with answers to the scientific questions.

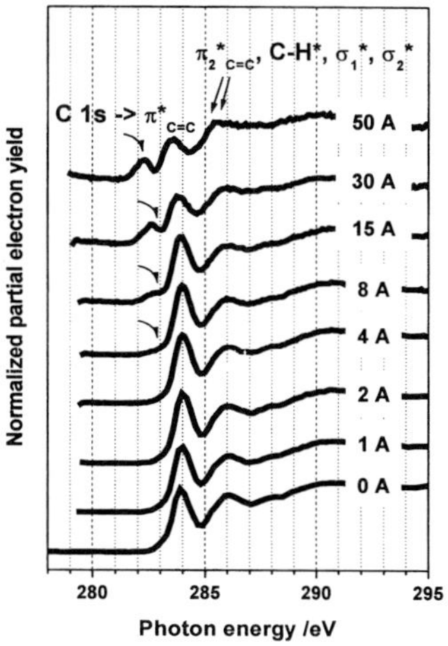

Figure 1. Evolution of NEXAFS C 1s peak spectra during Ca deposition on the film surface of CzEh-PPV.

NEXAFS Spectroscopy and XPS

We studied how NEXAFS spectrum[13, 14] changes on deposition of calcium metal on a thin film of CzEh-PPV and the spectral changes are shown in Figure 1. The figure shows that CzEh-PPV in the virgin state exhibits the lowest energy peak for $C_{1s} \rightarrow \pi^*$ electronic transition at 283.9 eV. The NEXAFS spectrum, however, shows appearance of a new peak within the leading edge of π^* resonance when the thickness of the calcium layer reaches 8Å. The new peak grows in intensity as more calcium is deposited. It also is noted that the original peak position is shifted to slightly lower energy side, 283.5 eV. And the energy difference between the new peak and the original one is 1.3 eV. Although the NEXAFS spectra for MEh-PPV are not given, formation of new intra gap states was observed, just as described for CzEh-PPV.

As one can see from Table 1, the turn-on electric field of LED devices is in the order of MEh-PPV $\approx$ CaEh-PPV < PPV. This order appears to correlate well with the polymers' ability to form new valence states within the band gap when contacted to the calcium metal. The formation of the intragap states is expected to enhance the electron injection into the emitting polymer layer from the calcium electrode by providing additional, lower energy states for electron transfer from near the Fermi level of the cathode metal.

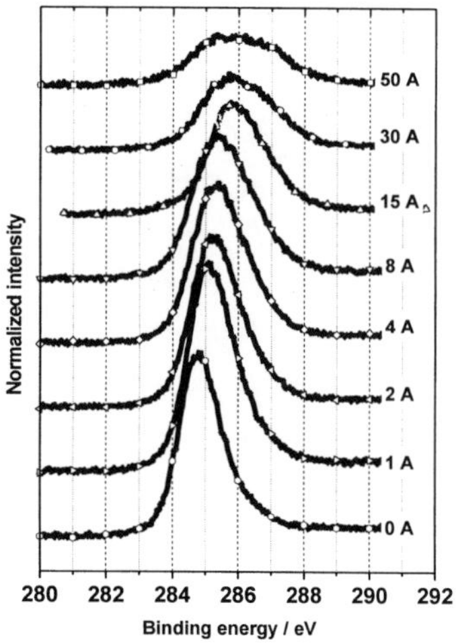

Figure 2. Evolution of XPS C 1s spectra during Ca depositon on CzEh-PPV.

We further studied the interface formation between calcium and CzEh-PPV by the XPS method. We collected the XPS spectra (Figure 2.) of C_{1s} electrons with increasing calcium coverage. We clearly note from Figure 2 that the C_{1s} peak steadily shift to the higher energy side as the thickness of deposited calcium layer increased. This is interpreted as originated from energy level bending induced by the calcium layer. The energy level bending amounts to about 0.7 eV when the thickness of calcium layer reach 30 Å. It, however, should be noted that the bending occurs even when Ca atoms are scarcely deposited, indicating that this bending does not require for the Ca coverage to be metallic. It is conjectured that the individual Ca atoms are involved in electron transfer that results in polaron formation. These polarons formed in the backbone π-electron system will eventually form new states as discussed about in the NEXAFS studies. In other words, the new gap states created near the surface of the polymer layer due to the electron transfer from Ca to the polymer causes the energy level shift resulting in band bending. There are some theoretical studies supporting this supposition.[13, 15]

Balance in Mobilities of Carriers

One of the major reasons for the poor device performance of PPV LEDs is ascribed to the much higher mobility of positive holes than that of electrons.[16] Such an imbalance will result in annihilation of most of the injected holes on the surface of the cathode before they are captured by electrons to form excitons.[17, 18] In order to reduce the undesirable waste of holes, one may apply an electron-transporting / hole-blocking layer between the cathode and the PPV layer. Instead, we have attached the carbazole pendants in CzEh-PPV and wanted to study their influence on the mobilities of carriers and balance in their mobilities. The pendants are directly attached to the phenylene rings of the PPV backbone.

Table 2. Comparison of carrier mobilities*

Polymer	Time of Flight		
	μ_h	μ_e	μ_h/μ_e
PPV	7.0×10^{-4}	3.0×10^{-6}	230
MEh-PPV	1.2×10^{-6}	$\sim 10^{-8}$	$>10^2$
CzEh-PPV	4.0×10^{-5}	7.5×10^{-6}	5.3

*in $cm^2/V \cdot S$

Table 2 summarizes the carrier mobilities of the three polymers measured by the time-of-flight method. [19-21] It is observed that, indeed, the hole mobility in PPV is higher than two hundred times the electron mobility. Surprisingly enough, a similar imbalance is observed for MEh-PPV. We, however, understand that LED devices prepared from MEh-PPV perform much better than those from PPV. This is attributed to mainly three factors: easier electron injection as predicted by the MEh-PPV's LUMO level, formation of intragap states on contact with Ca, and increased interchain distance owing to the presence of two alkoxy pendants one of which is fairly bulky. Increased interchain distance will hinder the formation of interchain polaron pairs[22, 23] that are expected to reduce the radiative relaxation. In contrast, CzEh-PPV shows impressively balanced mobilities of the two opposite carriers, although mobilities of both carriers are much diminished when compared with those of PPV and MEh-PPV. As one can see from Table 2, the ratio of hole mobility to electron mobility is 5.3 for CzEh-PPV.

Let's first pay our attention to the mobilities of the two carriers in CzEh-PPV. We note that, compared to PPV, the hole mobility is reduced to a greater extent than the electron mobility. In fact, the electron mobility remains about the same. This is rather surprising in the light of the fact that the polymer bears the hole-transporting carbazole pendants. It is our present conjecture that the positive holes traveling through the backbone π-electron system are stabilized by the carbazole pendant, which, in turn, is expected to diminish the hole mobility. This conjecture is supported by optically detected magnetic resonance studies of the polymer.[24]

Other Pertinent Experimental Results

Construction of the LED devices requires spin coating of polymers either on the surface of ITO-coated glass or on the surface of electrically conducting PEDOT clad on the ITO electrode. We became interested in the surface composition of the polymers after being coated on the ITO anode. We determined the surface composition by XPS and found that the concentration of the carbazole structures, due to their polar nature, are much less on the surface than in bulk. Since air is hydrophobic, it is expected that film surface is richer in hydrophobic moieties than in polar structures. In other words, the carbazole groups tend to orient themselves in such a way that they are more in contact to the ITO surface. Such an orientation of pendant groups will facilitate hole injection in CzEh-PPV. This may be one of the reasons for the significantly lower turn-on electric field for CzEh-PPV than for PPV. In short, surface composition, not the bulk composition, must be playing a very important role in interface formation and carriers' transfer between the electrode and the polymer.

We also observed that surface composition depends strongly on the polarity of solvent used in spin-coating. In addition, surface smoothness or morphology, strongly dependent on the nature of solvent is critical in achieving high device performance, which emphasizes the importance of better contact between the emitting polymer layer and electrodes.

Experimental

Device fabrication and characterization: The patterned ITO-coated glass slides were cleaned by sequential ultrasonication in acetone, 2-propanol and distilled water for 13 minutes, respectively, and finally cleaned by ultraviolet - ozone for 13 minutes. The conducting polymer solution of poly(3,4-ethylenedioxy-2,5-thiopene) doped with polystyrene sulphonate (PEDOT : PSS) (Bayer) was spin coated with 3300 rpm for 1min and dried at 150 ^{o}C for 30 minutes. The electrical conductivity of this film measured by the four line probe method was 10 Scm^{-1}. The solution (1wt %) of polymer in purified 1,1,2,2-tetrachloroethane was spin coated at 1000 rpm for 1 min. And the film was subjected to thermal treatment at 150 ^{o}C for 1h under vacuum. The Ca cathode 2000 Å thick was vacuum deposited from the tungsten boat at deposition rates of 2Å/s under the pressure of 3.0×10^{-7} torr. An Al capping layer was then evaporated to protect the Ca cathode at a deposition ratio of 4 Å/s under the same pressure. The active area of the EL devices was 4 mm^2. Photoluminescence (PL) and

Electroluminescence (EL) spectra were recorded on a PC1 photon counting spectrofluorometer (ISS Inc., USA). Current-voltage (I-V) characteristics and the intensities of EL emission were simultaneously measured with a Keithley 238 SMU electrometer and a BM7 luminance meter (Topcon Technologies, Inc., USA). Device fabrication and all the measurements were performed in a dry argon filled glove box without exposing to air.

Time-of-flight measurement: Sandwich-type structures of ITO/ polymer (3 um)/ Al (30 nm) were fabricated for the measurement. For carrier generation, an Nd-YAG laser of emitting wavelength of 355 nm and pulse with (FWHM) of 7 nm was used as a light source, and the polymer films were directly exposed to irradiation of the laser. The laser delay time was 150 ns, and an electric field of 20 kV/cm was applied for duration between 5 and 10 ms.

XPS: X-ray photoelectron spectroscopy (UPS) data were acquired at room temperature with a VG ESCALab 220i spectrometer (UK) with a Mg Kα line (1253.6 eV). The base pressure of the analysis chamber was lower than 1 x 10^{-11} Torr and the combined instrumental resolution was about 1.0 eV. Electron energy analyzer has a hemispherical type.

UPS: Ultraviolet photoelectron spectroscopy (UPS) data were acquired at room temperature with VG ESCALab 220i spectrometer (UK) with a VG UV lamp. UPS analysis was performed using He I (21.2 eV) photons. In the case of UPS, the combined instrumental resolution was about 0.1 eV.

Near-Edge X-Ray Absorption Fine Structure (NEXAFS): NEXAFS experiment was done on the 2B1 spherical grating monochromator beamline at the Pohang Synchrotron Lab. The base pressure of the experimental chamber was 1 x 10^{-10} Torr. For the detection of the output signal as a form of electron or photoelectron, photocurrent was measured, i.e., ‘total electron yield’ method was used.

Acknowledgement

This research was supported by the Korea Science and Engineering Foundation through the Center for Electro- and Photo-Responsive Molecules, Korea University. K. Kim wishes to acknowledge the support by the Ministry of Education through the Brain Korea 21 program. NEXAFS experiments at PLS were supported by the Ministry of Science and Technology and Pohang Steel Company. We also gratefully thank Ms. K. J. Kim and Dr. B. S. Kim in Pohang Accelerator Laboratory (2B1 beam line) for carrying out NEXAFS experiments.

[1] J. H. Burroughes, D. D. C. Bradley, A. R. Brown, R. N. Marks, K. Mackay, R. H. Friend, P. L. Burn, A. B. Holmes, *Nature* **1990**, *347*, 539.
[2] S.-J. Chung, J.-I. Jin, K. K. Kim, *Adv.Mater.* **1997**, *9*, 551.
[3] S.-J. Chung, K.-Y. Kwon, S.-W. Lee, J.-I. Jin, C. H. Lee, C. E. Lee, Y. Park, *Adv.Mater.* **1998**, *10*, 1112.
[4] D. Braun, A. J. Heeger, *Appl. Phys. Lett.* **1991**, *58*, 1982.
[5] A. Kraft, A. C. Grimsdale, A. B. Holmes, *Angew. Chem., Int. Ed. Engl.* **1998**, *37*, 402.
[6] T. Zyung, D.-H. Hwang, I.-N. Kang, H.-K. Shim, W.-Y. Hwang, J.-J. Kim, *Chem. Mater.* **1995**, *7*, 1499.
[7] D. W. Lee, K. Y. Kwon, J. I. Jin, Y. Park, Y. R. Kim, I. W. Hwang, *Chem. Mater.* **2001**, *13*, 565.
[8] H.-K. Shim, J.-I. Jin, *Adv. Polym. Sci.* **2002**, *158*, 193.
[9] K. Kim, Y. R. Hong, S. W. Lee, J. I. Jin, Y. Park, B. H. Sohn, W. H. Kim, J. K. Park, *J. Mater. Chem.* **2001**, *11*, 3023.
[10] H. G. Gilch, W. L. Weelwright, *J. Polym. Sci., Part A: Polym. Chem.* **1966**, *4*, 1337.
[11] A. Schmidit, M. L. Anderson, N. R. Armstrong, *J. Appl. Phys.* **1995**, *78*, 5619.
[12] I. D. Parker, *J. Appl. Phys.* **1994**, *75*, 1656.
[13] V. E. Choong, Y. Park, B. R. Hsieh, Y. Gao, in *Photonic Polymer Systems-'Fundamentals, Methods, and Applications'*, (Eds. D. L. Wise, G. E. Wnek, D. J. Trantolo, T. M. Cooper, J. D. Greesser), Marcel Dekker, New York, **1998**, Chap. 5.
[14] J. Stöhr, *NEXAFS Spectroscopy,* 1st ed., Springer-Verlag, New York, **1992**.
[15] K. Konstadinidis, F. Papadimitrakopoulos, M. Galvin, R. L. Opila, *J. Appl. Phys.* **1995**, *77*, 5642.
[16] C. Adachi, T. Tsutsui, S. Saito, *Appl. Phys. Lett.* **1990**, *56*, 799.
[17] A. Kraft, P. L. Burn, A. B. Holmes, D. D. C. Bradley, R. H. Friend, J. H. F. Martens, *Synth. Met.* **1993**, *57*, 4163.
[18] N. C. Greenham, R. H. Friend, A. R. Brown, D. D. C. Bradley, K. Pitchler, P. L. Burn, A. Kraft, A. B. Holmes, *Proc. SPIE-Int. Soc. Opt. Eng.* **1993**, *1910*, 84.
[19] Z. Burshtein, D. F. Williams, *Phys. Rev. B: Condens. Matter* **1977**, *15*, 5769.
[20] W. D. Gill, *J. Appl. Phys.* **1972**, *43*, 5033.
[21] H. Sher, E. W. Montroll, *Phys. Rev. B: Condens. Matter* **1975**, *12*, 2455.
[22] R. H. Friend, D. D. C. Bradley, P. D. Townsend, *J. Phys. D: Appl. Phys.* **1987**, *20*, 1367.
[23] M. Yan, L. J. Rothberg, F. Papadimitrakopoulos, M. Galvin, T. M. Miller, *Phys. Rev. Lett.* **1994**, *72*, 1104.
[24] C.-H. Kim, J. Shinar, D.-W. Lee, Y.-R. Hong, J.-I. Jin, in *Annual APS March Meeting 2002,* Indianapolis, Indiana, **2002**, Session B11.

Core-Shell Particles: Building Blocks for Advanced Polymer Materials

*Hung H. Pham, Eugenia Kumacheva**

Department of Chemistry, University of Toronto, 80 St. George Street, Toronto, Ontario, Canada, M5S 3H6

Summary: Polymer particles with submicrometer dimensions show promising applications in "bottom-to-top approach" to fabrication of materials with periodic structure, function, and composition. A novel approach to producing such materials is proposed, which employs core-shell particles with specific structures and compositions. We report on the synthesis of core-shell particles using interfacial polymerization and heterocoagulation techniques. The compositions of core-forming material and/or the shell-forming polymers were selectively controlled to be make the cores or the shells rigid or fluid, fluorescent or non-fluorescent, organic or inorganic. Several potential applications of nanocomposite materials obtained from these particles are demonstrated, including three-dimensional optical data storage and optical limiting and switching.

Introduction

Recently, nanocomposite materials with periodic structures have stimulated great interest in Materials Science because of their potential applications as chemical sensors, optical data storage devices, and optical limiters and switches.[1] In the "bottom-to-top" approach to fabrication of such materials functionalized particles with a very narrow size distribution and dimensions varying from several hundred nanometers to several microns can be viewed as the building blocks or structural units. Furthermore, the utilization of submicrometer particles with a core-shell or multilayer morphology allows one to obtain complex compositional and structural patterns in the ultimate nanocomposite material.

Our group developed a "core-shell" strategy for fabrication of periodically nanostructured polymer-based materials.[2] An overview of the core-shell approach is given in Figure 1. The essential feature of the core-shell particles synthesized in Stage A is a specific relation between the glass transition temperatures, T_g, of the core-forming polymer (CFP) and the shell-forming polymer (SFP): the glass transition temperature of the SFP should be substantially lower than that of the CFP, that is, $T_{g,SFP} < T_{g,CFP}$. The core-shell microspheres are assembles in a periodic array in Stage B and annealed at the temperature $T_{g,SFP} < T_{annealing} < T_{g,CFP}$ (Stage C), at which the SFP softens and flows, forms ultimately a continuous matrix,

 CCC 1022-1360/00/$ 17,50+.50/0

while the CFP remains intact and produces the disperse phase of the nanocomposite material. This strategy provides several levels of control over the structure and function of the nanostructured material, that is, over particle dimensions, number density, and the composition of the particles and the matrix. First, core-shell particles can contain organic or inorganic polymers in the core and/or in the shell. Second, a combination of inorganic cores and polymer shells can be used. Finally, core- and shell-forming polymers can be physically doped or chemically functionalized with different species tailoring interesting novel properties to the ultimate material. In particular, when the core and the shell have different optical properties, e.g., refractive index, nonlinearity, or fluorescence, the ultimate periodically structured material can perform as a photonic crystal.

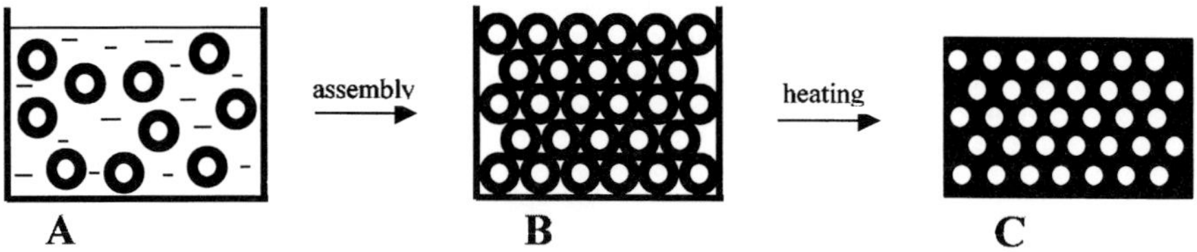

Figure 1: Schematics of the "core-shell" approach to producing polymer-based nanostructured materials. A: synthesis of core-shell particles; B: assembly of core-shell particles in periodic arrays; C: heat processing of core-shell particles to produce nanocomposite material.

Three different methods used in the preparation of core-shell particles are shown in Fig. 2. Interfacial polymerization of the shell on the surface of a polymeric or an inorganic core (Fig. 2a) is preferred when the core-forming material (CFM) and the shell-forming polymer (SFP) have a reasonably high compatibility. When the CFM and the SFP have a high interfacial tension, electrostatic interaction approach is the method of choice.[3] In this approach, dispersions of oppositely charged large particles of the CFM and small particles of the SFP are mixed and annealed, as shown in Fig. 2b. Upon mixing, electrostatic attraction between the beads leads to the formation of individual monodisperse heterocoagulates. Annealing them at the temperature exceeding T_g of the SFP results in the formation of core-shell particles.

Finally, organic or inorganic cores can be encapsulated with a polymeric shell by controlled phase separation technique, as shown in Fig. 2c.[4] In this approach, particles of the CFM are dispersed in the solution of the SFP. Controlled phase separation of the SFP solution is induced by either changing temperature or adding a non-solvent. Precipitation of the SFP on the core surface, accompanied by heat processing results in the formation of core-shell particles.

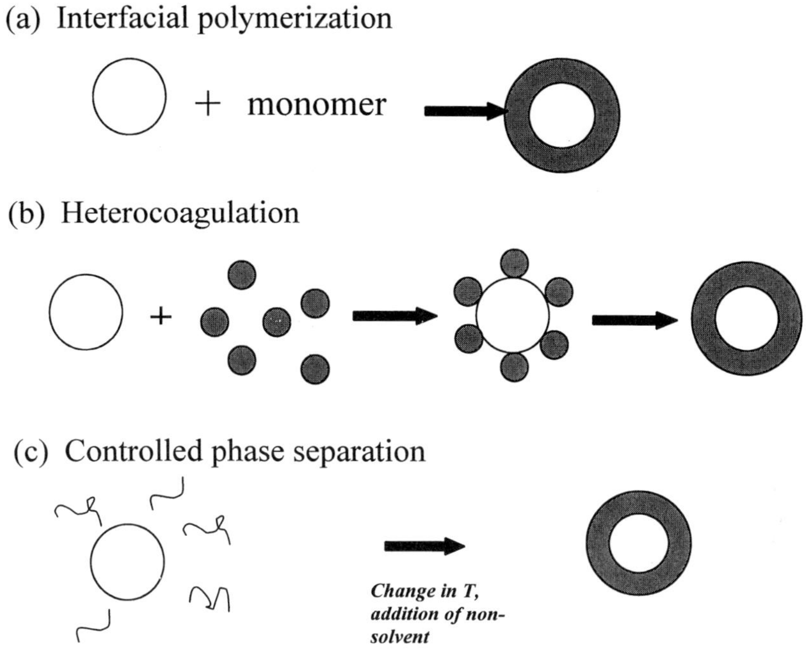

Figure 2: Methods used for preparation of core-shell particles.

In this paper we focus on interfacial polymerization and controlled heterocoagulation which have been extensively utilizing in our laboratory to produce core-shell particles. In addition, we also demonstrate and discuss several potential applications of these particles in fabrication of polymer materials with advanced optical properties.

Polymeric Core-Shell Particles Obtained by Interfacial Polymerization

Emulsion polymerization is a common technique used in production of organic core-shell particles. Extensive studies have been carried out to examine the effect of polymer/water and polymer/polymer interfacial tension,[5-7] the effect of initiator,[8,9] the mode of monomer addition (batch *vs*. semi-batch),[10-12] and the effect of cross-linking agents[13] on morphology of the composite latex particles.

Here, we modified the approach introduced by Rudin *et al.*[14] to prepare monodisperse core-shell particles with the size varying from ca. 200 nm to 1.5 μm. Core-shell particles with a hard core and a softer shell were synthesized to produce optically responsive polymer nanocomposite material with a periodic structure shown in Fig. 1c.[15] The core forming polymer, CFP, was a homopolymer of poly(methyl methacrylate), PMMA, and the SFP was a random copolymer of poly(methyl methacrylate) and poly(butyl methacrylate), P(MMA-co-BMA).[16] The CFP was functionalized with a small amount of 2-[methyl-(7-nitro-2,1,3-benzooxadiazol-4-ylamino]ethyl-2-methyl methyl methacrylate (NBD-MMA), a fluorescent dye-labelled comonomer. Ethylene glycol dimethacrylate (EGDMA) was used as a cross-linking agent to suppress diffusion of dye-labelled molecules from the core to the shell.

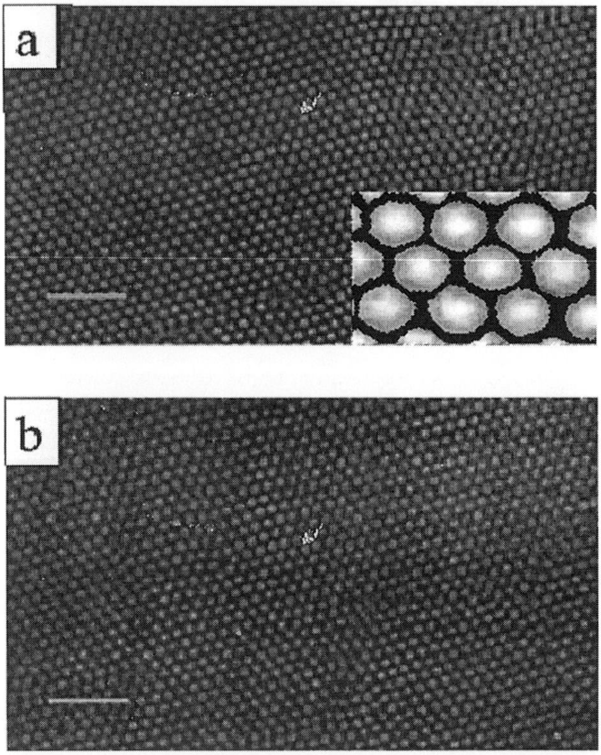

Figure 3. Laser confocal fluorescent microscopy (LCFM) image of the nanostructured material prepared from core-shell particles with rigid 650 nm PMMA cores and 205 nm P(MMA-co-BMA) shells. (a) surface; (b) 30 μm below the surface. Scale bar is 10 μm.

It was found that in order to produce a void-free material, the minimum thickness of the shell had to be at least $0.2r_c$, where r_c is the core radius. The glass transition temperature of the

P(MMA-co-BMA) shell was tuned to 90 °C by adjusting the MMA/BMA ratio to 2/1, while the glass transition temperature of the core was 134 °C. A continuous nanocomposite film was formed by slowly settling dispersion of core-shell particles at room temperature and then annealing the array at 110 °C, that is, about 20 °C above $T_{g,CFP}$. Figure 3 shows the surface and bulk morphologies of the nanocomposite, examined by laser confocal fluorescence microscopy (LCFM). The core has a diameter of 640 nm and the shell has a thickness of 205 nm. As shown in the images, the SFP forms a continuous optically inert phase (dark background), whereas fluorescent CFP particles remain intact (bright domains). The distance between the fluorescent cores is controlled by the thickness of the shell.

The core-shell approach to fabrication of periodically nanostructured materials was further modified by using three-layer latex particles with fluid cores.[17] Latex particles with a fluid core, rigid shell 1, and matrix-forming shell 2 were used as the building blocks in the fabrication of polymer nanostructured materials with liquid inclusions.[18] This work was motivated by the fact that the speed of photochemical reactions of dyes and chromophores could be significantly increased at temperatures exceeding the T_g of the host polymer.[19] In our work, fluid core particles having $T_g = -1.6$ °C and an average diameter of 300 nm were synthesized by copolymerizing butyl acrylate (BA), and MMA in the weight ratio of 1.5. The CFP were copolymerized with a small amount of NBD-MA and cross-linked with 1.5 mol % of EGDMA. In the second stage, the fluid cores were encapsulated with a thin rigid shell by copolymerizing BA and MMA in the BA/MMA weight ratio of 0.05. The challenge in synthesis of these fluid core-hard shell particles was to suppress rupture of shell 1, which released the fluorescent CFP in the surroundings. The rupture phenomenon occurred during the synthesis of the rigid shell stage at 80 °C and/or during cooling of the core-shell particles from 80 °C to room temperature. This feature was caused by the difference in thermoexpansion coefficients of the CFP and SFP-1, which generated pressure gradients in shell 1. It was found that rapture was closely related to the extent of cross-linking of the SFP-1. An increase in the amount of EGDMA to 9 mol % resulted in an increase of Young modulus of the SFP-1 sufficient enough to suppress shell rupture. The second shell (SFP-2) was consequently synthesized on the surface of the bilayer particles by copolymerizing MMA and BMA in the weight ratio of 1/1. The specific relationship between the glass transition temperatures, T_g's, of the polymers incorporated into the core-shell particles was $T_{g,CFP} < T_{g,,SFP\text{-}2} < T_{g,SFP\text{-}1}$. Nanocomposite polymeric films were obtained by heat processing of the arrays of three-layer particles at temperature, $T_{gSFP\text{-}2} < T_{ann} < T_{g,SFP\text{-}1}$. The morphology of the nanostructured material was studied by examining the distribution of the fluorescent dye in the polymer film.

Figure 4a shows LCFM image of the nanocomposite material obtained from the three-layer particles with fluid cores, while Fig. 4b illustrates the morphology of the control sample, prepared from the dispersion of the two-layer core-shell particles with rigid fluorescent cores. The control sample obtained from the core-shell particles with rigid PMMA cores and P(MMA-co-BMA) shells showed a higher optical contrast between the particles and the matrix, as shown in Fig. 4b. However, the film obtained from the three-layer particles with 9 mol % EGDMA in SFP-1 also had a reasonably high contrast, indicating that rupture of the SFP-1 was essentially suppressed.

In the next step, the distribution of the fluorescence species between the CFP, SFP-1, and SFP-2 was examined quantitatively. All LCFM images were taken at the same magnification at depth 10 μm below the surface of the sample using the same microscope settings (brightness, contrast, intensity of sample irradiation and the number of scans). The spatial variations of fluorescence intensity in the film prepared from the particles with fluid cores and in the control sample are shown in Fig. 4c and d, respectively. In fluorescence line profiles, the distribution of the fluorescent dye between the different phases in the films was characterized by the peak-to-well height, the width of the peaks, and the average brightness of the fluorescent domains. A contrast parameter, CP, calculated as the average of 50 values of the peak-to-well heights was equivalent to the signal-to-noise ratio in the nanocomposite film, and it characterized the distribution of the fluorescent molecules in the material. The values of the CP were 152 and 135 arb.u. for the films obtained from the control latex and from the three-layer particles, respectively. Thus, the CP value for the film of interest was ca. 89% of the CP of the control film. The average peak widths measured at their half-height were 0.42 and 0.52 μm for the control film and for the films prepared from three-layer particles, respectively. Comparison of these values with the diameter of the core particles of 0.3 μm showed a somewhat larger size (23 % larger) of the fluorescent domains in the films with liquid inclusions than that in the control film. Finally, the average brightness of the nanostructured films was found by measuring the average fluorescent intensity of the material. The fluorescent intensities of 500 experimental points on the fluorescence line profiles in Fig. 4c were summed up, and the total intensity was divided by the number of points. Since the rigid cores of the control microspheres and the fluid cores of the three-layer microbeads contained the same amount of fluorescent NBD molecules, it was anticipated that the total fluorescence intensity from the individual polymeric films would be very close, irrespective of the distribution of dye molecules in the nanocomposite material. Indeed, the values of average brightness for the control sample and for the polymeric films obtained from the three-layer microbeads were 151 and 141 arb.u., respectively.

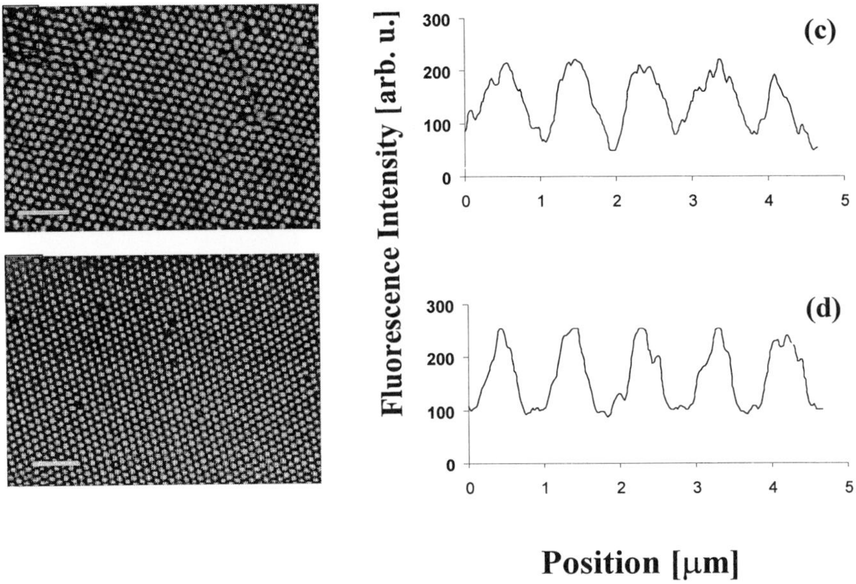

Figure 4. LCFM images of the nanostructured material obtained from particles with fluid cores (a) and core-shell particles with rigid cores (b). The images are taken 10 μm below the surface. Scale bar is 10 μm. Fluorescence line profiles: (c) films obtained from three-layer particles; (d) film obtained from control sample.

Thus the use of three-layer core-shell particles as the building blocks led to the formation of the periodically nanostructured material with liquid inclusions, in which the spatial variation in fluorescent properties was close to that in the control film built from the core-shell particles with rigid cores. The three-layer morphology of the particles, in particular, the existence of the intermediate highly cross-linked shell 1 was extremely important in the fabrication of such materials, since SFP-1 formed a rigid wall confining fluid "containers" and suppressing the release of the fluid fluorescent CFP into the surroundings.

Furthermore, it was found that in the core-shell approach covalent attachment of dyes or chromophores to the CFP or SFP accompanied by cross-linking of the corresponding polymer was vital for the preparation of the nanostructured material with a strong modulation in optical

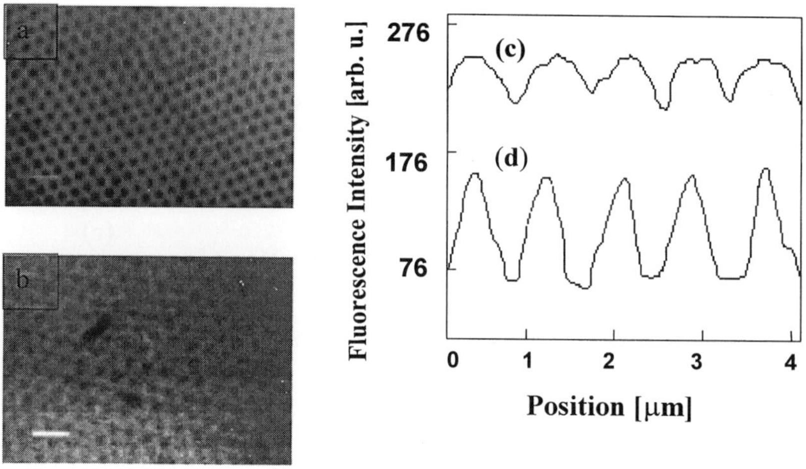

Figure 5: LCFM images of the nanostructured material obtained from core-shell particles with fluorescent matrix. (a) NBD dye is covalently attached to the crosslinked SFP; (b) NBD dye is mixed with the SFP. The scale bar is 2 μm. Fluorescence line profiles in (c) and (d) correspond to samples (b) and (a), respectively.

properties.[20] The effect of chemical attachment of dyes and polymer cross-linking versus physical incorporation of dyes in a polymer was demonstrated for materials with "inverse" structure. Figures 5a and b show the morphology of such materials obtained from the core-shell particles with fluorescent shells. When NBD-MA was copolymerized with P(MMA-co-BMA) shell-forming polymer and the SFP was cross-linked with EGDMA, high optical contrast between the fluorescent matrix and the particles was obtained, as shown in Fig. 5a. Blending of the dye with the same SFP (Fig. 5b) led to a substantial reduction in optical contrast, as a result of the dye-labeled molecules diffuse between the cores and the shells during synthesis and /or annealing. These results were further confirmed by analysis of the fluorescence line profiles as described above. Figure 5c shows the fluorescence line profiles obtained from image analysis of Fig. 5a and b. In the films obtained from the latex particles in which the dye molecules were covalently attached to the crosslinked SFP (bottom curve), the average value of CP was ca. 2.5 times higher than that obtained from the particles in which the dye molecules were physically mixed with the SFP (top curve). Covalent attachment of the NBD dyes to the non-crosslinked SFP led to the intermediate mean value of CP. To conclude, chemical attachment of dye molecules to the corresponding cross-linked polymer was critical in producing high quality nanostructured materials.

Polymeric and Hybrid Core-Shell Particles Obtained by Controlled Heterocoagulation

Heterocoagulation of large and small oppositely charged colloid particles, accompanied by spreading of small beads over the surface of large spheres (Fig. 2b) offers a promising alternative to synthesis of core-shell particles. This approach is particularly useful for the production of core-shell particles with polar cores and nonpolar shells since preparation of these particles using interfacial polymerization often results in nucleation and growth of secondary particles and/or in synthesis of particles with an "inverse" structure. In the first series of experiments, rigid cores were synthesized from polypyrrole (PPy),[21] a polar polymer with high surface energy, whereas in the second series, inorganic SiO_2-$TiOSO_4$ (SiO_2-TS) cores were used. Both types of particles carried a positive charge.

The choice of PPy as a core-forming polymer was motivated by the fact that core-shell particles with conductive cores and dielectric shells can be used as the building blocks in fabrication of nanostructured materials with periodically modulated conductive properties. In such materials an ordered array of monodispersed spheres synthesized from a conjugated polymer is dispersed in an insulating or a semi-conductive polymer matrix. Nanostructured materials produced in this fashion have potential applications in sensor and nanocapacitor devices, and in nano-resettable fuses. Inorganic SiO_2-TS particles were used to achieve a high refractive index modulation in the ultimate nanocomposite material.

The individual core particles (PPy or SiO_2-TS particles) were mixed with small negatively charged beads of PMMA, P(MMA-co-BMA), or PBMA. Heterocoagulation induced by electrostatic interaction, followed by heating the dispersion at temperature above the glass transition temperature of the polyacrylic particles generated core-shell particles.

PPy core particles were prepared by dispersion polymerization. Interfacial polymerization of PMMA on the surface of PPy using $K_2S_2O_8$, 2,2'-azobis(2-methyl propionitrile) and 4,4'-azobis(4-cyanovaleric acid) as the initiators proved to be problematic. In the first two cases polymer conversion was relatively low (<40%), whereas with the latter initiator, broad particle size distribution was often obtained. In contrast to interfacial polymerization, electrostactically-driven heterocoagulation proved to be an efficient way to the preparation of PPy core-polyacrylate shell particles. To obtain monodisperse core-shell particles with smooth polyacrylic shells several requirements had to be fulfilled. First, monodisperse heterocoagulate units could be obtained only at the optimum total concentration of particles in the dispersion of ca. 0.1 wt %. In more concentrated systems, notable particle aggregation was observed. Second, a critical number ratio of small-to-big particles, N_S/N_L, existed above which individual monodisperse heterocoagulates were obtained. Figure 6a shows the variation in dimensions of heterocoagulate units as a function of N_S/N_L for 204 nm PPy beads mixed with 48 nm PMMA particles. For N_S/N_L < 84, the mean size of the heterocoagulate units was substantially larger than ca. 300 nm anticipated for individual heterocoagulate units, which

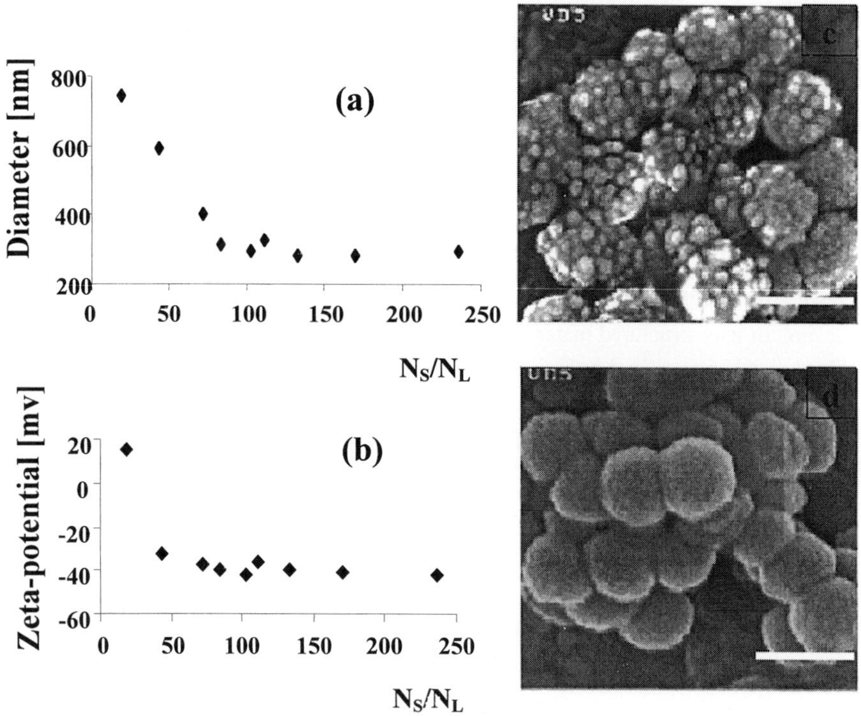

Figure 6. Effect of N_S/N_L on the variation in diameter (a) and electrokinetic potential (b) of heterocoagulate units formed by electrostatically-driven heterocoagulation of 204 nm PPy beads and 48 nm PMMA particles. SEM images of PPy/PMMA-co-PBMA composite particles prior to (c) and after annealing for 23 h annealing at 90 °C (d). Ns/N_L = 180. Scale bar 300 nm.

indicated strong uncontrolled particle aggregation. For 84 < N_S/N_L < 112, the average heterocagulate size gradually reduced to ca. 300 nm and for N_S/N_L > 112, no further change was observed. These results correlated with the variation in ζ-potential of heterocoagulates

(Fig. 6b). In Fig. 6b, the value of ζ-potential gradually decreased and for N_S/N_L= 84 it reached a constant value of – 40 mV, which was close to that of the individual PMMA beads. The dependence of heterocoagulate size and ζ- potential could be explained by bridging of several PPy beads through the small polyacrylic particles, that is, in excess of PPy particles.

The role of charge of cationic and anionic particles was two-fold. A large contrast between the charges of PPy and polyacrylic particles resulted in a strong driving force for heterocoagulation, however repulsion between the similarly charged small beads adsorbed on the core surface counteracted their dense packing. Low packing density of the small particles resulted in a raspberry structure of the shells.[22] Figure 6c and d show PPy/PMMA-co-PBMA heterocoagulate units imaged before and after heat processing. For optimized N_S/N_L after annealing the raspberry structure of the heterocoagulate units was substantially smeared.

A similar approach was used for the preparation of hybrid core-shell particles. SiO_2-TS particles encapsulated with P(MMA-co-BMA) copolymer. Monodisperse SiO_2 particles with dimensions varying from 300 to 800 nm were prepared using sol-gel process. These particles were then coated with TS using a modified approach of Matijevich.[23] The process of coating was driven by electrostatic attraction between the negatively charged silica particles and positively charged TS clusters at pH = 4; therefore TS coating had a raspberry structure.[24]After heterocoagulation, the SiO_2-TS beads carried a small positive charge. Figure 7a shows the results of Energy Dispersion X-Ray analysis of the composite SiO_2-TS microspheres. In this graph the weight ratio TS/SiO_2 in the particles is plotted as a function of ϕ, the ratio of the mass of TS in the system to the total surface area of silica particles. When ϕ increased, the fraction of TS in the composite particles grew due to the more complete coverage of the SiO_2 core. On the basis of the EDS measurements, for the complete coverage of the silica cores with TS achieved at ϕ = 0.085 g/cm^2, the weight ratio TS/SiO_2 reached ca. 35%, this ratio corresponded to ca. 26 wt % of TS in the composite particles.

Encapsulation of the composite SiO_2-TS cores with a polymeric shell was accomplished via electrostatic heterocoagulation of the cationic SiO_2-TS particles and anionic P(MMA-co-BMA) microbeads. The requirements for controlled heterocoagulation were similar to those described for PPy-polyacrylic particles. Upon heating SiO_2-TS/P(MMA-co-BMA) heterocoagulates obtained at sufficiently high N_S/N_L formed core-shell particles with reasonably smooth shells. Figures 7b and c show the SEM images of the SiO_2-TS/P(MMA-co-BMA) heterocoagulates before and after annealing, respectively. A near complete polymer spreading was observed and the surface roughness of the hybrid particles was almost smoothed over. Some voids remaining in the P(MMA-co-BMA) shells resulted presumably from the polymer flow and penetration in the gaps between the TS clusters adhering to the surface of silica cores.

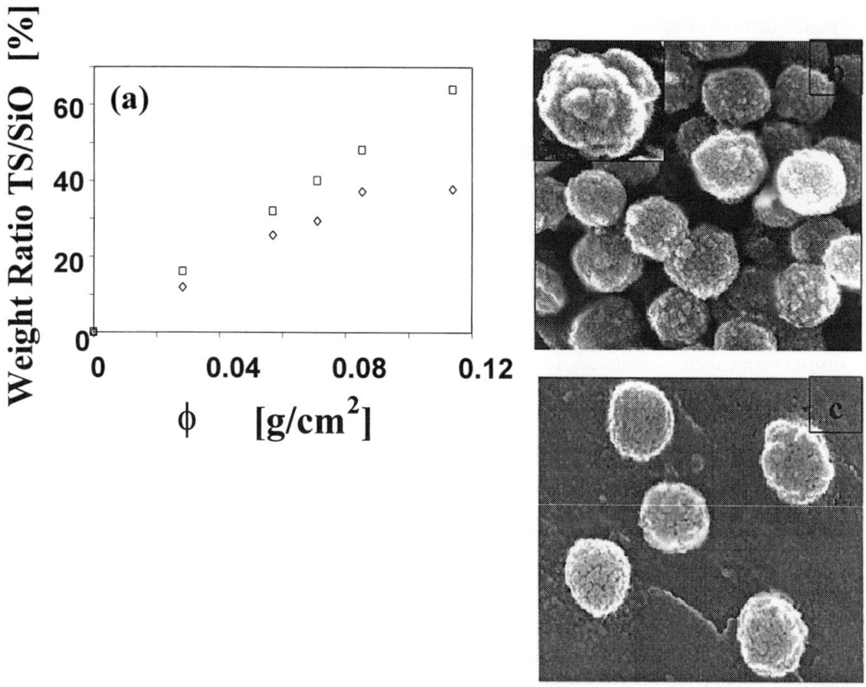

Figure 7. (a) Experimental (◊) and calculated (□) weight ratios of TS/SiO_2 in th composite particles plotted as a function of the weight ratio of TS/surface area of silica. (b) SEM images of SiO_2-TS/PMMA-PBMA hybrid particles obtained at $N_S/N_L = 119$ (c) after heat processing.

Optical Applications of Polymer-Based Materials Produced from Core-Shell Particles

Three-dimensional optical data storage. Further progress in information technologies depends critically on the development of new materials for high-density optical memory storage. As an alternative to two-dimensional (2D) data storage, 3D media promise a dramatic increase in memory capacity since the storage density scales as $1/\lambda^2$ and $1/\lambda^3$, where λ is the wavelength of the reading beam, for 2D and 3D optical memories, respectively. Nanostructured materials with a periodic modulation in optical properties (photonic crystals) show great promise as a medium for high-density 3D optical data storage.[25]

Polymer nanocomposite material with a periodic close-packed array of fluorescent particles periodically embedded in an optically inert matrix (Fig. 3) was examined with respect to 3D optical writing/reading. In the two-photon recording experiments, we employed a Biomedical Photometrics confocal microscope equipped with a home-built femtosecond laser. Optical recording in the polymer nanocomposite was achieved by addressing and photobleaching the individual fluorescent particles. Pulses were typically 100 fs centered at 844 nm. Reading out of the resulting bit pattern was accomplished with the same two-photon confocal microscope using 0.5 mW average power laser fluences. Figure 8a shows a two-dimensional plane of the nanostructured material located at the depth of ca. 200 mm, in which optical recording was carried out by photobleaching fluorescent particles for approximately 150 ms at 5 mW of average power. A photobleached fluorescent bead appears as a black spot in the periodic structure of the nanocomposite. Since the absorption peak of the dye is 476 nm and the polymer is transparent in the visible and near IR range, absorptive coupling of the laser light occurred through two-photon absorption by the NBD dye. The photobleaching of the NBD

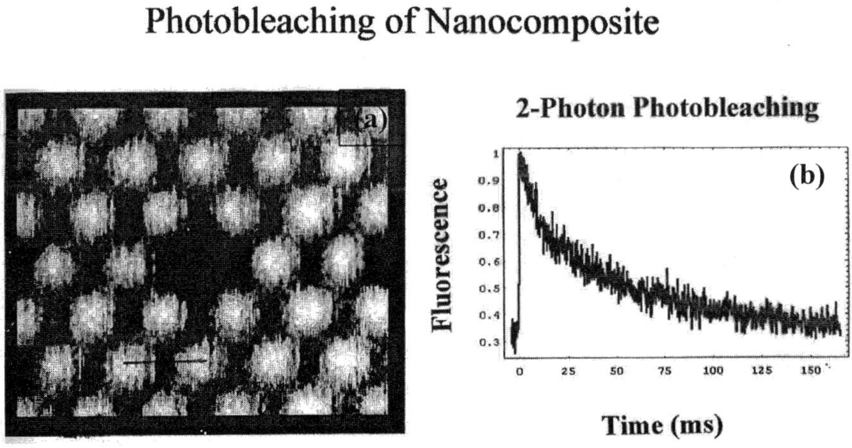

Figure 8: Two-photon writing in a 3D nanostructured material induced by local photobleaching of the fluorescent dye incorporated into the core particles. (a) Structure of the addressed plane located 200 μm from the top surface. The center-to-center particle distance is 1 μm. (b) Fluorescence decay *vs.* time for the addressed particle.

dye caused by photooxidation was irreversible, providing the basis for permanent data storage.

As can be seen from Fig. 8b showing a decay of fluorescence in time, a contrast ratio of 2:1 in fluorescence intensity of unbleached to bleached particles, respectively, was achieved after an exposure time of 20 ms (Fig. 8b).

In these experiments, core-shell particles with rigid fluorescent cores were used as the building blocks of the nanostructured recording media. Preliminary results obtained in our laboratory showed that NBD dye was photobleached substantially faster when the nanostructured material was fabricated from the three-layer particles with a fluid fluorescent core.

Optical limiters and switches. Recently, Sargent *et al*[26] have demonstrated that nanostructured materials with periodically modulated nonlinear refractive index have potential applications in optical limiting and switching devices. In such materials, the linear refractive index of the particles and the matrix are very close, whereas their nonlinear refractive indices are different and opposite in sign. This feature can be accomplished by incorporating nonlinear optics species in the core and in the shell of the core-shell particles used as the structural units in the fabrication of such materials. Covalent attachment of organic species, e.g., azobenzenes to the CFP or SFP or doping of the CFP or SFP with inorganic nanocrystals (quantum dots) proved to be a very efficient method in tuning optical limiting properties of nanocomposite films. Currently, these experiments are conducted in our laboratory in collaboration with Sargent's group.

Conclusion

The core-shell strategy shows a broad avenue to the fabrication of advanced polymer materials. A large number of possible combinations of the core-forming materials and the shell-forming polymers allows for production of nanostructured materials with various structural and compositional patterns. A particular method used for the preparation of core-shell particles depends on the corresponding polymers and should be chosen with caution. Core-shell particles with tunable optical properties great potential applications as high density three-dimensional optical data storage and optical limiting.

[1] Lee, K., Asher, S.A. *J. Am. Chem. Soc.* **2000**, *122*, 9534; Siwick, B.J., Kalinina, O. Kumacheva, E., Miller, R.J.D., Noolandi, J. *J. Appl. Phys.* **2001**, *90*, 5328; Brzozowski, L., Sargent, E.H. *IEEE J. Quant. Eletron.* **2000**, *36*, 550.

[2] Kumacheva E., Kalinina O., Lilge L. *Adv. Mater.* **1999**, *11*, 231.

[3] Okubo M., Lu Y. *Col. Surf.* **1996**, *109*, 49; Ottewell R.H. Schofield A.B., Waters J.A., Williams N.S. *J. Col.*

Polymer Sci. **1997**, *275*, 274; Furusawa K., Velev O.D. *Col. Surf. A* **1999**, *159*, 359.
[4] Dudnik V., Sukhorukov G.B., Radtchenko I.L., Mohwald H. *Macromolecules* **2001**, *34*, 2329.
[5] Daniel J.C. *Makromol. Chem. Suppl.* **1985**, 10/11, 359.
[6] Sutterlin N. *Makromol. Chem. Suppl.* **1985**, 10/11, 403.
[7] Min T.I., Klein A., El-Aasser M.S., Wanderhoff J.W. *J. Polym. Sci., Polym. Chem.* **1983**, *21*, 2845.
[8] Chen Y-C., Dimonie V.L., El-Aasser M.S. *J. Polym. Sci., Polym. Chem.* **1991**, *42*, 1049.
[9] Jonsson J-E., Hassander H., Hansson L.H., Tornel B. *Macromolecules* **1994**, *27*, 1932.
[10] Okubo M., Yamada Y., Matumoto T. *J. Polym. Sci., Polym. Chem.* **1980**, *18*, 3219.
[11] Lovell P.A., McDonald J., Saunders D.E.J., Young R.J. *Polymer* **1993**, *34*, 61.
[12] Jonsson J-E., Hassander H., Hansson L.H., Tornel B. *Macromolecules* **1991**, *24*, 126.
[13] Hu R., Dimonie V.D., El-Aasser M.S. *J. Appl. Polym. Sci.* **1995**, *58*, 375.
[14] O'Callagham K.Y., Paine A.Y., Rudin A. *J. Appl. Polym. Sci.* **1995**, *58*, 2047; O'Callagham K.Y., Paine A.Y., Rudin A. *J. Appl. Polym. Sci.* **1995**, *33*, 1849.
[15] Kalinina O., Kumacheva E. *Macromolecules* **1999**, *32*, 4122.
[16] Kalinina O., Kumacheva E. *Macromolecules* **2001**, *34*, 6380.
[17] Kalinina O., Kumacheva E. *Macromolecules* **2002**, *35*, 3675.
[18] It was found that core-shell particles with low T_g cores could not be used in the fabrication of nanocomposite material with liquid inclusions due to the release of the fluorescent CFP into the matrix.
[19] Talhavini M., Atvars T.D.Z. *Photochem. Photobiol.A: Chem.* **1998**, 50, 65.
[20] Kalinina O., Kumacheva. E. *Chem. Mater.* **2001,** *13,* 35.
[21] Li H., MSc. thesis, University of Toronto, 2002.
[22] Strictly speaking, a critical packing density required for smooth shell formation could be predicted on the basis of values of interfacial tension between the CFP and the SFP and the size of small particles. For a particular system the distance between the small particles on the surface of large core could be controlled by varying ionic strength of the dispersion medium.
[23] Matijevic E., Hsu W.P., Yu R. *J. Colloid Interface Sci.* **1993**, *156*, 56.
[24] Han J., Kumacheva E. *Langmuir* **2001,** *17,* 7912.
[25] Koroteev N. I., Magnitskii S.A., Tarasishin A.V., Zheltikov A.M. *Laser Physics* **1999**, *9(6)*, 1253.
[26] Pelinovsky D., Sears J., Brzozowski L., Sargent E.H. *J. Opt. Soc. Amer. B: Opt. Phys.* **2002**, *19*, 43

Macromol. Symp. **192**, 207–216 (2003)

Effects of External Fields on Macromolecular Cholesteric Phase

Linge Wang,[a] *Yong Huang**

Institute of Chemistry, Chinese Academy of Sciences, Beijing 100080, China
Laboratory of Cellulose and Lignocellulosics Chemistry, Guangzhou Institute of Chemistry, Chinese Academy of Sciences, Guangzhou 510650, China

Summary: Effects of external fields such as magnetic field and boundary conditions on supramolecular structure, molecular arrangement and the texture of the ethyl-cyanoethyl cellulose cholesteric liquid crystalline solutions were investigated. It was found that the molecules of (E-CE)C are oriented to perpendicular to the magnetic field and the diamagnetic anisotropy of (E-CE)C is negative. With homeotropic anchoring boundary condition, the molecules are aggregated with focal-conics arrangement the molecules are aggregated with planar arrangement with homogeneous anchoring boundary condition. The effects of the external field on the orientation of the cholesteric phase were influenced by the concentration of the solution because the twist power of the cholesteric was varied with the concentration. And the effects are also restrains by the surface tension of the interphase.

Introduction

Cellulose and many of its derivatives can form cholesteric liquid crystals in the appropriate solvents[1]. Ethyl-cyanoethyl cellulose [(E-CE)C], which is a cellulose derivative with two different ether groups, can form cholesteric liquid crystals in many organic solvents[2], such as dichloroacetic acid (DCA) and acrylic acid (AA). Observing by an optical polarizing microscope, (E-CE)C cholesteric liquid crystalline solutions can show multi-texture behavior with variation of concentration[3]. When concentration is in the region of 25-35 wt%, there are both cholesteric and isotropic phases in the (E-CE)C/DCA cholesteric solutions. The mesophase generally shows the disk-like texture, the fingerprint texture and the mosaic texture with concentration increase. When concentration is above 35 wt%, there is an uniform cholesteric phase in the (E-CE)C/DCA solution, and the mesophase generally shows the oily-streaks texture and the planar texture which are characterized by the optical reflective

[a] Ph.D. student in Graduate School of Chinese Academy of Sciences.

 CCC 1022-1360/00/$ 17,50+.50/0

properties.

Liquid crystals are always composed of anisotropic constituents. In the presence of a magnetic field, the arrangement of liquid crystalline molecules may be varied according to its different diamagnetic anisotropy along different body axis. And liquid crystalline molecules may be influenced by different boundary conditions, because of the small molecular elastic constants. Generally, with the slides imposing a homeotropic anchoring condition, the cholesteric phase tends to form fingerprint texture in which molecules aggregate with focal-conics arrangement[4-6]. The homogeneous anchoring on the slides is beneficial to the planar arrangement of liquid crystalline phase[7,8]. In the paper, the variations of the (E-CE)C/DCA mesophase textures under external fields, a magnetic field and different boundary conditions, were investigated. And the effects the external fields on the cholesteric LC supramolecular structure were also discussed.

Experimental

Materials and Sample Preparation

The (E-CE)C was prepared by reaction of ethyl cellulose (from Luzhou Chemical Plant, China) and acrylonitrile[2] and its molecular structure is shown in Figure 1. The degree of substitution for ethyl was about 2.1 and for cyanoethyl was about 0.33, calculated from the content of nitrogen of (E-CE)C that was determined by elementary analysis (Heraeus, CHN-O-RAPID, Heraeus, Germany). The molecular weight of (E-CE)C, M_n, measured by a gel permeation chromatograph (GPC) (Waters-ALC-244-GPC) and calibrated by standard polystyrene, was 7×10^4. DCA was chemically pure reagents. The (E-CE)C was mixed with DCA at room temperature and the solution was sealed in a test-tube. The solution was laid aside at room temperature for over 20 days to achieve an equilibrium solution. Then, the solution was sandwiched between two glass slides to form the solution film with the thickness of about 20 μm and sealed with solid wax. The mesophase texture of the solution was observed by a polarizing optical microscope (POM) (ORTHOPLAN-POL, Leitz, Germany).

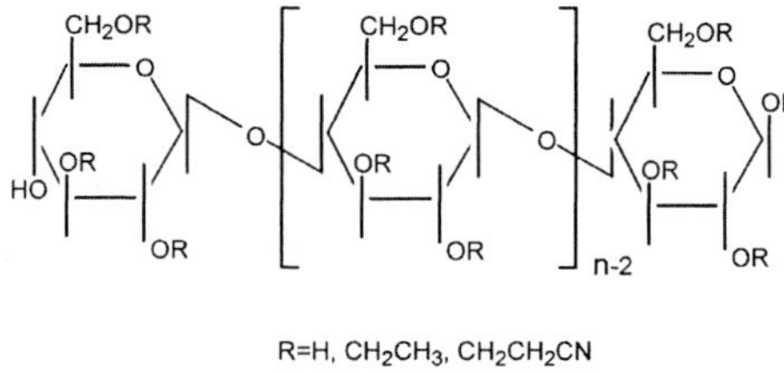

Figure 1. Molecular formula of (E-CE)C.

Magnetic Field Processing

The specimen was treated in a magnetic field with an intensity of 9.4 T that got from a Superconducting-Magnet NMR Spectrometer (DRX-400 MHz, BRUKER, Germany and Switzerland). Then, the specimen was observed by POM.

Boundary Conditions

Two glass microscope slides were filled with a lecithin/ethylether solution in order to achieve homeotropic anchoring condition. And the glass slides surface was treated with a polyvinyl alcohol aqueous solution to achieve homogeneous anchoring condition. The solvents were evaporated before the liquid crystalline solution was sandwiched between the slides.

Results and Discussion

Effects of Magnetic Field on the Fingerprint Texture

When the concentration is 29 wt%, there are both anisotropic and isotropic phases in the (E-CE)C/DCA cholesteric solutions and the mesophase generally shows the fingerprint texture. Some parallel equidistant dark and bright alternative striations are observed in the mesophase and the width of the striations is equal to a half pitch. The direction of the helix axis is aligned perpendicular to the striations and parallel to the slide surface. Without effect of the magnetic field, the direction of the helix axis of the fingerprint texture is random in different domains (Figure 2a). The striations, however, are formed a zigzag pattern and the helix axis are aligned almost parallel to the direction of the magnetic field after the solutions is treated in the magnetic field (Figure 2b). It is indicated that the direction of the molecular orientation is almost normal to the direction of the field and the diamagnetic anisotropy (χ_{α}) of (E-CE)C

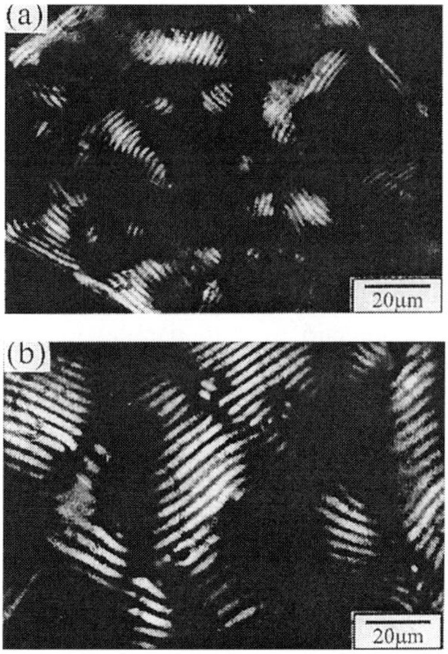

Figure 2. Effect of magnetic field on the fingerprint texture: (a) absence of the magnetic field; (b) after treated in the magnetic field.

is negative[9]. As we know, the diamagnetic anisotropy of molecules is correlated to the molecular structure, and the negative χ_α of the (E-CE)C is attributed to its conformation of the β-D-glucose backbone and the side cyanoethyl groups[9].

Effects of Magnetic Field on the Oily-Streaks Texture

The pitch of the (E-CE)C/DCA cholesteric phase decreases with increasing the concentration. When the concentration is about 40 wt%, there is an uniform cholesteric phase in the (E-CE)C/DCA solutions and the mesophase generally shows the planar texture and the oily-streaks texture (Figure 3a). In the mesophase, the helix axis is mostly aligned perpendicular to the slide surface for the planar texture but aligned parallel or oblique to the slide surface in the oily-streaks texture. No periodical birefringent striations can be observed by POM in the mesophase, because the pitch is very small ($P <$ 1μm). From Figure 3b, it can be observed that the mesophase texture is nearly unchanged after the solution is treated in the magnetic field. If the magnetic field can effect on the planar texture and the oily-streaks texture, in the case of $\chi_\alpha<0$, the helix axis will be aligned from perpendicular to the magnetic field to parallel to it and the mesophase will be changed to the fingerprint texture. The results indicate that the magnetic field with the intensity of 9.4 T cannot influence the cholesteric structure of the planar and the oily-steaks textures. The variation of the pitch with the critical intensity of the magnetic field, in which the cholesteric order is changed, can be described by the equation derived from the de Gennes's theory[10]:

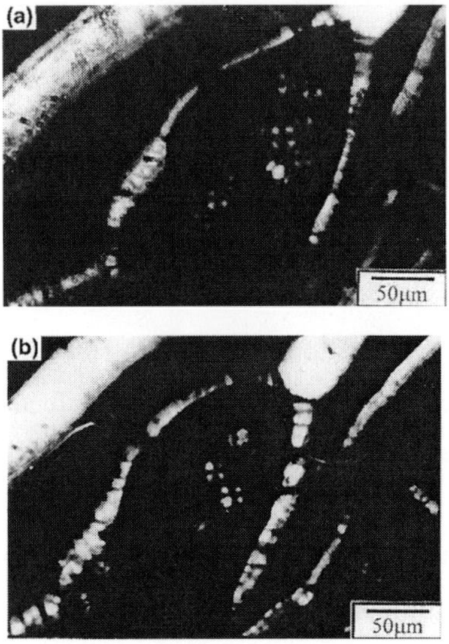

Figure 3. Effect of magnetic field on the oily-streaks texture: (a) in absence of the magnetic field; (b) after treated in the magnetic field; the morphology of the texture was remained.

$$H_C = (\pi^2/2)(k_{22}/\chi_\alpha)^{1/2}(1/p_0) \qquad (1)$$

Where H_C is the critical intensity of the magnetic field, P_0 is the pitch in the absence of the magnetic field, k_{22} is the twist elastic constant. According to the equation (1), the value of H_C depends on the k_{22}/χ_α and P_0. H_C is proportional to the $1/P_0$, because the parameter of k_{22}/χ_α of the oily-steaks texture is equal to that of the fingerprint texture in the (E-CE)C/DCA liquid crystalline solutions. The P_0 of the fingerprint texture is about 4-6 μm, which is about 4 times larger than that of the oily-streak texture. Therefore, the H_C of the oily-streak texture must be 4 times larger than that of the fingerprint texture if the influence of the magnetic field on the planar texture and the oily-streaks texture is the same as on the fingerprint texture. In the magnetic field with the same intensity, the effect on the fingerprint texture will be larger than on the planar and the oily-streaks textures. This result is different from that of the magnetic field effect on cholesteric liquid crystalline DNA systems[11]. The diamagnetic anisotropy of the DNA molecules is negative and the helix axis in the planar texture were aligned from perpendicular to parallel to the direction of the magnetic field, and the pitch of about 2.2 μm. We believe the orientation is cause by the pitch of the cholesteric DNA is bigger than that of (E-CE)C/DCA.

Effects of Magnetic Field on the Disk-Like Texture

At a low concentration, 26 wt%, the cholesteric phase is dispersed in the isotropic phase and the mesophase generally shows the disk-like texture in the (E-CE)C/DCA liquid crystalline solutions. Due to the surface tension of interphase between the mesophase and the isotropic phase, the cholesteric phase generally forms "round" shape for minimizing the surface and the free energy. The disk-like texture has also been observed in many macromolecular cholesteric phases[12-14], such as DNA fragment[13] and polypeptide[14]. The diameter of "disk" general is ranged between 20-50 μm and some periodical concentric extinction rings are observed in the (E-CE)C/DCA disk-like texture (Figure 4a). When the concentration is increased, the "dike" will be aggregated and merged to fingerprint texture. The distance between the neighboring concentric rings in the disk-like texture is equal to the half pitch, which is about 4-6 μm and is approached to that in the fingerprint texture. The helix axis is aligned along the radial directions of the "disk" and parallel to the slide surface. Figure 4 shows that the effect of the field on the "disk" seems to be dependent on its diameter. When the diameter of the "dike" is about 28μm, the shape and the morphology are nearly unchanged and the concentric rings are still shown the round shape (see Figure 4a) after the solution is treated in the magnetic field. It is interesting that when the diameter increased to 42μm, the shape of the "disk" is changed to the oval form and the helix axis tends to be aligned to parallel to the direction of the magnetic field (Figure 4b). Candau and co-workers[12] have reported that the critical intensity of the magnetic field is inversely varied with the diameter of the spherulites. The magnetic field with the intensity of 12 T has been used to effect the orientation of the cholesteric phase in the spherulite texture with the diameter of 122 μm. It is clear that the effect of the magnetic field on the disk-like texture with large diameter is large but on one with small diameter is small. The same phenomenon has also been found in the effect of the magnetic field on PBLG cholesteric solutions[14], in which the diameter of the "spherulites" is ranged between 600-1000μm and the pitch is 42.6μm.

The results mentioned above indicate that the effect of the magnetic field on the cholesteric phase with the fingerprint texture is more effective than that with the disk-like texture. The pitch of the mesophase with the fingerprint texture and the disk-like texture is almost the same and the surface tension of the cholesteric phase in the fingerprint texture is different from that in the in the disk-like texture. It seems that there is high surface tension in the disk-like texture and there exists a competition between the effect of the surface tension and the magnetic field when the solution is treated in the field. The surface tension of the mesophase with the disk-like texture increases with decreasing the diameter of the "disk". So, the effect of magnetic field may be restrained by the effect of the surface tension when the diameter of the "disk" is small (Figure 4a). When the diameter is increased to big enough, the effect of the magnetic field will be larger than the effect of the surface tension and the helix axis are tended to be aligned parallel to the magnetic field (Figure 4b).

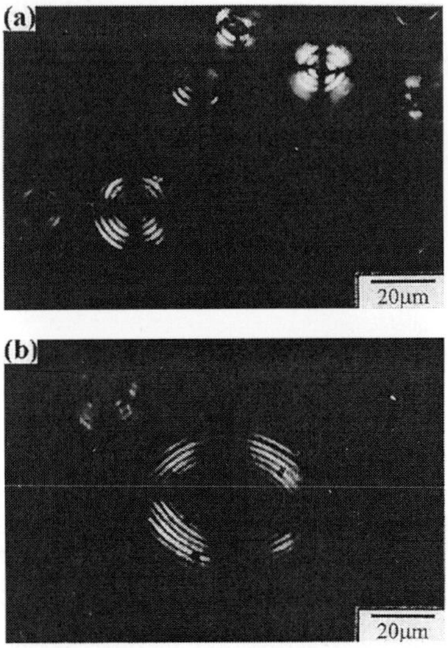

Figure 4. Effect of the magnetic field on the (E-CE)C/DCA disk-like texture: (a) unchanged disk, diameter ≈ 28μm; (b) changed disk, diameter≈42μm.

Effects of Different Boundary Conditions on the Fingerprint Texture

Without any effects of anchoring power, the width of the striations in the (E-CE)C/DCA fingerprint texture is about 2.6 μm (Figure 5a). With the slides imposing homeotropic anchoring condition, the width of the striations decreases to 1.8 μm (Figure 5b). With the slides imposing homogeneous anchoring condition, the striations almost disappeared, in which only some narrow strips can be observed in the solution and the mesophase shows the planar texture and oily-streaks texture (Figure 5c). In each domain, molecules are spontaneously twisted and aggregated with focal-conics arrangement in which the helix axis aligned along the radial direction and paralleled to the slide surface (Figure 6a). With the slides imposing homeotropic anchoring condition, molecules may be induced to align perpendicular to the slide surface and the twisting power of cholesteric phase competes with

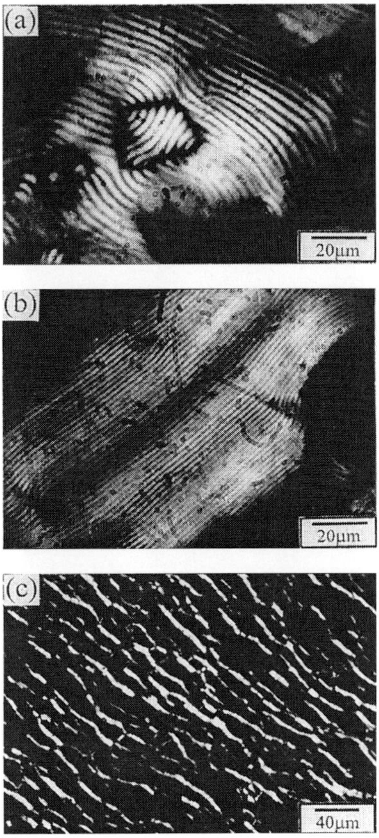

Figure 5. Effect of different boundary conditions on the fingerprint texture: (a) no anchoring; (b) with homeotropic anchoring; (c) with homogeneous anchoring.

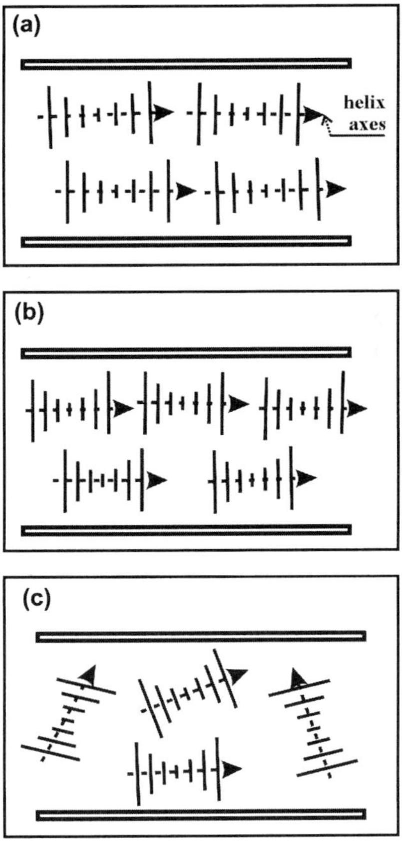

Figure 6. Scheme of the effect of boundary conditions on molecular focal-conics arrangement: (a) no anchoring; (b) with homeotropic anchoring; (c) with homogeneous anchoring.

the anchoring power of the boundary. The combination of both twisting and anchoring power results in the molecule orientation perpendicular to the slide surface but the cholesteric supramolecular structure is retained and the cholesteric pitch decreases (Figure 6b). With the slides imposing homogeneous anchoring condition, molecules may be induced to align parallel to the slide surface in which the helix axis is aligned perpendicular to the slide surface. And molecules are ordered according to the planar texture arrangement (Figure 6c), and the mesophase shows the planar texture and the oily-streaks texture.

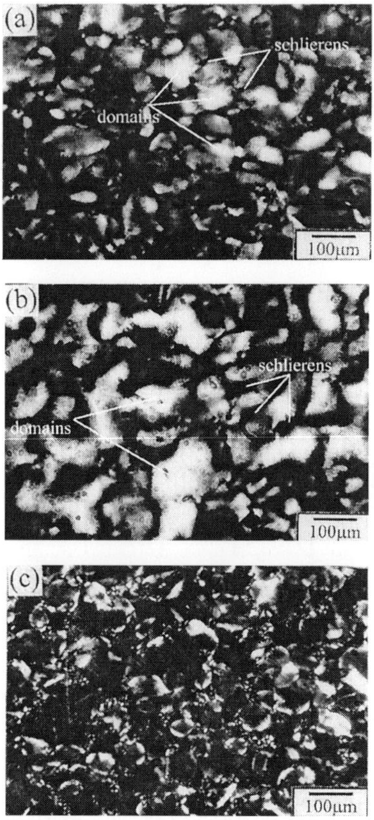

Figure 7. Effect of different boundary conditions on the mosaic texture: (a) no anchoring; (b) with homeotropic anchoring; (c) with homogeneous anchoring.

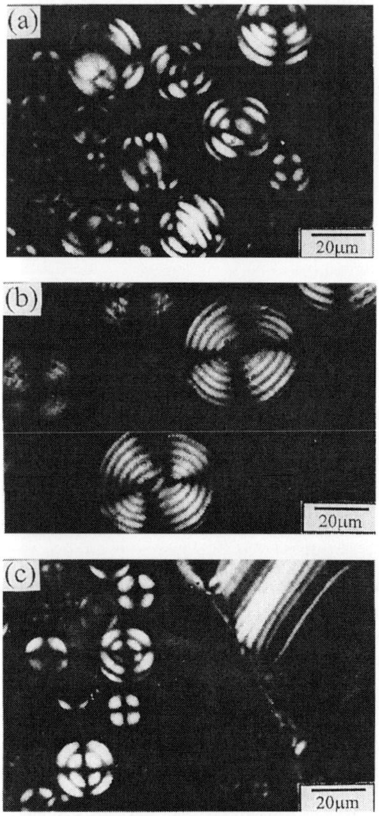

Figure 8. Effect of different boundary conditions on the disk-like texture of (E-CE)C/DCA solution: (a) no anchoring; (b) with homeotropic anchoring; (c) with homogeneous anchoring.

Effects of Different Boundary Conditions on the Mosaic Texture

When the concentration is 32 wt%, the cholesteric phase becomes the continuous phase and the isotropic phase is dispersed in the cholesteric phase. The mesophase generally shows the mosaic texture in the (E-CE)C/DCA cholesteric liquid crystalline solution. In the mosaic texture, some extinction schlierens are observed in the boundary of the domains and they generally show hyperbolic shapes (Figure 7a). With the slides imposing homeotropic anchoring condition, the area of each cholesteric domain and the length of the schlierens increases (Figure 7b). With the slides imposing homogeneous anchoring condition, the area of each cholesteric domain decreases the intensity of birefringence and the length of the schlierens are also decreased (Figure 7c).

In the mosaic texture, the helix axes are located in a certain inclination angle with the slide surface and it is very similar to the fan texture of smectic phase. In each domain, the helix axis aligns along with same direction molecules are also aggregated with focal-conics arrangement. The growth and aggregation of each domain are restrained by other domains and the direction of the helix axis is discontinued in the boundary of the domains. The disclination lines generally show hyperbolic shapes because the growth and aggregation of each domain are different. The anchoring power is restrained by the strong twisting power. With the slides imposing homeotropic anchoring condition, the combination of the twisting and the anchoring power induces the increase of each domains. The length of disclination lines is also increased. With the slides imposing homogeneous anchoring condition, the aggregation of the mosaic texture mesophase is restrained by the anchoring power, which results in the decrease of both the domain in size and the length of disclination lines. Parts of the domains are changed to the planar texture in which the helix axis aligns perpendicular to the slide. Due to the intensity of twisting power enhance with the increasing of concentration of cholesteric phase, the effect of boundary conditions on the mosaic texture is smaller than that on the fingerprint texture.

Effect of Different Boundary Conditions on the Disk-Like Texture

Without any effects of anchoring power, the distance between neighboring concentric rings is about 3.9 μm (Figure 8a). With the slides imposing homeotropic anchoring condition, the distance between neighboring concentric rings decreases to 3.2 μm, see Figure 8b. With the slides imposing homogeneous anchoring condition, the distance between neighboring concentric rings increases to 5.2 μm (Figure 8c).

With the slides imposing homeotropic anchoring condition, molecules may be induced to align perpendicular to the slide surface, the pitch of the disk-like texture is decreased. With the slides imposing homogeneous anchoring condition, molecules may be induced to align parallel to the slide surface, but the mesophase retains the disk-like texture. The effect of anchoring power is restrained due to the surface tension of interphase is strong. Most of molecules are still aggregated with focal-conics arrangement and some of theirs helix axes are oblique aligned to the slides. The homogeneous anchoring results in the loosen arrangement in polymer chains packing and the increase of cholesteric pitch, because part of molecules are aligned parallel to the slide surface.

Conclusion

The supramolecular structure of (E-CE)C/DCA cholesteric LC solutions depends on the twisting power, the external field effects and the surface tension of interphase. The twisting power increases with increasing the concentration and results in the pitch decreasing. Without the external field, the cholesteric texture is changed with the variation of the concentration. Under the effects of the external fields such as the magnetic field and different boundary conditions, the molecules are aligned to the field and the structure will be influenced by the competition between the cholesteric twisting power and the orientation power. The effect of the external field on the cholesteric order is restrained when the concentration is high, because the twist power is increased with increasing the concentration. Moreover, the surface tension of interphase is powerful in the low concentration and restrains the effect of the external fields on the cholesteric order.

Acknowledgment. Financial support by National Natural Science Foundation of China (Grant No. 29925411) is greatly appreciated.

[1] D. G, Gray, *Appl. Polym. Symp.* **1983**, *37*, 179.
[2] Y. Huang, M. C. Chen, L. S. Li, *Acta Chimica Sinica (Chi)*, **1988**, *46*, 367.
[3] Y. Huang, *J. Macro. Sci. Phys.* **1989**, *B28(1)*, 131.
[4] P. Ribiere, S. Pirkl, P. Oswald, *Liq. Cryst.*, **1994**, *16(2)*, 203.
[5] S. Pirkl, *Cryst. Res. & Tech.*, **1993**, *28(7)*, 1027.
[6] S. Wen, M. M. Labes, *Mol. Cryst. Liq. Cryst.*, **1994**, *239*, 55.
[7] H. –S. Kitzerow, B. Liu, F. Xu, P. P. Crooker, *Phys. Rev. E*, **1996**, *54(1)*, 568.
[8] W. D. John, Z. J. Lu, J. W. Doane, *J. App. Phys.*, **1995**, *78(9)*, 5253.
[9] L.Wang, Y. Huang, *Macromolecules*, **2000**, *33(19)*, 7062.
[10] P. G. de Gennes, *Solid State Commun.*, **1968**, *6*, 163.
[11] D. H. Van Winkle, A. Chatterjee, R. Link, R. L. Rill, *Phys. Rev. E.*, **1997**, *55*, 4354.
[12] S. Candau, P. Le Roy, F. Debeauvais, *Mol. Cryst. Liq. Cryst.*, **1973**, *23*, 283.
[13] D. H. Van Winkle, M. W. Davidson, W. X. Chen, R. L. Rill, , *Macromolecules,* **1990**, *23*, 4140.
[14] D. L. Patel, D. B. Dupré, *J. Polym. Sci.: Polym. Phys. Ed.*, **1980**, *18*, 1599.

Topological Polymer Chemistry: Classification and Construction of Nonlinear Polymer Architectures

Yasuyuki Tezuka

Department of Organic and Polymeric Materials, Tokyo Institute of Technology, O-okayama, Meguro-ku, Tokyo 152-8552, Japan

Summary: A systematic classification of nonlinear polymer architectures is formulated by reference to constitutional isomerism in a series of alkanes (C_nH_{2n+2}), monocycloalkanes (C_nH_{2n}) and polycycloalkanes (C_nH_{2n-2}, C_nH_{2n-4}, etc). And a variety of topologically unique polymers, such as monocyclic and polycyclic polymers, cyclic telechelics (kyklo-telechelics), cyclic macromonomers and polymeric topological isomers have been constructed by means of an "electrostatic self-assembly and covalent fixation" strategy. In this process, new telechelic polymer precursors having moderately strained cyclic onium salt groups carrying multifunctional carboxylate counteranions have been prepared as key polymer precursors. The unique electrostatic self-assembly, formed particularly in a dilute solution, was covalently fixed through the ring-opening reaction by heating, to produce a variety of nonlinear polymer architectures in high efficiency.

Introduction

The precise control in polymer architectures has been an ongoing challenge in synthetic polymer chemistry, since new polymer topologies will realize unprecedented properties and functions in polymeric materials. A variety of branched polymer structures have so far been constructed, including star polymers, H-shaped polymers, super H-shaped polymers, pom-pom shaped polymers, as well as comb polymer and dendrimers. Moreover, cyclic, multicyclic and cyclic-branched combined polymers, such as ring polymers, 8-shaped polymers as well as tadpole polymers have recently gained increasing interests.[1]

We show first a systematic classification of nonlinear polymer topologies by reference to constitutional isomerism in a series of alkanes (C_nH_{2n+2}), monocycloalkanes (C_nH_{2n}) and polycycloalkanes (C_nH_{2n-2}, C_nH_{2n-4}, etc).[2] Subsequently, a novel methodology has been proposed to produce various non-linear polymer architectures in high efficiency, by exploiting the self-assembly principle to preorganize linear polymer precursors, prior to convert them into covalently linked permanent structures. Thus, we have utilized Coulombic self-assembly of hydrophobic macromolecules having moderately strained cyclic onium salt groups carrying

 CCC 1022-1360/00/$ 17,50+.50/0

multifunctional carboxylates as counteranions.[3,4]

Systematic Classification of Nonlinear Polymer Topologies

A rational classification of nonlinear polymer topologies will indicate their structural relationships, and eventually provide their convenient synthetic pathways. There have been, however, few attempts on the systematic classification of nonlinear, and in particular cyclic and multicyclic, polymer architectures composed of sufficiently long and thus flexible segment components.

We have formulated a systematic classification of a series of well-defined cyclic and branched polymer architectures by reference to constitutional isomerism in alkanes (C_nH_{2n+2}) and in mono- and polycycloalkanes (C_nH_{2n}, C_nH_{2n-2}, etc).[2] Each alkane and (poly)cycloalkane molecules is taken as a source to generate a unique topological construction. The total number of termini (chain ends) and of junctions (branch points) are taken as invariant (constant) geometric parameters. The total number of branches at each junction and the connectivity of each junction are maintained as invariant parameters as well. On the other hand, such Euclidian geometric properties as the distance between two adjacent junctions and that between the junction and terminus are taken as variant parameters. This conforms to the flexible nature of the randomly-coiled polymer segments. Furthermore, topological constructions having five or more branches at one junction is allowed while the relevant isomers having the corresponding molecular formula are absent.

In Table 1-A, alkane molecules of generic molecular formula of C_nH_{2n+2} with n = 3–7 and their relevant topological constructions are listed. A point from methane (CH_4) and a line construction from ethane (C_2H_6) are not included, since the former is not significant with respect to polymer topology and the latter and propane (C_3H_8) produce an equivalent line construction. Two butane isomers produce a linear and a three-armed star construction, respectively. Likewise, pentane isomers produce a four-armed star construction in addition to the two produced already from butane isomers. From five hexane isomers, two new constructions of an H-shaped and of a five-armed star architecture are produced. And further, heptane isomers produce the two new constructions of a super H-shaped and a six-armed star architectures. Accordingly, a series of branched polymer topologies are hierarchically ranked as shown in Table 1-A.

Topological constructions from monocycloalkane molecules of C_nH_{2n} with up to n = 6 are collected in Table 1-I. Cyclopropane (C_3H_6) produces a simple cyclic topology. From the two

Table 1. Systematic Classification of Nonlinear Polymer Topologies. (Reprinted with permission from ref [2]. Copyright, 2001 by American Chemical Society)

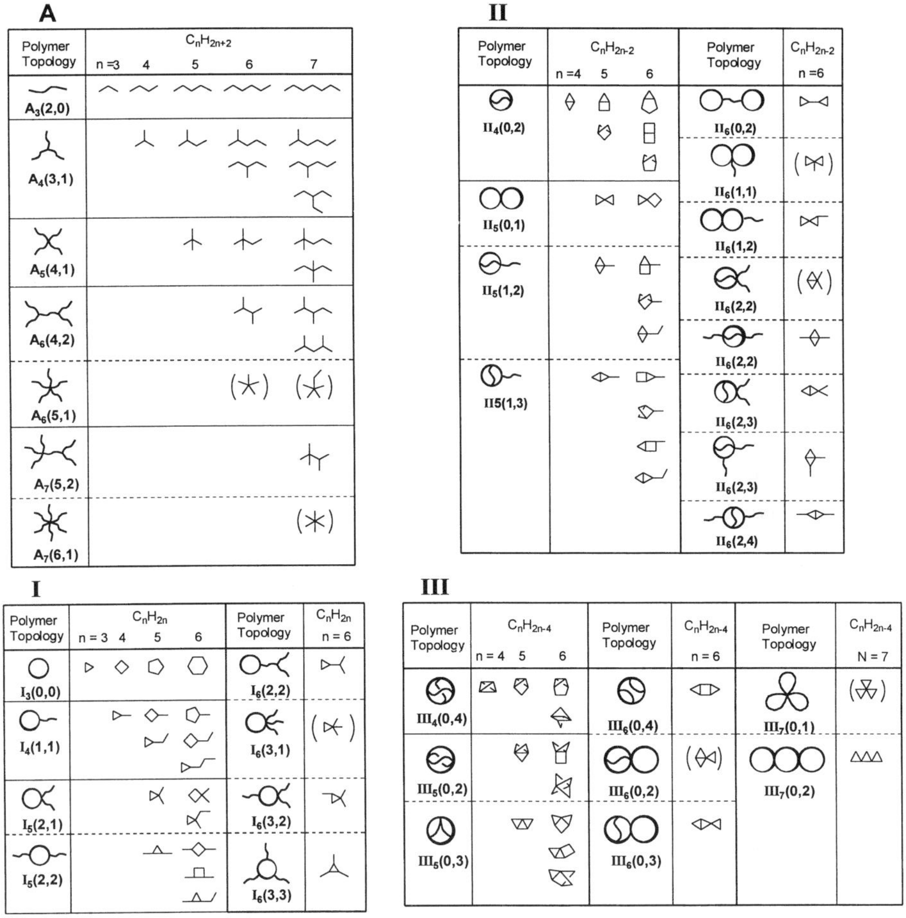

isomers of C_4H_8, a tadpole and a simple ring structures are produced. Two constructions are newly produced from C_5H_{10} isomers, and are distinguished from each other by their junction and branch structures, i.e., one has two outward branches at one common junction in the ring unit, while the other has two outward branches located at two separate junctions in the ring unit. Further, four new topological constructions are produced from C_6H_{12} isomers; one having five branches at one junction is hypothetical and is therefore shown in parentheses in Table 1-I. A series of "a ring with branches" constructions have been thus classified as summarized in Table 1-I.

Topological constructions from bicycloalkanes of C_nH_{2n-2} with up to n = 6 are listed in Table

1-II, where basic three bicyclic constructions free of outward branches are included. Those are θ-ring, 8-ring and manacle-ring, respectively. First, a θ construction is produced from bicyclo[1,1,0]butane, C_4H_6. From the five bicycloalkane isomers of C_5H_8, three new constructions are produced. One of them is an 8-shaped construction from spiro[2,2]pentane. By reference to bicyclohexane (C_6H_{10}) isomers, moreover, eight new constructions are produced. All constructions but one from bi(cyclopropane) possess outward branches emanated from either θ-ring or 8-ring shown above. In this manner, a series of internally and externally linked double cyclic topological constructions can be classified into the corresponding categories by reference to the bicycloalkane isomers.

Selected topological constructions produced from tricycloalkanes of C_nH_{2n-4} are also shown in Tables 1-III. A large number of "rings with branches" constructions are produced in this group. Eight topological constructions are derived from θ- or 8-forms. The six of them are produced directly from the corresponding tricycloalkanes, and the additional two from hypothetical molecular formula possessing five or six branches at a junction as given also in Table 1-III. Moreover, the seven topological constructions of externally linked (or "bridged") rings with a simple cyclic, θ- or 8-forms are obtainable in reference to the relevant tricycloalkane isomers.

Single Cyclic and Multicyclic Polymers

We have developed an "electrostatic self-assembly and covalent fixation" strategy to construct a variety of topologically unique, nonlinear polymer architectures, where new telechelic polymer precursors having a moderately strained cyclic onium salt group carrying multifunctional carboxylate counteranions have been employed as key polymer intermediates. The unique electrostatic self-assembly, formed particularly in a dilute solution, has been subsequently converted to permanent covalent product through the ring-opening reaction by heating, to produce a variety of nonlinear polymer architectures in high efficiency. [3,4](Scheme 1)

Thus, a polymer cyclization was carried out using a telechelic poly(THF) having N-phenylpyrrolidinium salt groups carrying a dicarboxylate counteranion.[5-7] Typically, the reaction proceeded by heating under dilution at the concentration of 0.2 [g/L] in as high as 93 % yield, and the cyclic poly(THF) of more than 98 % purity was obtained in 74 % yield after the purification with preparative thin-layer-chromatography (TLC) technique.[5]

Bicyclic and tricyclic polymer topologies were also constructed by using the telechelic poly(THF) having N-phenylpyrrolidinium salt groups, and carrying a tetra and a hexacarboxylate counteranion, respectively.[4] The balance of the charges between the ammonium salt groups in the polymer precursor and the carboxylate anions was maintained,

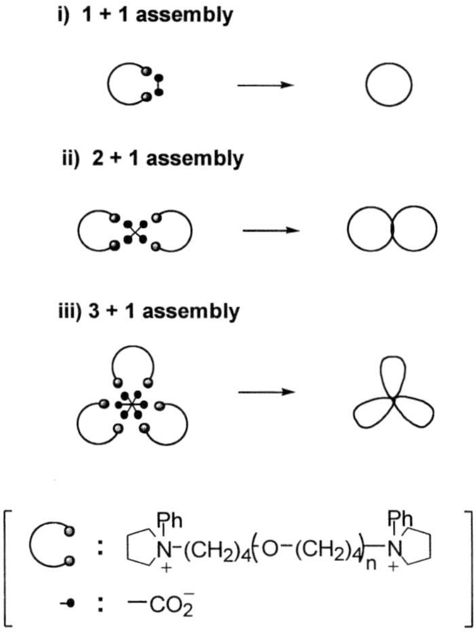

Scheme 1

(Reprinted with permission from ref [3]. Copyright, 2001 by Wiley-VCH)

corresponding to the molar ratio of 2:1 and 3:1, respectively. (Scheme 1) These polymer precursors were then subjected to the heat treatment under appropriate dilution. The bi- and tricyclic poly(THF)s were readily obtained after the preparative thin layer chromatography. These results indicate that the electrostatic polymer assemblies formed under dilution are comprised of two units of the polymer precursor and one unit of tetracarboxylate, and of three units of the polymer precursor and one unit of hexacarboxylate, respectively. Subsequent heat treatment could lead to bicyclic and tricyclic polymer products, through the covalent fixation by the ring-opening reaction of pyrrolidinium salt groups.

Kyklo-Telechelics

Cyclic polymer precursors having functional groups at the designated positions, termed *kyklo*-

telechelics (the Greek "*kyklos*" means cyclic), will be of a potential importance as a macromolecular building block to construct topologically unique and complex macromolecular architectures containing cyclic polymer units. The linear telechelic poly(THF) having N-phenylpyrrolidinium salt groups, and optionally with an additional functional group at the center position of the polymer chain (termed *kentro*-telechelics) has been utilized to design *kyklo*-telechelics having a single or two functional groups.[8] (Scheme 2) The "electrostatic self-assembly and covalent fixation" process was applied for an efficient polymer cyclization process by the linear telechelic poly(THF)s, where a unique self-assembly is formed under dilution by polymer precursors carrying specific counteranions, while balancing the charges between cations and anions.

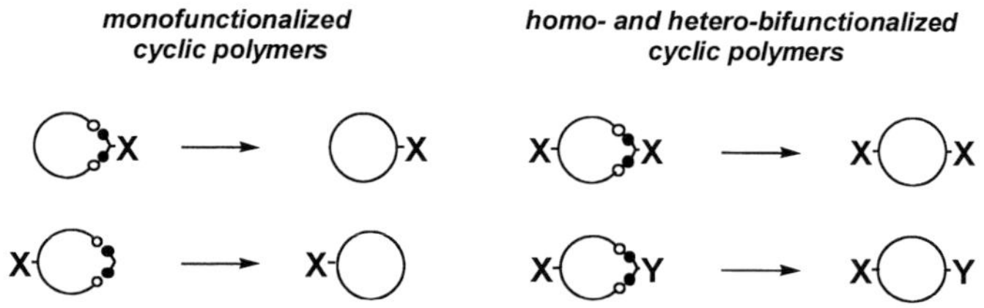

Scheme 2

(Reprinted with permission from ref [3]. Copyright, 2001 by Wiley-VCH)

Thus, a telechelic poly(THF) carrying a 5-hydroxyisophthalate counteranion (a hydroxy-functionalized dicarboxylate) was prepared. This ionically-linked polymer precursor was subjected to the heat treatment under dilution to produce a cyclic poly(THF) having a hydroxy group, derived from the dicarboxylate counteranion. Homo- and hetero-bifunctionalized cyclic poly(THF)s having either two identical or two different functional groups at the opposite positions, respectively, have subsequently been synthesized by using the relevant hydroxy-functionalized cyclic poly(THF) obtained alternatively from an α,ω,*kentro*-telechelic poly(THF), which carries not only N-phenylpyrrolidinium salt groups at both chain ends but also an additional hydroxy group at the center position of the polymer chain. Thus, the α,ω,*kentro*-telechelic polymer precursor carrying a 5-hydroxyisophthalate counteranion was subjected to the heat treatment under dilution to give a homo-bifunctionalized cyclic poly(THF) having two hydroxy groups at the opposite positions. Similarly, an α,ω,*kentro*-telechelic poly(THF) carrying a 5-allyloxyisophthalate counteranion was prepared as a precursor to produce a hetero-bifunctionalized cyclic poly(THF) having a hydroxy and an allyloxy group at the opposite positions.

Cyclic Macromonomer

Cyclic polymers will be utilized to construct physically (non-covalently) cross-linked polymer architectures based on its loop topology. However, a tailored non-covalent cross-linking process by threading through flexible and large size cyclic polymers has scarcely been achieved. Since a long and flexible polymer chain tends to assume a randomly coiled and hence constrained three-dimensional structure, the threading by another polymer chain through such a large cyclic polymer unit is unwilling to proceed. This is contrastive with a variety of medium size macrocyclic compounds up to around 100-membered ring, including 30~60 membered crown ethers, bipyridinium-based cyclophanes, cyclodextrins and macrocyclic amides. They assume stiff and extended cyclic conformation and are used for the synthesis of rotaxanes and catenanes.

We have succeeded in synthesizing a novel functional cyclic polymer, i.e., a cyclic macromonomer of 280-membered ring, to design an unusual polymer network structure

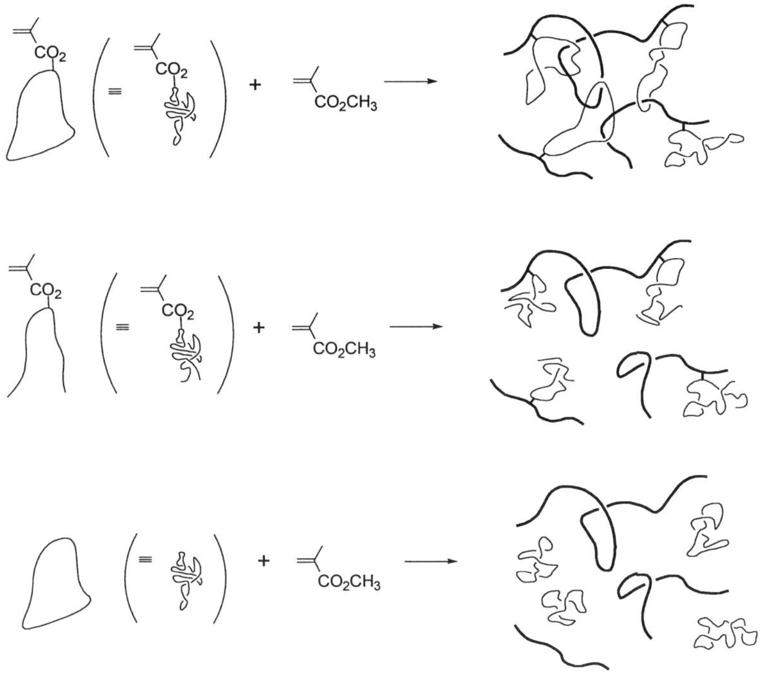

Scheme 3

(Reprinted with permission from ref [3]. Copyright, 2001 by Wiley-VCH)

having both covalent and physical linkages. This network is formed by a chain threading of the propagating polymer segment through large cyclic polymer units attached to the backbone polymer segments.[9] (Scheme 3)

A cyclic macromonomer having a methacryl group with the degree of polymerization of 70, corresponding to 280-membered ring (Mn = 5200), was obtained from the hydroxy-functionalized cyclic poly(THF), in a quantitative yield. A free radical copolymerization with methyl methacrylate (MMA) to a complete-conversion resulted in a gel product. In contrast, no gelation took place either in the relevant quantitative-conversion copolymerization of MMA with an open-chain poly(MMA) macromonomer, or in a quantitative-conversion homopolymerization of MMA in the presence of a simple cyclic poly(THF) of the relevant ring size. (Scheme 3)

These results indicate that the gelation took place through the physical cross-linking, i.e., the threading by the propagating chain through the pendant cyclic branches attached to the polymer backbone. And interestingly, this chain threading through large polymer loops occurred only when they are covalently attached to the polymer backbone.

Topological Isomers

The covalent conversion of an electrostatically self-assembled precursor, obtained from the bifunctional poly(THF) having N-phenylpyrrolidinium salt groups carrying tricarboxylate (trimesate) counteranions, was also performed under dilution.[4](Scheme 4) The reversed-phase chromatography (RPC) showed the presence of two components in the covalent fixation product. The ^{1}H NMR and IR of the two fractions isolated by RPC fractionation were identical each other, and showed the quantitative ring-opening reaction of pyrrolidinium salt groups by carboxylate groups in the tricarboxylate. The molecular weights of the product (as a mixture) determined by the VPO and by the ^{1}H NMR (assuming quantitative chemical conversion of polymer end groups) coincided, within an experimental error, with that of the three times of its linear precursor analogue. The SEC showed, however, that one fraction is larger in its size than another.

These results indicate that a pair of topological isomers, i.e. θ- and manacle-constructions, were formed simultaneously from three bifunctional polymer precursors and two trifunctional end-linking reagents.[1,10] (Scheme 4) Since they are produced from the identical precursors by a common chemical reaction, their chemical compositions are identical each other, possessing an identical molecular weight. They are, on the other hand, topologically distinctive each other. The random combination of cations and anions will produce the two constructions of manacle-ring and θ-ring in the ratio of 3:2. The size, i.e., hydrodynamic volume of the manacle-form isomer is considered to be larger than that of the θ-form counterpart. The observed isomer ratio (78:22) indicates that the intramolecular process to produce the manacle-type isomer is slightly favored in this covalent fixation process.

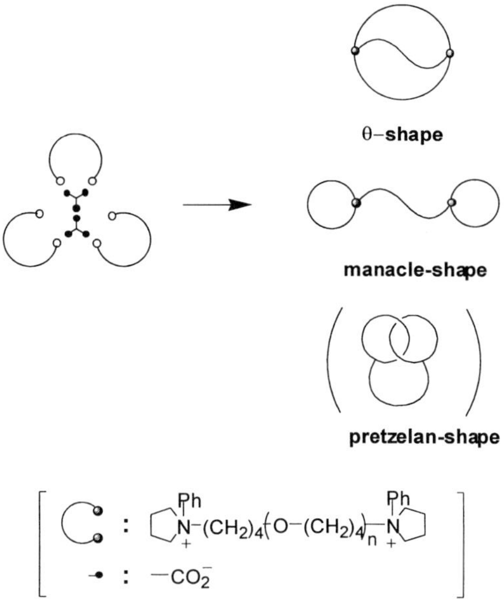

Scheme 4

Conclusions

We have formulated a systematic classification of nonlinear polymer topologies. It provides useful insights for the relationship between different polymer topologies, and eventually for the rational synthetic pathway. Thus, an "electrostatic self-assembly and covalent fixation" process has been devised for constructing various non-linear polymer topologies, including monocyclic and polycyclic polymers, cyclic telechelics (*kyklo*-telechelics), cyclic macromonomers and a pair of polymeric topological isomers. New telechelic prepolymers having moderately strained cyclic ammonium salt group were prepared as key intermediates. The electrostatic self-assemblies, formed in particular under dilution, were subsequently converted into the permanent covalent products by the selective ring-opening reaction.

Acknowledgments

The author is grateful to Dr. S. Corona-Galvan and Professor R. P. Quirk, the organizers of the microsymposium "Recent Advances in Anionic Living Polymerizations" held in the occasion of the 7th Pacific Polymer Conference in Oaxaca, Mexico, for the arrangement of the travel grant. The symposium was sponsored by BASF Mexicana, Bridgeston Corporation, Bridgeston/Firestone, Inc., Chemetall Foote Corporation, Chevron Phillips, Dow Elastomers, Dynasol Elastomers, Goodyear Tire & Rubber Company, Korea Kumho Petrochemical Company, Polymer Source, Sartomer, Synthomer, and Waters S.A. Financial supports from The Iwatani Naoji Foundation's Research Grant and from NEDO Nanostructure Polymer Project grant (2001-2008) are gratefully acknowledged. This study was also partly supported from the Ministry of Science and Education, Japan (13450377).

[1] Y. Tezuka, H. Oike, *Prog. Polym. Sci.*, **2002**, *27*, 1069.
[2] Y. Tezuka, H. Oike, *J. Am. Chem. Soc.*, **2001**, *123*, 11570.
[3] Y. Tezuka, H. Oike, *Macromol. Rapid Commun.*, **2001**, *22*, 1017.
[4] H. Oike, H. Imaizumi, T. Mouri, Y. Yoshioka, A. Uchibori, Y. Tezuka, *J. Am. Chem. Soc.*, **2000**, *122*, 9592 .
[5] H. Oike, T. Mouri, Y. Tezuka, *Macromolecules,* **2001**, *34,* 6592.
[6] H. Oike, M. Hamada, S. Eguchi, Y. Danda, Y. Tezuka, *Macromolecules,* **2001**, *34*, 2776.
[7] Y. Tezuka, K. Mori, H. Oike, *Macromolecules* in press.
[8] H. Oike, S. Kobayashi, T. Mouri, Y. Tezuka, *Macromolecules,* **2001**, *34*, 2742.
[9] H. Oike, T. Mouri, Y. Tezuka, *Macromolecules,* **2001**. *34*, 6229.
[10] D. M. Walba, *Tetrahedron,* **1985**, *41*, 3161.

Structural Evidence of Grafting in Electron Beam Irradiated Starch-Allylurea Blends

*A. Olivier, J. Luna-Carlos, F. Cazaux, X. Coqueret**

Laboratoire de chimie macromoléculaire – UPRESA CNRS 8009
Université des Sciences et Technologies de Lille
F-59655 Villeneuve d'Ascq Cedex, France

Summary: The effectiveness of covalent grafting in mixtures of N-allylurea (AU) and amorphized starch submitted to electron beam irradiation was studied for elucidating the chemical changes at the origin of the physical stabilization observed after radiation processing. FTIR and nmr analysis were used to quantify the progress of AU conversion upon irradiation of the mixtures and to gain information on the structure of the products. The grafting yield was determined by gravimetry after selective solubilization that extracts AU monomer and its homopolymer from the covalently modified polysaccharide. MALDI-TOF spectrometry applied to irradiated mixtures of AU and maltotriose, as a model compound, affords unambiguous evidence of covalent grafting. The obtained data allow to estimate the size of the AU grafts and to propose mechanistic pathways for radiation grafting.

Introduction

The development of biodegradable materials exhibiting most of the use properties of the conventional thermoplastics is still a major challenge. Starch-based materials are particularly attractive because they can be produced from a cheap and annually available resource.[1] For obtaining thermoplastic starch materials, native granular starch is amorphized at temperatures above 100°C and plasticized with the aid of additives of relatively low molecular weight.[2] We have recently reported that the electron beam (EB) irradiation of amorphous blends of potato starch and N-allylurea (AU) efficiently impedes the spontaneous blooming of mixtures including a large wt-fraction of the low molecular weight additive.[3] The physical instability of untreated blends was evidenced not only by the occurrence of blooming at film surface, but also by internal phase separation and by recrystallization of the starch and AU constituents monitored by X-ray diffraction.[4] The observed physical stabilization of the blends obtained after irradiation was confirmed by thermomechanical analysis of films submitted to various aging conditions.[3,4]

 CCC 1022-1360/00/$ 17,50+.50/0

AU was purposely selected as a reactive additive, claimed to undergo radiation induced polymerization.[5] Thus, the physical changes observed after irradiation of the blended materials possibly originate (i) from AU conversion into free polymer, (ii) from AU grafting polymerization onto the anhydroglucose units of starch (AGU), (iii) from polysaccharide chains degradation or (iv) from various combinations of these three basic chemical reactions (see Scheme 1).

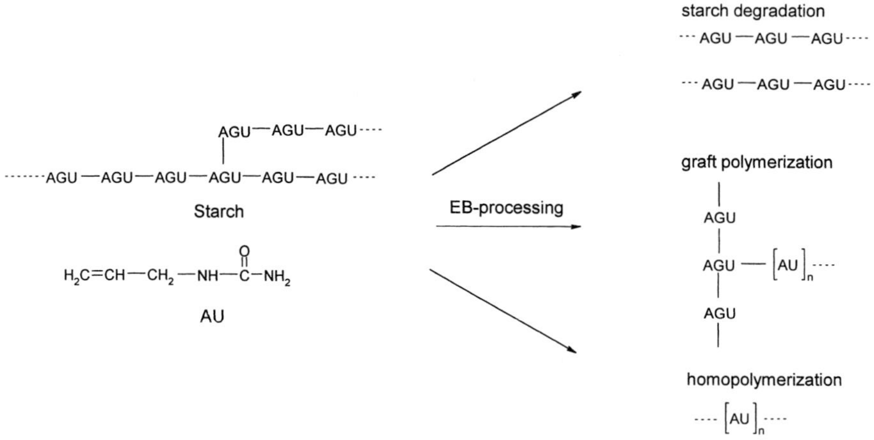

Scheme 1 - Simplified representation of the reactions induced by the EB-treatment in starch - allyllurea (AU) blends [AGU for anhydroglucose unit].

A detailed understanding of the chemical effects induced by high energy radiation into the starch - AU blends conditions the optimization of formulations and of processing with the view of meeting a set of desired requirements. The present report is devoted to the effectiveness of AU grafting polymerization in the amorphous blends with starch, with the objective to determine the grafting yield to get structural and mechanistic information from MALDI-TOF mass spectrometry.

Experimental Part

Native potato starch (NPS) was obtained from Roquette Frères (Lestrem, France). N-allylurea (reagent grade) was purchased from Sigma-Aldrich. Thermoplastic starch (TPS) materials were prepared in an internal Brabender mixer at 125°C with a screw rotation speed of 50 rpm for 15 min. The blends were chopped into small cubes and quenched at –17°C immediately after processing in order to avoid physical evolution during storage. TPS films of thickness ranging between 20 and 40 μm were prepared by grinding the material into powder and by compression molding at 130°C (6 MPa for 5 min). The materials named SAU20 and SAU50 were prepared from NPS (100 parts) and AU, 20 or 50 parts (w/w), respectively. The desired water content in the pressed materials was achieved by conditioning the films in cabinets where a constant relative humidity was maintained (58 % RH at R.T. over saturated aqueous NaBr solution).

Irradiation of blend films or of crystalline AU was carried out under the EB of an Electrocurtain CB150 generator (Energy Sciences Inc.) operating at 175 kV. The radiation was performed in air at a dose rate of 200 $kGy.s^{-1}$. Current and conveyor speed were adjusted to obtain 25 or 50 kGy per pass. Higher doses were obtained by increasing the number of passes in the chamber of the EB processor.

Thermogravimetric analyses were performed with a Shimadzu TGA apparatus in an air flow, with heating rate of 10°$C.min^{-1}$. Infrared and nmr spectroscopic measurements were performed as described elsewhere.[6] For grafting yield determination, the procedure was as follows. Irradiated films (100 mg) were stirred in a methanol / water mixture (v/v 75/25, 20 mL). After stirring for 24 h, the turbid suspension was filtered with a Büchner funnel with a paper filter (Whatman No 1005 070). The liquid fraction was concentrated in vacuo. The solid residues of both fractions were then allowed to dry for 1 week at room temperature (constant weight) before analysis by ^{1}H nmr spectroscopy. Matrix-assisted laser desorption/ionization time of flight (MALDI-TOF) mass spectra were obtained on a Vision 2000, Finnigan Mat Spectrometer. The matrix was prepared by dissolving 2,5-dihydroxybenzoic acid (10 mg) in a 50:50 v/v mixture of water and methanol. Aqueous NaCl was added so as to obtain a 5 mM salt concentration. Equal volumes of the matrix and of the sample solution (0.1 wt-% in water) were mixed before filling the target pit. SEC analyses were performed with a Waters apparatus equipped with Ultrahydrogel columns (1 linear, 1 DP) and a refractometric detector. Water was used as the eluent with a flow rate of 1$mL.min^{-1}$.

Results and Discussion

Dose Dependence of AU Conversion

The polymerizability of allylurea under ionizing radiation is strongly dependent on the physical state of the monomer. It has been reported to undergo polymerization in the crystalline state with formation of long chain polymer.[5] Its reactivity in solution is believed to be poor, on the basis of low polymer yield after precipitation.[7] Various effects can be expected from the differences in molecular organization (crystal *vs.* amorphous blend) and in chemical purity of irradiated medium (recrystallized monomer *vs.* intimate mixture with the polysaccharide) that exist between the various forms of AU we have to consider. We have monitored by infrared spectroscopy the progress of AU conversion, irradiated under various solid states at room temperature with increasing doses of accelerated electrons. The two allotropic crystalline forms of AU were recrystallized in the appropriate solvents, so as to obtain the monoclinic and tetragonal crystals with high purity, from methanol and water solutions, respectively.

The conversion profiles shown in Figure 1 were deduced from infrared absorbance measurements at 929 cm^{-1}. The lines corresponding to the two allotropic forms of crystalline AU indicate that the polymerization proceeds with similar rate. This only suggests that the molecular organization within the crystal has no overall effect on polymerization. Keeping in mind the possible differences of behavior in each step of the chain addition mechanism, a more detailed study would be necessary to further comment on the compared reactivity of the crystalline forms of AU. The point of interest for the present discussion seems to be the apparently higher reactivity observed when AU (20 wt-parts) is blended with destructurized starch (80 wt-parts), if reasoning on fractional conversion.

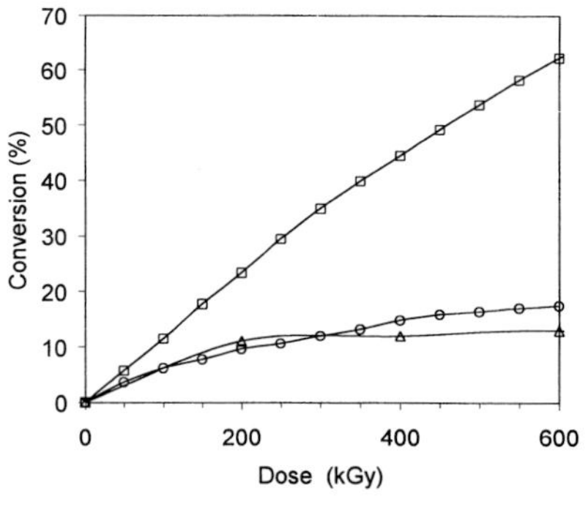

Figure 1. Progress of fractional conversion of AU as a function of EB-irradiation dose, in a SAU20 blend (□) and in the two crystalline forms of AU (O, monoclinic; Δ: tetragonal).

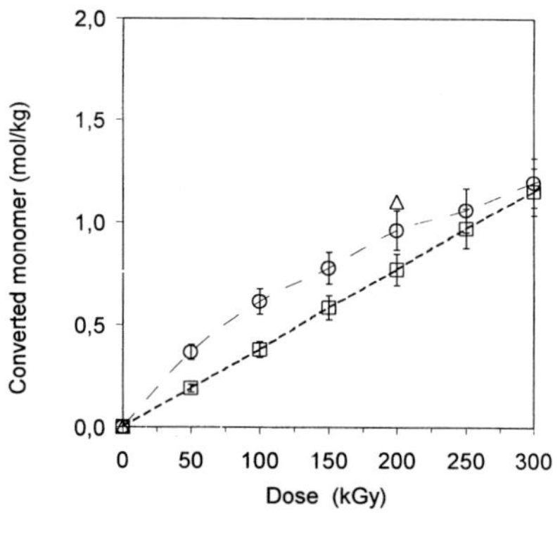

Figure 2. Progress of the amount of converted AU as a function of EB-irradiation dose, in a SAU20 blend (□) and in the two crystalline forms of AU (O, monoclinic; Δ: tetragonal).

However, since AU is diluted in the solid starch blend, the absolute amount of converted monomer is the pertinent criterion for the comparison of monomer reactivity. The plots of Figure 2 indicate that the actual polymerization profiles are not so different. The topochemical control acting strongly over some polymerization conducted under radiation in the solid state does not seem to operate. One could have expected that the presence of starch in the amorphous and homogeneous blends could lead to the generation of additional initiating centers favoring graft polymerization. Thus, this study of AU reactivity does not afford any conclusive information nor any indication on the possibility of graft formation.

Thermogravimetric Analysis

One of the claimed characteristic effects of grafting polymerization onto starch is the modification of the material sensitivity to thermal degradation in air.[8] Several examples concern the γ-radiation induced grafting polymerization of acrylonitrile. The degradative behavior in the high temperature range (300°C-700°C) strongly modified, as an expected consequence of polyacrylonitrile grafts rearrangement. The changes in the low temperature range (50°C-300°C) are said to constitute a strong evidence for grafting.[9] We have recorded the thermograms of native starch for comparison with that of EB-irradiated starch-AU blends.

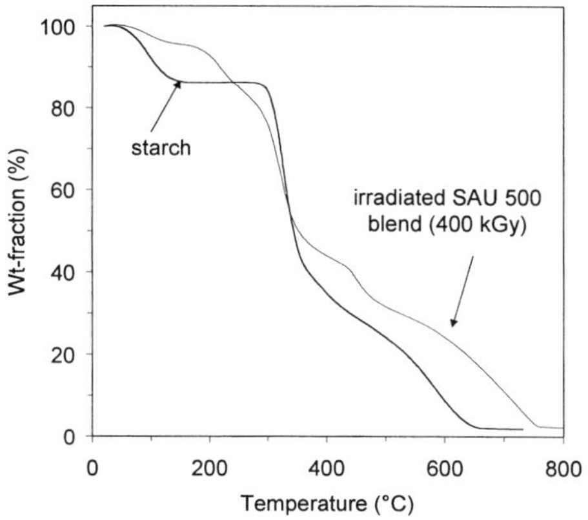

Figure 3. Plots representing the weight loss of native starch and irradiated SAU50 blend submitted to thermogravimetric analysis in air at a heating rate of 10°C.min^{-1}.

The profiles we observe with our samples are quite similar to that described for PAN modified starch, in which covalent grafting has been demonstrated by independent methods. Though we consider the argument to be indirect and questionable in a number of examples, our material exhibits the typical behavior of other grafted starches.

NMR Analysis of Grafting

At the molecular level, the changes in spectroscopic properties related to the sites modified by the covalent attachment of grafts may afford unambiguous information. We have tried to obtain such arguments by using ^{13}C and ^{1}H nmr spectroscopies. The anomeric carbon atom (C_1) of the AGU unit is a most plausible site for a free radical generation on starch.[10] Owing to the presence of the overlapping broad signal due to the NH_2 group of monomeric and polymeric AU in the ^{1}H nmr spectra, there is no possibility to determine, even qualitatively, the effectiveness of grafting from changes in the intensity for the AGU anomeric protons at δ = 5.5 ppm. The signals assigned to poly(AU) units, irrespective to their free state in the blend or of their grafting onto starch can be used to determine accurately AU conversion. It was not possible to obtain conclusive information from ^{13}C nmr spectra, in spite of careful examination for the presence of quaternary C atoms, as the probable grafting sites.[6]

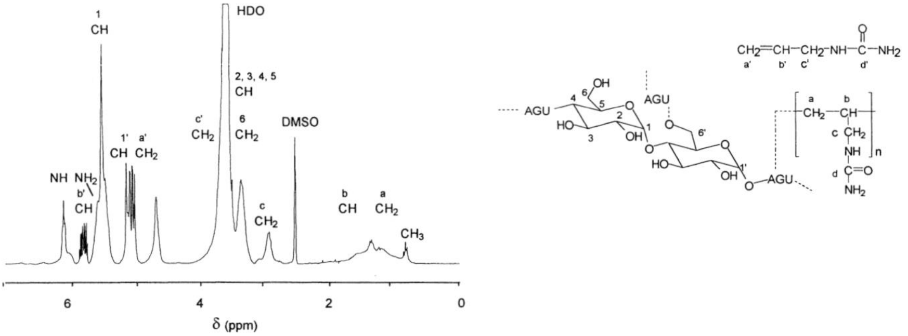

Figure 4. ^{1}H nmr spectrum (300 MHz, d-DMSO) of an SAU50 blend after EB processing with a 750 kGy dose.

Selective Solubilization of Free Poly(AU)

In order to quantify the relative extent of grafting and of AU homopolymerization, we have developed a procedure for achieving the selective solubilization of the different constituents. The solubility of amorphized starch and of AU - poly(AU) mixtures were tested separately at 1 wt-% concentration in water – methanol solutions. It was possible to solubilize efficiently AU together with free poly(AU) in water-methanol mixtures of composition ranging from 25:75 to 34:66 v/v, with only 2 wt-% of dry residue, whereas starch was not dissolved in similar conditions, as confirmed by gravimetry of the dried filtrate, indicating 100 % recovery.

From the ^{1}H nmr spectra of the crude irradiated samples and of the separated constituents it was possible to determine the overall conversion of AU as well as the fraction of polymerized AU present in grafts. The spectrum in Figure 5 shows the presence of poly(AU) grafts that can be quantified by integration of appropriate signals.

The results collected in Table 1 indicate that grafting is in competition with the formation of free homopolymer. The relative grafting yield is higher for the SAU20 films, in which AU is present in amounts lower than the limiting solubility.

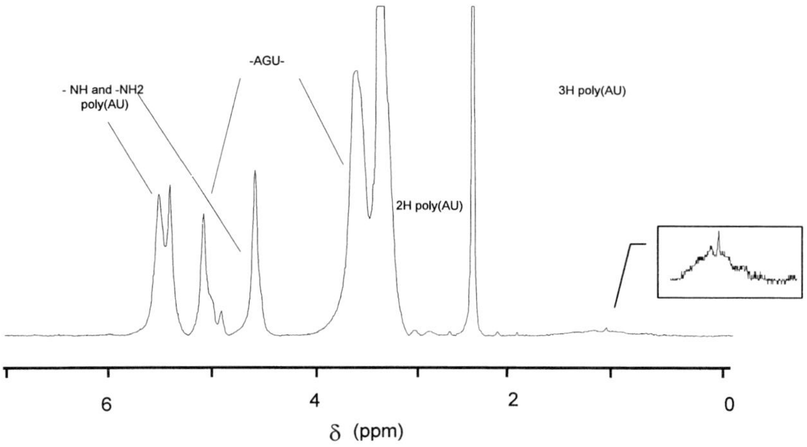

Figure 5. ^{1}H nmr spectrum (300 MHz, d-DMSO) of the water - methanol insoluble fraction of an irradiated SAU50 treated with a 400 kGy EB dose.

Table 1. Graft yield in SAU20 and SAU50 samples treated by a 400 kGy EB dose.

Sample	f_{AU} a)	$f_{poly(AU)}$ a)	f_{graft} a)	$[AU]_0$ b) mol.kg^{-1}	poly(AU) content c) mol.kg^{-1}	graft content c) mol.kg^{-1}
SAU20	0.25	0.15	0.60	1.6	0.24 (20%)	0.96 (80 %)
SAU50	0.70	0.175	0.125	3.3	0.58 (59%)	0.41 (41%)

a) Fractional amount of unreacted AU, of free poly(AU) and of grafted AU in the blend after 400 kGy irradiation.
b) Initial AU content in the blends.
c) Amount of AU in the form of free or of grafted polymer after 400 kGy irradiation.

We see from these data that reacted AU is essentially under the form of graft polymer in the SAU20 sample, whereas less than one half of the reacted AU has been coupled to starch in the SAU50 film. However, it has been shown that, from the viewpoint of physical properties, a

400 kGy treatment applied to SAU50 samples induces sufficient stabilization.[3,4]

Maldi-TOF Mass Spectrometry

In order to gain a deeper insight into the grafting mechanism, amorphous mixtures of various maltodextrines and AU, as well as pure crystalline AU, were irradiated with accelerated electrons and subsequently analyzed by MALDI-TOF mass spectrometry.

The obtained mass spectrograms reveal the presence of short chain AU oligomers with DP < 15. A detailed analysis of the mass of the various series of homologous oligomers is underway that will allow us to propose different mechanisms for initiation and termination, and to discuss the particular issue of transfer reactions.[12] We will focus here on some selected results obtained from an irradiated blend of maltotriose and AU. A mixture including equal weight amounts of the two components was melted to obtain an homogeneous blend that was quenched over an aluminum plate cooled at a temperature of –15°C. The resulting solid film was then exposed to EB-radiation so as to obtain a dose of 435 kGy and was prepared for MALDI-TOF analysis. The obtained spectra reveal, in addition to the presence of unreacted starting materials, the formation of six oligomer families, recognized by peaks separated by a mass of 100 that corresponds to AU molecular weight. Interestingly, one series that was not found in the spectra recorded from irradiated pure AU, corresponds to maltotriose - AU adducts, with the possible structure depicted in Scheme 2.

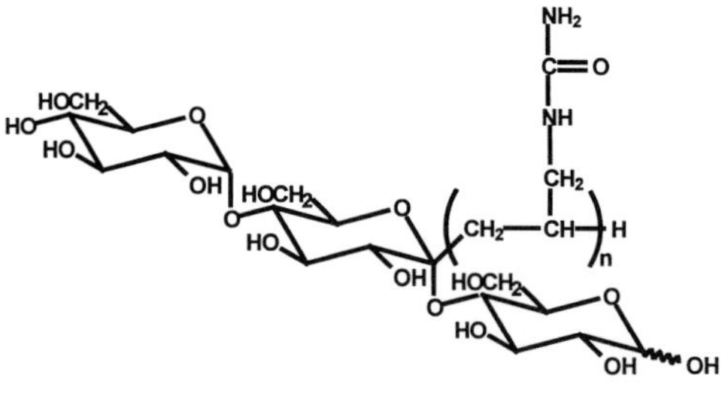

Scheme 2. Possible structure for the maltotriose – AU adducts detected by MALDI-TOF MS.

This series was identified with values for the number of grafted AU units ranging from 1 to 7, with maximum of the distribution for 3 repeat units (Table 2).

A possible sequence of reactions yielding the observed adduct is depicted in Scheme 3. It includes maltotriose free radical generation, likely on one of the two anomeric C atoms at the glucosidic junctions, by direct interaction with the energetic beam or by hydrogen abstraction caused by some free radical. Chain growth would result from repeated addition of the propagating free radical onto AU propenyl function. Finally abstraction of some hydrogen atom from monomer or saccharide would stop this growth by a transfer process.

Table 2. Main characteristics of the MS peaks assigned to maltotriose (MT) AU adducts.

m/z (observed)	Cation structure	Relative intensity	m/z (theoretical)
527.4	MT + Na^+	30	527.4
627.7	MT + 1AU+ Na^+	3.3	627.6
727.6	MT + 2AU+ Na^+	3.9	727.7
827.7	MT + 3AU+ Na^+	4.2	827.8
927.7	MT + 4AU+ Na^+	3.4	927.9
1027.7	MT + 5AU+ Na^+	3	1028.0

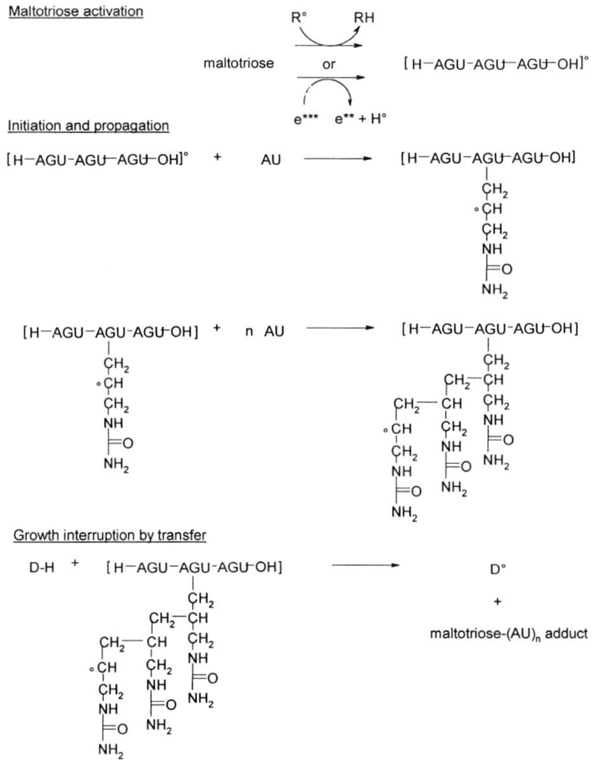

Scheme 3. Proposed mechanism for the graft polymerization of AU onto maltotriose under EB irradiation.

The evidence of grafting supported on a molecular basis by MALDI-TOF mass spectrometry now provides a reasonable explanation for the increase of SAU50 material solubility in water with higher processing doses (Figure 6).[3,5] We have suspected that polysaccharide degradation upon treatment with high energy radiation could induce chain scission at the origin of the higher soluble fraction. However, control measurements by SEC analysis performed with model maltodextrins of high solubility in water, blended with AU and EB treated in conditions similar to that used for starch blends, did not show any sign of degradation. The SEC traces of Figure 7 indeed indicate no significant modification of the distributions at large retention volumes that correspond to low molecular weights.

The change in water solubility observed with the starch AU blends therefore seems to be due to some effect impeding aggregate formation from the polysaccharide chains, as observed with gelatinized starch. The presence of short poly(AU) grafts along the polysaccharide chains reasonably explains this effect. It contributes also to the long term stability of the material, by avoiding retrogradation.

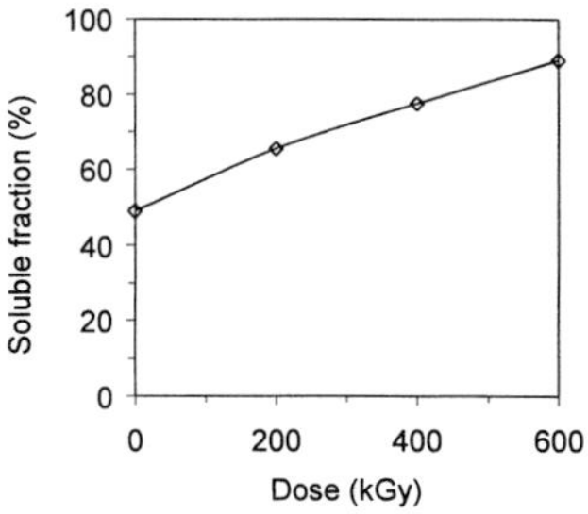

Figure 6. Variations of the soluble fraction of material in water, after irradiation of SAU50 blends.

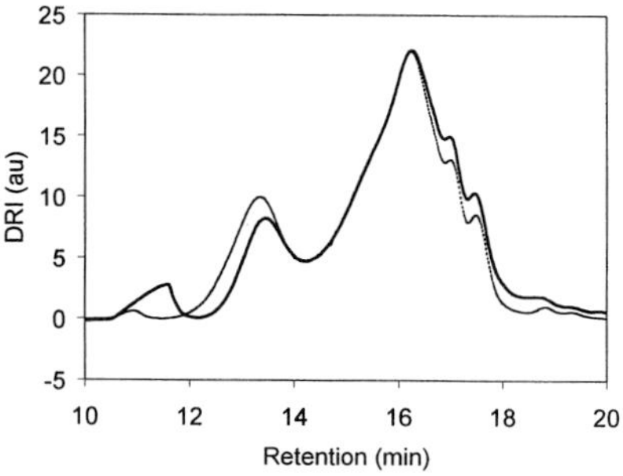

Figure 7. SEC traces of a mixture of maltodextrine (100 wt-parts) and AU (50 wt-parts) mixture before (---) and after (—) 400 kGy irradiation in the solid state.

Conclusion

The chemical modifications induced in starch-AU blends submitted to electron beam irradiation have been examined by various methods for addressing the important issue of the contribution of grafting in the observed physical effects. The selective solubilization in water - methanol mixtures of AU and free poly(AU) from irradiated blends allowed us to assess the extent of grafting onto starch that appeared dependent on the AU content. MALDI-TOF mass spectrometry affords unambiguous evidence of grafting that was suggested by other indirect observations or measurements. The mechanism of AU polymerization under high energy radiation is currently investigated in more details by the informative MS analytical method.

Acknowledgements

The authors wish to thank Dr. U. Maschke for assistance during EB irradiations and Dr. G. Ricart for MALDI-TOF MS analyses. The Electron Beam Project conducted in Lille is supported by the Centre National de la Recherche Scientifique, the Conseil Régional Nord Pas de Calais and the EU program FEDER.

[1] W. M. Doane , C. L. Swanson, G. F. Fanta, in "Emerging technologies for materials and chemicals from biomass", A.C.S. Symposium Series No 476, p. 197, 1992.
[2] K. Poutanen, P. Forssell, *Trends Polym. Sci.* **1996**, 4, 128.
[3] D. Ruckert; F. Cazaux, X. Coqueret, *J. Appl. Polym. Sci.* **1999,** 73, 409.
[4] A. Olivier, F. Cazaux, C. Gors, X. Coqueret, *Biomacromolecules* **2000,** 1, 282.
[5] A. Usanmaz, O. Yilmaz, *Eur. Polym. J.* **1986** 22, 657-660.
[6] A. Olivier, F. Cazaux, C. Gors, X. Coqueret, *Biomacromolecules* **2001,** 2,1260.
[7] A. S. Urun, A. Usanmaz, *METU J. Pure Appl. Sci* **1981**, 14, 1
[8] V. Iyer, P. V. Varadarajan, K. H. Sawakhande, N. D. Nacahane, *J. Appl. Polym. Sci.* **1990**, 39, 2259.
[9] B. K. Patel, V. K. Sinha, C. P. Patel, B. N. Misra, H. C. Trivedi, V. Vidyanagar, G. State, *Starch/Stärke* **1995**, 47, 151.
[10] A. M. Henderson, A. Rudin, *J. Polym. Sci., Polymer Chemistry Edition* **1981**, 19, 1721.
[11] A. Olivier, Ph.D. Dissertation, Université des Sciences et Technologies de Lille, 2000.
[12] J. Luna-Carlos, A. Olivier, F. Cazaux, X. Coqueret, *in preparation.*

Nemematic Conformation of Oxyethylene Spacers Involved in Main-Chain Liquid Crystals

Hidemine Furuya,[1] *Hiroshi Iwanaga,*[2] *Takeshi Nakajima,*[3] *Akihiro Abe*[4*]

[1]Tokyo Institute of Technology, Department of Organic and Polymeric Materials, 2-12-1 Ookayama, Meguro-ku, Tokyo 152-8552, Japan
[2]Fuji Photo Film Co. Ltd., Ashigara Research Laboratories, 210 Nakamura, Minamiashigara 250-0193, Japan
[3]SunAllomer Ltd., Kawasaki Development Center, 2-3-2 Yako, Kawasaki-ku, Kawasaki 210-8548, Japan
[4]Tokyo Institute of Polytechnics, Department of Applied Chemistry, 1583 Iiyama, Atsugi 243-0297, Japan

Summary: ^{2}HNMR measurements were performed on main-chain dimer and polymer liquid crystals (LC) having oxyethylene (OE) spacers $-(OCH_2CH_2)_xO-$ (x=2, 3). The orientational as well as conformational characteristics of these molecules have been investigated in bulk and in a nematic solution. The OE spacer was found to take spatial arrangements characteristics of the nematic phase. The nematic conformation of the spacer remains nearly invariant over a wide range of temperature and concentration. In these analyses, the ratio of the deuterium quadrupolar splittings $\Delta\nu/\Delta\nu^R$ ($\Delta\nu$: spacer and $\Delta\nu^R$: mesogenic unit) provided an important information regarding the spatial configuration of molecules in the LC state. The results obtained in this study are consistent with our previous conclusion drawn on a series of main-chain LC oligomers and polymers comprising n-alkane spacers $-O(CH_2)_nO-$ (n=9, 10).

Introduction

Flexible polymers often exhibit crystallization or liquid crystallization. In these first-order transitions, changes in the spatial arrangement of polymer chains are considered to be a major factor governing the transition entropy. Thermodynamic significance of the conformation entropy in these examples has been extensively studied on the basis of the rotational isomeric state (RIS) treatment.

In a series of papers,[1-8] we have reported the results of ^{2}HNMR and pressure-volume-temperature (*PVT*) studies of dimer and trimer liquid crystals (LC) having structures such as

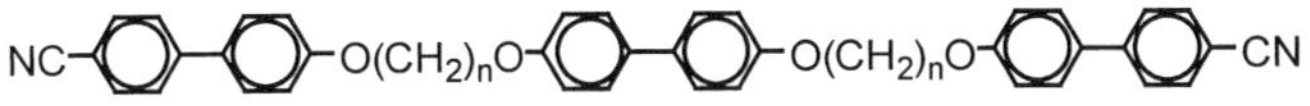

CBA-Tn (n = 9 and 10)

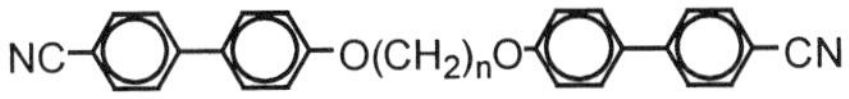

CBA-n (n = 9 and 10)

 CCC 1022-1360/00/$ 17,50+.50/0

where n designates the number of carbon atoms in the spacer. In these compounds, mesogenic cores (cyanobiphenyl groups) are linked with n-alkane spacers via the ether linkage. CBA-n and CBA-Tn exhibit two well-defined phase transitions as a function of temperature: crystal (C) - nematic LC (N) - isotropic melt (I). These ether-type LCs are known to exhibit a pronounced odd-even oscillation in various thermodynamic quantities at the NI phase boundaries,[9] indicating that the spatial configuration of the spacer plays a significant role.[10] The values of the constant-volume transition entropies $(\Delta S_{tr})_v$ derived from *PVT* studies were favorably compared with the conformational entropies of the spacer calculated from the spatial arrangements elucidated spectroscopically in the two relevant states.[3]

The results of the ^{2}HNMR analysis on these dimer and trimer LCs may be summarized as follows: (1) Under a uniaxial potential field, both spacer and mesogenic units at the terminals tend to align along the domain axis. Consequently the individual mesogenic cores inevitably incline to some extent with respect to the direction of molecular extension, giving rise to a moderate value of the orientational order parameter of mesogenic cores.[6] (2) While the orientational fluctuation of the entire molecule varies as a function of temperature, the nematic conformation of the spacer remain nearly invariant over the entire range of the LC state.[3,6] (3) It has been confirmed by the ^{2}HNMR and *PVT* studies that 50 - 60% of the transition entropy $(\Delta S_{tr})_p$ arises from the variation in the conformational distribution of the spacer at the phase boundary.[2,8]

The lowest energy state of n-alkane or polymethylene chains are the all-trans planar arrangement, which could be readily accommodated in the nematic environment. In contrast, polyoxyethylene (POE) is known as one of the most flexible polymers, and the chain crystallizes in a helical (7/2) arrangement. The OC-CO moiety of the POE chain prefers to take a gauche form in the unconstrained state. The purpose of the present work is to investigate the conformational characteristics of the OE spacer $-(OCH_2CH_2)_xO-$ under the influence of the nematic potential field. In the following, we report the nematic conformation of OE spacers revealed by the ^{2}HNMR analysis.

Experimental

Preparation of Samples

Twin compounds having various mesogenic units on both terminals of the OE spacers were prepared according to the scheme[11-13] listed below:

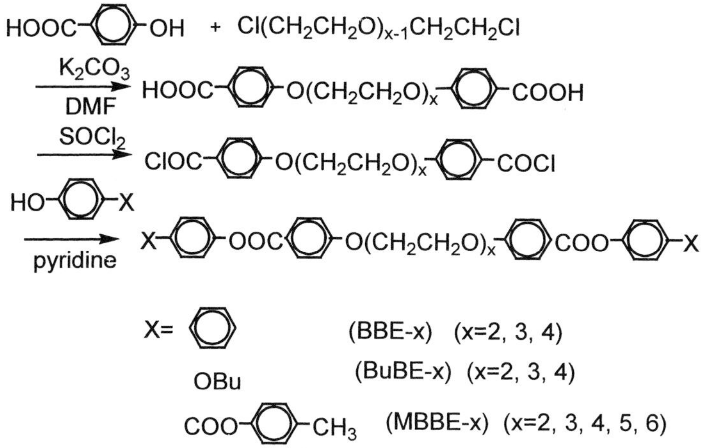

Abbreviated names of the compounds are indicated in parentheses: x denotes the number of OE units in the spacer.

Polymer samples having similar spacers were prepared by the conventional polycondensation method:[14-16]

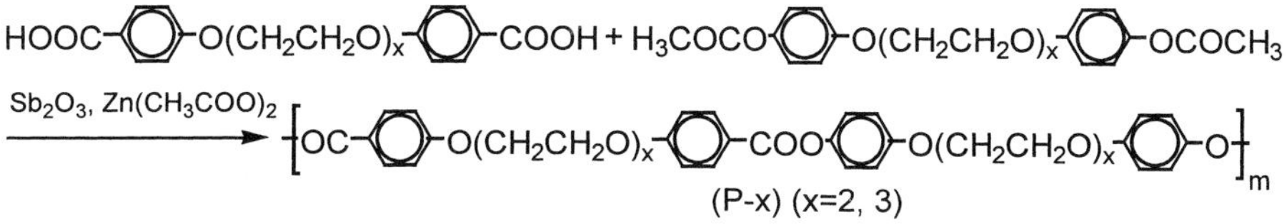

(P-x) (x=2, 3)

Polymers having partially deuterated spacers and mesogenic units were also prepared following the procedure described in the literature.[17-19] These polymer samples are hereafter represented by P-x.

Measurements

Thermal transitions were measured with a Perkin-Elmer differential scanning calorimeter (DSC) calibrated with indium. Heating rates of 10 deg min^{-1} were employed in most of the DSC scans. The phase transitions were observed by using an Olympus polarizing microscope equipped with a Mettler FP82 hot stage. The ^{2}HNMR spectra were recorded on a JEOL JNM-GSX-500 spectrometer operating at a 76.65 MHz deuterium resonance frequency. Measurements were carried out under a nonspinning mode. The molecular weights of polymer samples were obtained by gel permeation chromatography (GPC) with a SHOWDEX K-80M column using chloroform as the eluent: P-2, M_n=1.3x10^4, M_w=3.6x10^4;

P-3, M_n=3.0x10^3, M_w=1.1x10^4.

Results and Discussion

Thermal Analysis and Polarizing Microscope Observation of the Meso Phase

Phase transition temperatures and associated thermodynamic quantities obtained by using DSC are summarized in Table 1, where T_m, T_c, and T_{NI}, respectively, indicate the melting temperature of the crystal, the crystallization temperature, and the transition temperature from the isotropic to nematic phase. ΔH and ΔS are respectively the enthalpy and entropy changes at the phase transition as specified by the appended suffix, values being observed on heating unless otherwise noted. Most of the samples prepared exhibited meso phases, either enantiotropic or monotropic, the textures of the phase being identified as nematic under the polarizing microscope.

In consideration of the guaranteed temperature limit (200°C) of our standard ^{2}HNMR probe, BuBEs were chosen as the most appropriate dimer candidate to study the conformational characteristics of the OE spacer in the nematic state. With this reason, BuBE-x were extensively deuterated and used in the following analysis (see Table 1).

Table 1. Summary of DSC Data

	x	T_m °C	T_c °C	T_{NI} °C	ΔH_m kcal mol^{-1}	ΔH_{NI} kcal mol^{-1}	ΔS_m cal mol^{-1} K^{-1}	ΔS_{NI} cal mol^{-1} K^{-1}
BBE	2	188.9	169.0	191.9	18.1	0.59	39.2	1.28
	3	195.4	133.6	162.1[a]	17.9	0.62[a]	37.6	1.42[a]
	4	156.3	108.1	130.0[a]	14.8	0.23[a]	34.5	0.58[a]
BuBE	2	124.6	110.8	132.2	12.9	0.80	32.5	1.97
	3	99.0	68.9	103.0	12.6	0.76	33.8	2.03
	4	98.7	56.0	76.0[a]	13.5	0.54[a]	36.3	1.55[a]
MBBE	2	189.1	136.1	300.6	14.1	1.00	30.6	1.74
	3	163.9	121.3	271.0	13.3	0.95	30.5	1.75
	4	145.4	100.4	241.7	16.3	0.64	38.8	1.24
	5	127.9	91.9	215.5	6.1	0.28	15.3	0.58
	6	97.8	----	186.0	2.1	0.22	5.8	0.47
P	2	175.0	154.0	183.0[a]	----[b]	0.72[a,c]	----[b]	1.56[a,c]
	3	113.0	----	114.0[a]	----[b]	0.76[a,c]	----[b]	1.84[a,c]

[a]Estimated on cooling. [b]Values are not available due to an extensive overlap of the two endothermic peaks. [c]Expressed in terms of 'mol. of repeat unit'.

^{2}HNMR Observation

(1) BuBE-x (x = 2,3) and polymers in bulk

Examples of ^{2}HNMR spectra obtained for deuterated BuBE-x samples in the LC state are shown in Figs.1a and 2a. The deuterium atoms substituted on the aromatic cores of the mesogenic units exhibit splittings due to the dipolar couplings (D^{R}_{HD}) with the neighboring hydrogen in addition to the quadrupolar splittings ($\Delta\nu^{R}$). Here the values of $\Delta\nu$ and D_{HD} due to the mesogenic core are distinguished by the superscript R.

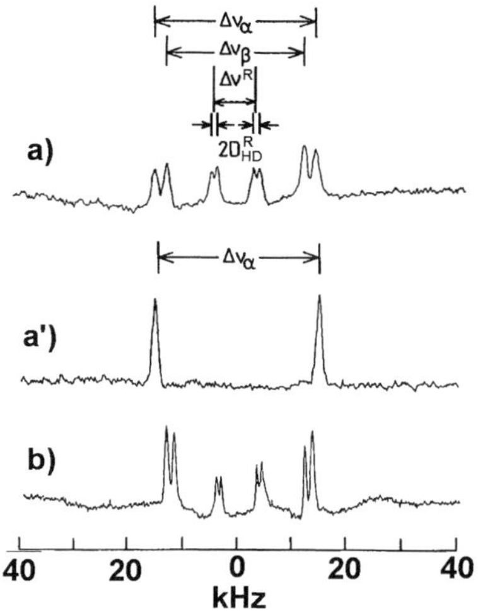

Figure 1. Examples of ^{2}HNMR spectra observed for (a) BuBE-2, (a') α-deuterated BuBE-2, and (b) BuBE-2 in PAA.

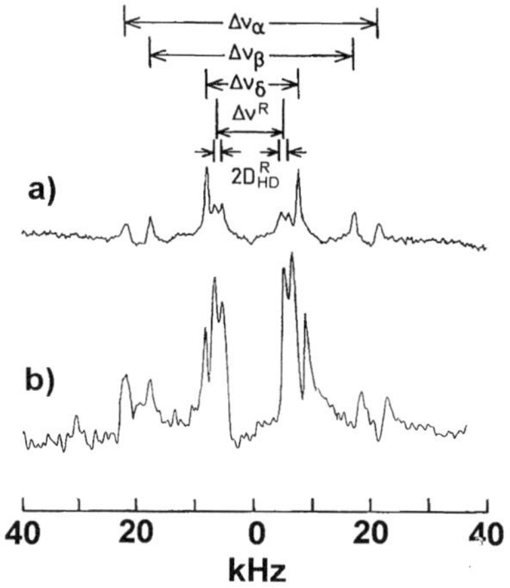

Figure 2. Examples of ^{2}HNMR spectra observed for BuBE-3 (a) in bulk and (b) in PAA.

Table 2. Observed Values of Quadrupolar and Dipolar Splittings for BuBE-x Samples in Bulk

Compound	T_r	$\Delta\nu^{R}$ / kHz	$2D^{R}_{HD}$ / kHz	$\Delta\nu_{\alpha}$ / kHz	$\Delta\nu_{\beta}$ / kHz	$\Delta\nu_{\delta}$ / kHz
	1.0	-7.13	-0.86	-27.97	-23.72	
BuBE-2	0.999	-7.10	-0.97	-27.65	-23.76	
	0.991	-9.09	-1.19	-34.60	-29.68	
	0.981	-10.21	-1.27	-38.04	-32.55	
	0.971	-10.86	-1.32	-40.75	-35.09	
	1.0	-8.96	-1.10	-35.13	-28.42	-12.43
BuBE-3	0.995	-10.36	-1.25	-39.70	-32.43	-14.28
	0.990	-11.26	-1.30	-45.53	-35.10	-15.48
	0.984	-11.89	-1.38	-45.39	-36.69	-16.31

For an unambiguous assignment of the peaks, a BuBE-2 sample deuterated specifically at the α-position of the OE unit of the spacer was prepared. Comparison shown in Figs.1a and a' indicates that the outermost peaks originate from the α-CD bonds. In the ^{2}HNMR spectrum for the BuBE-3 sample (Fig.2a), the degree of deuteration helps to deduce a proper assignment: the deuterium content at the δ-position is about twice of that at the α-position along the spacer. The three doublet peaks of the spacer were thus identified as α, β, and δ in the order from the terminal carbon. Subscripts are appended to the individual $\Delta\nu$-values to distinguish the position of the CD bond along the chain. In Table 2, the observed values of quadrupolar and dipolar splittings are summarized, where the temperatures are expressed relative to that of the NI transition point: $T_r = T/T_{NI}$. At lower temperatures, the resonance peaks tend to be broader.

The present study has been extended to include polymers comprising the same spacers. As-made samples were adopted for the spectroscopic examination. The deuterium NMR studies were carried out for deuterated samples P-2 and P-3 (see Figs. 3a and 4a). While the

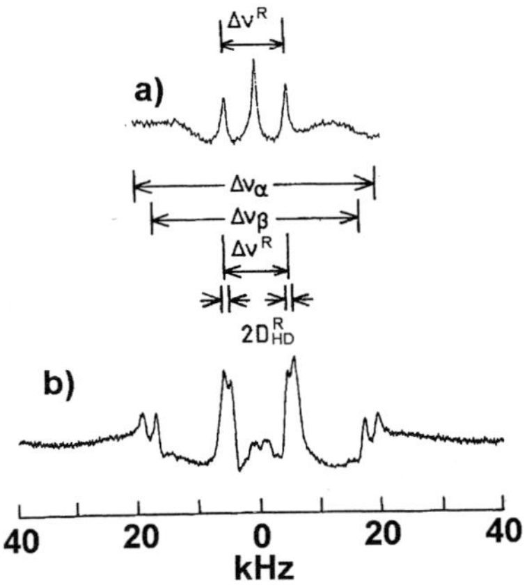

Figure 3. ^{2}HNMR spectra of (a)P-2 in bulk and (b) P-2 in PAA.

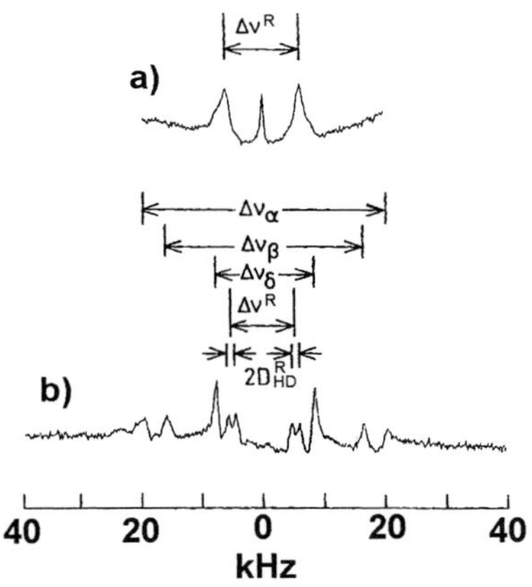

Figure 4. ^{2}HNMR spectra of (a) P-3 in bulk and (b) P-3 in PAA.

mesogenic cores exhibit broad resonance peaks, the signals from the CD bonds of the spacer are smeared out and almost indistinguishable. In all cases, the intensity of the central signals due to the isotropic fraction remains to be substantial throughout the LC state, indicating that the transformation to the nematic phase is incomplete. The origin of such an isotropic fraction is not clear at this moment. By judging from the molecular weight estimated by GPC, the samples should contain a sizable amount of low molecular weight fraction. The values of the observed $\Delta\nu^R$ are listed in Table 3.

Table 3. Observed Values of Quadrupolar Splittings for Mesogenic Cores of Polymers in Bulk

Compound	T_r	$\Delta\nu^R$ / kHz
P-2	1.0	-7.98
	0.992	-9.02
	0.982	-9.97
	0.974	-10.85
	0.964	-11.44
P-3	1.0	-11.09
	0.995	-11.92
	0.991	-11.94
	0.986	-12.27
	0.984	-12.50
	0.978	-12.05

Table 4. Observed Values of Quadrupolar and Dipolar Splittings for the Dimer and Polymer Samples in PAA

Compound	T_r	$\Delta\nu^R$ / kHz	$2D^R_{HD}$ / kHz	$\Delta\nu_\alpha$ / kHz	$\Delta\nu_\beta$ / kHz	$\Delta\nu_\delta$ / kHz
BuBE-2	1.0	-7.45	-0.88	-26.85	-24.07	
	0.999	-7.79	-1.00	-27.80	-24.97	
	0.994	-8.71	-1.04	-30.89	-27.74	
	0.986	-9.62	-1.10	-34.06	-30.39	
	0.979	-10.31	-1.20	-36.28	-32.25	
	0.969	-11.02	-1.26	-38.69	-34.30	
BuBE-3	1.0	-7.54	-0.95	-29.23	-23.69	-10.89
	0.998	-8.47	-0.97	-31.79	-26.02	-11.82
	0.993	-9.24	-1.16	-35.20	-28.39	-13.31
	0.988	-10.05	-1.21	-37.77	-30.25	-14.94
	0.980	-10.95	-1.24	-40.16	-32.69	-15.48
	0.968	-12.10	-1.36	-43.80	-36.08	-17.09
P-2	1.0	-10.01	-1.10	-37.94	-33.71	
	0.995	-10.45	-1.10	-38.64	-34.40	
	0.985	-10.92	-1.23	-40.70	-35.92	
	0.970	-11.61	-1.35	-43.63	-37.99	
P-3	1.0	-9.17	-1.18	-37.35	-29.54	-14.45
	0.995	-9.72	-1.20	-38.84	-30.71	-15.09
	0.990	-10.35	-1.19	-40.24	-32.42	-15.90
	0.982	-11.04	-1.20	-41.77	-33.86	-16.55

(2) Dimers and polymers hosted in a nematic solvent.

To obtain a better picture for the nematic conformation, P-x samples were dissolved in *p*-

azoxyanisole (PAA), which is a nematic low-molar-mass LC.[20] The ^{2}HNMR spectra obtained for a mixture containing 20wt.% P-x are included in the corresponding figures (see Figs.3b and 4b). From the phase diagram,[21] the eutectic composition of the mixture is estimated to be about 20-30wt.%. The temperature range (T_{NI}-T_c) of the nematic phase tends to be most expanded around this composition.

To facilitate comparison with polymers, the analysis was also carried out for the dimer sample BuBE-x dissolved in a nematic solvent, PAA. The spectrum obtained for a diluted sample (20wt.%) at T_r = 0.97 is shown in Figs.1b and 2b. The spectrum remains nearly identical irrespective of the dilution, suggesting that the conformation of the spacer is invariant as long as molecules are kept under the nematic field. The values of the observed splittings are summarized in Table 4.

Orientational Characteristics and the Nematic Conformation in the LC State

The order parameters of the mesogenic core axis were estimated from the observed $D_{\mathrm{HD}}^{\mathrm{R}}$ and $\Delta\nu^{\mathrm{R}}$ according to the procedure previously described.[22,23]

$$D_{\mathrm{HD}}^{\mathrm{R}} = -\gamma_{\mathrm{H}}\gamma_{\mathrm{D}}h/(4\pi^2 r_{\mathrm{HD}}{}^3)S_{\mathrm{ZZ}}^{\mathrm{R}} \qquad (1)$$

$$\Delta\nu^{\mathrm{R}} = (3/2)[S_{\mathrm{ZZ}}^{\mathrm{R}}\cdot q_{\mathrm{ZZ}} + (1/3)(S_{\mathrm{XX}}^{\mathrm{R}} - S_{\mathrm{YY}}^{\mathrm{R}})(q_{\mathrm{XX}} - q_{\mathrm{YY}})] \qquad (2)$$

Numerical values of the parameters required in these expressions were adopted from literature[22]: γ_H = 2.6752x10^{8} kg^{-1}sA, γ_D = 4.1065x10^{7} kg^{-1}sA, h = 6.6262x10^{-34} Js, r_{HD} = 2.48 A, q_{ZZ} = -18.45 kHz, q_{XX} = 113.85 kHz, and q_{YY} = -95.4kHz.

The order parameters $S_{\mathrm{ZZ}}^{\mathrm{R}}$ and the associated biaxiality terms $S_{\mathrm{XX}}^{\mathrm{R}} - S_{\mathrm{YY}}^{\mathrm{R}}$ are tabulated in Tables 5 and 6. Since the splittings due to dipolar couplings D_{HD} could not be obtained for polymer samples in bulk, $S_{\mathrm{ZZ}}^{\mathrm{R}}$ values were estimated from $\Delta\nu^{\mathrm{R}}$, assuming that contribution of the $S_{\mathrm{XX}}^{\mathrm{R}} - S_{\mathrm{YY}}^{\mathrm{R}}$ term is comparable to that of the corresponding dimer LCs. The values obtained both in bulk and in the PAA solution are compared in these tables for given spacers. Although the actual phase transition temperatures T_{NI} are quite different between the bulk and solution, the ordering behaviors become quite similar when the temperature is expressed relative to T_{NI} (i.e., T_r). In all cases, the magnitude of the biaxiality term is small (<0.05) as is usually the case for the conventional nematic LCs.

In our previous studies,[3] the conformation of the spacer has been analyzed on the basis of the deuterium quadrupolar splitting data collected for the constituent CD bonds of the chain. In the RIS simulation scheme previously established,[3] the following expression was used for the

Table 5. Order Parameters for BuBE-2 and P-2 in Bulk and in PAA solution (20wt.%)

Compound	T_r	S^R_{ZZ}	$S^R_{XX} - S^R_{YY}$
	1.0	0.343	0.023
BuBE-2	0.999	0.396	0.037
	0.991	0.497	0.044
	0.981	0.517	0.039
	0.971	0.517	0.033
	1.0	0.375	
P-2	0.992	0.470	
	0.982	0.503	
	0.974	0.535	
	0.964	0.557	
	1.0	0.355	0.023
BuBE-2/PAA	0.999	0.403	0.032
	0.994	0.419	0.028
	0.986	0.444	0.025
	0.979	0.484	0.030
	0.969	0.508	0.029
	1.0	0.444	0.022
P-2/PAA	0.995	0.444	0.018
	0.985	0.496	0.027
	0.970	0.544	0.033

Table 6. Order Parameters for BuBE-3 and P-3 in Bulk and in PAA solution (20wt.%)

Compound	T_r	S^R_{ZZ}	$S^R_{XX} - S^R_{YY}$
	1.0	0.444	0.032
BuBE-3	0.995	0.505	0.034
	0.990	0.525	0.031
	0.984	0.557	0.034
	1.0	0.521	
P-3	0.995	0.555	
	0.991	0.556	
	0.986	0.568	
	0.984	0.576	
	0.978	0.560	
	1.0	0.383	0.029
BuBE-3/PAA	0.998	0.389	0.022
	0.993	0.468	0.035
	0.988	0.488	0.033
	0.980	0.498	0.027
	0.968	0.546	0.029
	1.0	0.476	0.038
P-3/PAA	0.995	0.484	0.035
	0.990	0.480	0.028
	0.982	0.484	0.023

deuterium quadrupolar splittings ($\Delta\nu_i$) of the spacer:

$$\Delta\nu_i = (3/2)(e^2qQ/h)S_{ZZ}\left(3\langle\cos^2\phi_i\rangle-1\right)/2 \qquad (3)$$

where the quadrupolar coupling constant for the aliphatic CD bond is taken to be (e^2qQ/h) = 174 kHz.[24,25] Here we assume that the molecular axis (z) lies in the direction parallel to the line connecting the centers of the two neighboring mesogenic cores, and the molecules are approximately axially symmetric around the z-axis, and thus the orientation of these anisotropic molecules can be described by a single order parameter S_{ZZ}, the biaxiality of the system S_{XX} - S_{YY} being ignored for simplicity. In the equation above, ϕ_i is the angle between the ith CD bond and the molecular axis. The bracket indicates statistical mechanical averages taken over all allowed conformations in the system. In this model, the same value of S_{ZZ} is assumed to be applicable to all conformers in the mesophase. The order parameter S_{ZZ} may then be approximately related to S^R_{ZZ} of the mesogenic core by

$$S^R_{ZZ} = S_{ZZ}\left(3\langle\cos^2\psi\rangle-1\right)/2 \qquad (4)$$

where ψ denotes the disorientation of the mesogenic core axis with respect to the molecular axis. The analysis of the experimental results may be facilitated by taking ratios such as

$$\Delta\nu_i/\Delta\nu^R = \text{Const}\cdot\left(3\langle\cos^2\phi_i\rangle-1\right)/\left(3\langle\cos^2\psi\rangle-1\right) \qquad (5)$$

or

$$\Delta\nu_i/\Delta\nu_j = \text{Const}\cdot\left(3\langle\cos^2\phi_i\rangle-1\right)/\left(3\langle\cos^2\phi_j\rangle-1\right) \quad (i\neq j) \qquad (6)$$

Here contribution from the biaxiality term to the observed $\Delta\nu^R$ is assumed to be negligible (cf. equation (2)). These ratios should solely depend on the spacer conformation, being free from the orientational order of the molecular axis S_{ZZ}.

The values of $\Delta\nu/\Delta\nu^R$ estimated for given temperatures (T_r) are plotted in Figs.5 and 6, respectively, for x=2 and 3. In these figures, the dimer and polymer carrying the same spacer are compared. For the dimer, the ratios obtained both in bulk and in the PAA solution are accommodated in the same figure. The values tend to decrease slightly with lowering temperature. The $\Delta\nu/\Delta\nu^R$ ratios fall nearly in the same range for given i's. Representative values taken from the figures are as follows: for spacers with x=2, $\Delta\nu_\alpha/\Delta\nu^R$=3.8, $\Delta\nu_\beta/\Delta\nu^R$=3.2; those for x=3, $\Delta\nu_\alpha/\Delta\nu^R$=3.9, $\Delta\nu_\beta/\Delta\nu^R$=3.1, $\Delta\nu_\delta/\Delta\nu^R$=1.4. These results immediately suggest that the conformational correlation along the spacer is controlled by the same principle throughout the nematic LC state (T_r = 1.0 ~ 0.97). It does not matter whether the spacer is a part of dimer or polymer. Since the flexible spacer maintains liquid-like characteristics in the nematic state, the conformation of the spacer should be affected by the temperature, leading to a small variation in the ratio over a certain temperature range.

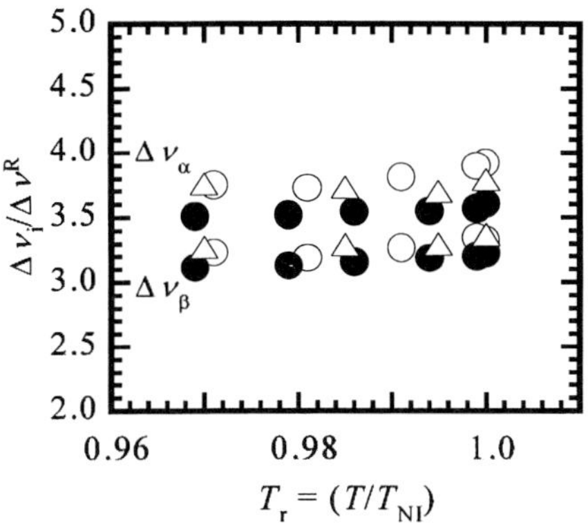

Figure 5. Quadrupolar splitting ratios $\Delta\nu_i/\Delta\nu^R$ plotted against temperature: BuBE-2 (open circles), BuBE-2 in PAA (filled circles), and P-2 in PAA (triangles).

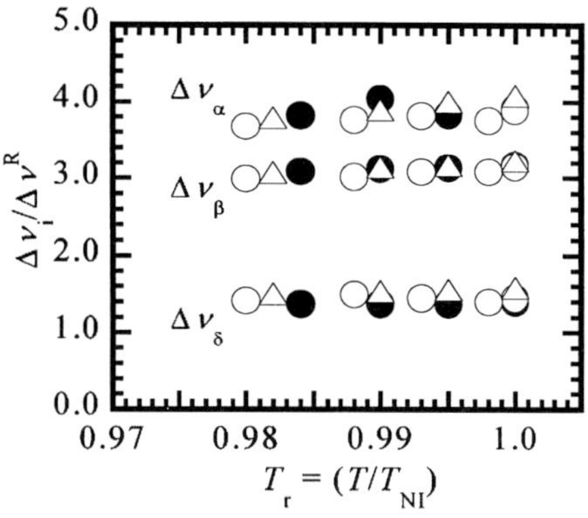

Figure 6. Quadrupolar splitting ratios $\Delta\nu_i/\Delta\nu^R$ plotted against temperature: BuBE-3 (open circles), BuBE-3 in PAA (filled circles), and P-3 in PAA (triangles).

Concluding Remarks

It is rather surprising to find that the ratios $\Delta\nu_i/\Delta\nu^R$ remain nearly invariant for given spacers independent of its origin: i.e., those from the dimer or polymer, or those observed in bulk or in the mixture with a nematic solvent. The same conclusion can be derived from the plots of $\Delta\nu_i/\Delta\nu_j$ vs. T_r constructed according to eq.6: the corresponding figures are omitted here to avoid redundancy. An ensemble of conformers characteristic to the nematic state may be termed the nematic conformation. These observations seem to confirm our previous conclusion derived from the analysis of a series of main-chain LC oligomers and polymers comprising n-alkane spacers. Elucidation of the individual conformers of OE spacers has however encountered some difficulty due to the fact that the number of adjustable parameters exceeds that collected from the ^{2}HNMR measurements. Further elaboration is needed.[26]

In the liquid crystalline state, the spacers should be allowed to take various conformations in as much as they are compatible with the preferred orientation of the mesogenic cores. Under these circumstances, the spacers are comparatively extended, and thus participate to some extent in the anisotropic intermolecular interactions in the nematic field.[27] While the orientation of the molecular axis (i.e., the whole molecule) varies sizably as a function of temperature, the nematic conformation of the spacer is not much affected.

Some of the dimer compounds prepared in this study exhibit a nematic phase over a wider temperature range. These samples are well suited for the investigation of thermodynamic (*PVT*) properties of the nematic ensemble. Studies along this line are in progress.

[1] Abe, *Macromol. Symp.*, **1997**,*118*, 23.
[2] Abe, S. –Y. Nam, *Macromolecules*, **1995**, *28*, 90.
[3] Abe, H. Furuya, R. N. Shimizu, S. –Y. Nam, *Macromolecules*, **1995**, *28*, 96.
[4] Abe, H. Furuya, *Macromolecules*, **1989**, *22*, 2982; A. Abe, H. Furuya, *Mol. Cryst. Liq. Cryst.*, **1988**, *159*, 99; A. Abe, H. Furuya, D. Y. Yoon, *Mol. Cryst. Liq. Cryst.*, **1988**, *159*, 151.
[5] K. Inomata, Ph. D Thesis, Tokyo Institute of Technology, 1991.
[6] Abe, R. N. Shimizu, H. Furuya, "Ordering in Macromolecular Systems", A. Teramoto, M. Kobayashi, T. Norisuye, Ed., Springer, Berlin, 1994, 139.
[7] Y. Maeda, H. Furuya, A. Abe, *Liq. Cryst.*, **1996**, *21*, 365.
[8] Abe, T. Takeda, T. Hiejima, H. Furuya, *Polym. J.*, **1999**, *31*, 728.
[9] J. W. Emsley, G. R. Luckhurst, G. N. Shilstone, I. Sage, *Mol. Cryst. Liq. Cryst.*, **1984**, *102*, 223.
[10] Abe, *Macromolecules*, **1984**, *102*, 223.
[11] S. Tanimoto, R. Taniyasu, M. Okano, *Bull. Inst. Chem. Res.*, *Kyoto Univ.*, **1978**, *56*, 340.
[12] J. –I. Jin, C. Ober, *Bull. Korean Chem. Soc.*, **1983**, *4*, 143.
[13] J. –I. Jin, Y. –S. Chung, J. –S. Kang, *Mol. Cryst. Liq. Cryst.*, **1982**, *82*, 261.
[14] L. Strzlecki, D. Van Luyen, *Eur. Polym. J.*, **1980**, *16*, 299.
[15] D. Van Luyen, L. Strzlecki, *Eur. Polym. J.*, **1980**, *16*, 303.
[16] W. Volksen, D. Y. Yoon, P. M. Cotts, *Macromolecules*, **1989**, *22*, 3846.
[17] T. Ishiguro, S. Kato, Y. Akazawa, *Yakugaku Zasshi*, **1943**, *63*, 282.
[18] G. B. Butler, Y. C. Chu, *J. Polym. Sci.*, *Polym. Chem. Ed.*, **1979**, *17*, 859.
[19] S. Z. Perry, H. Hibbert, *Can. J. Res.*, **1936**, *14*, 77.
[20] F. Volino, M. M. Gauthier, A. M. Giroud-Godquin, R. B. Blumstein, *Macromolecules*, **1985**, *18*, 2620.
[21] T. Nakajima, Master Thesis, Tokyo Institute of Technology, 1991.
[22] J. W. Emsley, K. Hamilton, G. R. Luckhurst, F. Sundholm, B. A. Timimi, D. L. Turner, *Chem. Phys. Lett.*, **1984**, *104*, 136.
[23] N. Kimura, A. Abe, *Polym. Bull.*, **1992**, *28*, 81; A. Abe, N. Kimura, M. Nakamura, *Makromol. Chem.*, *Theory Simul.*, **1992**, *1*, 401.
[24] S. Bruckner, J. C. Scott, D. Y. Yoon, A. C. Griffin, *Macromolecules*, **1985**, *18*, 2709.
[25] J. C. Rowell, W. D. Phillips, L. R. Melby, M. Panar, *J. Chem. Phys.*, **1965**, *43*, 3442.
[26] H. Iwanaga, Master Thesis, Tokyo Institute of Technology, 1990.
[27] H. Furuya, S. Okamoto, A. Abe, G. Petekidis, G. Fytas, *J. Phys. Chem.*, **1995**, *99*, 6483.

Fluorescence Study for the Electrostatic Interaction and Aggregation in Dilute Polar Solution of Polyelectrolytes

*Qilong SUN, Biye REN, Xinxing LIU, Fang ZENG, Ping LIU, Zhen TONG**

Research Institute of Materials Science, South China University of Technology, Guangzhou 510640, China

Summary: Fluorescence decay and quenching of pyrene labels on copolymers of 2-acrylamido-2-methylpropanesulphonic acid (AMPS) and *N,N*-dimethylacrylamide (DMAA) were observed in dilute salt-free aqueous solutions as a function of the mole fraction F_{AMPS} of AMPS from 0 to 0.896. Monoexponential and biexponential decays were found for the samples of $F_{AMPS} < 0.35$ and samples of $F_{AMPS} > 0.35$, respectively. The fast decay component is 80% and the averaged lifetime $<\tau>$ and lifetime τ_1 of the fast decay decreased with increasing F_{AMPS}. Quenching efficiency of Cu^{2+}, CH_3NO_2, and dinitrobenzene to the pyrene label was investigated in the framework of Stern-Volmer plot. The quenching effects of Cu^{2+} included both of dynamic and static ones, the latter was due to the condensed Cu^{2+}. For the neutral quenchers, the quenching rate constant k_q increased when $F_{AMPS} < 0.449$ then decreased, showing a decline of accessibility to the pyrene label. I_1/I_3 value in salt-free dilute aqueous solution and in DMSO solution decreased obviously with an increase in F_{AMPS}, indicating that the labeled fluorophore experienced a decrease in polarity of its microenvironment with increasing charge density of the polymer. This I_1/I_3 decrease was enhanced with increasing the polymer concentration and adding salt NaCl up to 0.75 mol/L showed no effect on the appearance of this decrease. These results were interpreted consistently with the counterion condensation concept, where condensed counterions induced the "temporal" aggregation of less-polar in the polyelectrolyte solutions surrounding the pyrene labels.

Introduction

Polyelectrolyte attracts researches for more than a half century due to its significance in many industrial areas and in modeling the assembly structure of DNA and proteins. According to the traditional consideration for polyelectrolytes in dilute aqueous solutions, the ionizable residues on the polymer will dissociate into polyions and counterions. The dominant interaction governing the molecular behavior of polyelectrolytes is the electrostatic repulsion screened by the atmosphere of small ions.[1, 2]

Actually, for a vinyl polymer with a pendant ionizable group on each repeat unit, the linear distance between two neighboring ions on the chain is about 2.5 Å, independent of polymer

 CCC 1022-1360/00/$ 17,50+.50/0

concentration when the polyelectrolyte is completely dissociated. This charge density is even higher than that in 10 mol/L of aqueous NaCl solution if it exists actually, leading to an unstable system due to the high free energy. This means that there is no infinite dilute state for polyelectrolytes as the reference state used for normal solutions and that the effective charge density should be reduced to decrease the electrostatic free energy. Based on this consideration, Oosawa[3] and Manning[4] proposed the counterion condensation concept. When the linear charge density of polyelectrolyte is beyond a threshold value, the counterion will condense surrounding the polyion to reduce the actual charge density. Then, the condensed counterions may cause a new kind of aggregation domain in the solution as suggested by the simulation of Ray and Manning.[5-8]

On the other hand, there exist abnormal observations for the structure of polyelectrolyte in polar solutions at low ionic strength. One example is the existence of fast and slow diffusion modes revealed by dynamic light scattering.[9-14] The origin of this slow mode, usually attributed to so-called large-scale "heterogeneities" or "domains" in the solution, is still unclear. Another example is the appearance of a peak in the angular dependence of the scattered intensity along with a steep upturn at very low scattering angles as observed by small angle neutron and X-ray scattering on polyelectrolyte solutions at low ionic strength.[12, 15, 16] This unusual scattering is electrostatic nature and corresponds to "multichain" domains. These observations seem to suggest the possible existence of somewhat aggregated structure in polar solutions of polyelectrolytes at low ionic strength.

Fluorescence is a powerful tool to monitor the change in the microenvironment surrounding the chromophore in polymer solutions.[17, 18] The fluorophore covalently attached to polymer chains can directly reflect the local situation of labeled polymer, such as local polarity, local viscosity, local segment mobility, and accessibility of the label. Therefore, pyrene labeled strong electrolyte copolymers of 2-acrylamido-2-methylpropanesulphonic acid (AMPS) and *N,N*-dimethylacrylamide (DMAA) were chosen because the homopolymers of AMPS and DMAA were completely dissolvable in water. The pyrene content in the polymer has to be restricted as low as 0.1 mol% to avoid the association induced by the aggregated labels.

Experimental Section

Synthesis of pyrene labeled sulfonate polyelectrolyte samples has been described elsewhere[19] and a brief introduction is given below. Pyrene label monomer *N*-(1-pyrenylmethyl)-methacrylamide (PyMA) was prepared from 1-pyrenylmethylamine hydrochloride (Aldrich) according to the method of Ref. 20. 2-Acrylamido-2-methylpropanesulphonic acid (AMPS, Fluka), *N,N*-dimethylacrylamide (DMAA, Kohjin, Japan), and PyMA were initiated to copolymerize in DMSO/water mixture (2:3 in volume) at 70 °C by 0.2 mol% of ammonium persulfate. The total monomer concentration and PyMA content were kept at 0.80 mol/L and 0.1 mol%, respectively, while the monomer ratio of AMPS was changed to produce a series of samples with different charge densities. The refined products were analyzed with a Heraeus CHN-O apparatus and the F_{AMPS} value of AMPS mole fraction in the sample was evaluated. The weight averaged molecular weight M_w of samples was determined by gel permeation chromatography (GPC) of Waters using aqueous Na_2SO_4 solution of 0.1 mol/L as the eluent and poly(ethylene oxide) as the standard. The pyrene label content of 0.1 mol% in the sample was confirmed by UV absorbance. These samples were referred to as ADPy-x series; with x denoting the mole ratio of the AMPS monomer in the feed, and the composition and M_w are listed in Table 1.

Fluorescence lifetime was determined with a Horiba NABS-1100 time-resolved spectrofluoremeter excited at 340 nm. The aqueous solution with polymer concentration c_P = 0.05 g/L was prepared by dissolving the ADPy sample in highly pure water (resistance of 18×10^6 Ωcm by a Millipore purification apparatus) for one day with stirring. Before measurement, the solution was deaerated by bubbling with highly pure nitrogen for 30 min.

Table 1. Composition of ADPy samples*

Sample	AMPS/DMAA in feed (mol)	yield (wt%)	F_{AMPS}	$M_w \times 10^{-4}$	M_w/M_n
ADPy-0	0/100	78.55	0	-	-
ADPy-1	1/99	66.52	0.033	187	6.0
ADPy-5	5/95	80.95	0.147	72.2	4.8
ADPy-10	10/90	84.71	0.301	-	-
ADPy-20	20/80	63.31	0.449	40.1	3.1
ADPy-30	30/70	68.62	0.574	55.2	2.9
ADPy-45	45/55	84.58	0.711	84.1	2.6
ADPy-60	60/40	86.91	0.820	85.7	2.9
ADPy-75	75/25	89.50	0.896	80.8	2.8

*: Pyrene content is 0.1 mol% of total monomers.

The steady-state fluorescence emission spectrum was measured with a Hitachi F-4500 fluorescence spectrophotometer at 25 °C. The excitation wavelength was 343 nm with the exciting and emitting slit of 5.0 and 2.5 nm, respectively, which were required because of low emission intensity due to very low label content in the samples. For the quenching experiments, the quencher solution of nitromethane in water (0.6 mol/L) or dinitrobenzene in ethanol (0.1 mol/L) was added into the polymer solution with a microsyringe. The aqueous quencher solution of cupric sulfate (0.02 mol/L) was mixed with a polymer solution first and then diluted to desired concentrations.

Results

Fluorescence Decay

Figure 1 shows experimental fluorescence decay data (points) of pyrene labels on five ADPy samples in water at concentration of $c_P = 0.05$ g/L. The decay is biexponential of the form[21]

$$I(t) / I(0) = \alpha_1 \exp(-t / \tau_1) + \alpha_2 \exp(-t / \tau_2) \quad (1)$$

where $I(t)$ and $I(0)$ are the fluorescence intensities at time t and $t = 0$, respectively, α_i is the fraction of decay with the lifetime τ_i, and $\alpha_1 + \alpha_2 = 1$. The solid curves are the fit of eq (1) to the observed data, giving the parameters in Table 2. The monoexponential decay is observed for the samples ADPy-5 and ADPy-10 with lower F_{AMPS}, but the biexponential decay is the case for the other samples with higher F_{AMPS}. In the latter case, as F_{AMPS} is increased the lifetime τ_1 of the faster decay becomes shorter with constant α_1 and τ_2.

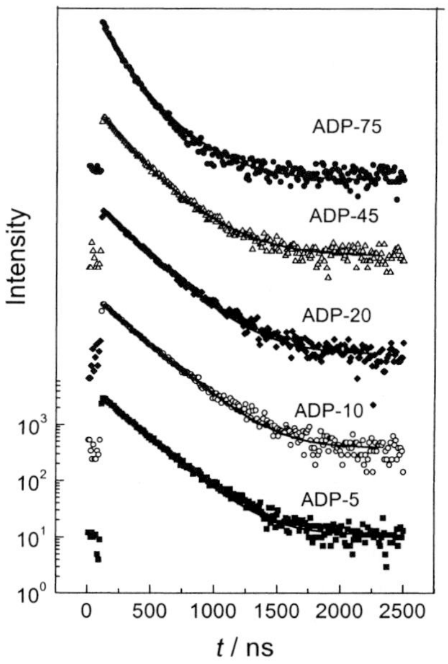

Figure 1. Fluorescence decay traces (points) of pyrene labels on indicated ADPy samples in aqueous solution at 25 °C with polymer concentration c_P of 0.05 g/L, the solid curves are the exponential fit with the parameters in Table 2.

The biexponential decay of fluorescence emission has been found from aqueous solution of polyelectrolytes. Hoyle et al. considered a dual environment for pyrene probes in water-soluble polyphosphazene based on the biexponential decay.[22] Morishima and colleagues observed the biexponential decay of pyrene labels on polyelectrolytes with long alkyl side chains for hydrophobic modification and correlated the fraction of slower decay component with the mole fraction of hydrophobic modifier.[23, 24]

Table 2. Fluorescence lifetime of pyrene label on ADP in water solutions

Sample	F_{AMPS}	ξ	α_1	τ_1 /ns	τ_2 /ns	$<\tau>$* /ns	for Cu^{2+} quenching		
							K_{SV} /Lmol^{-1}	**K**	α
ADPy-5	0.147	0.420	1.0	231	-	231	100	80	0
ADPy-10	0.301	0.860	1.0	231	-	231	-	-	-
ADPy-20	0.449	1.283	0.8	180	290	202	700	80	0.0008
ADPy-45	0.711	2.031	0.8	145	290	174	1450	80	0.0008
ADPy-75	0.896	2.560	0.8	135	290	166	1700	80	0.0008

*$<\tau>$: Average lifetime, $<\tau> = \alpha_1\tau_1 + \alpha_2\tau_2$

The present results seem to suggest that there would be two different microenvironments for pyrene labels on the strong polyelectrolyte in aqueous solution when its charge density is beyond a specific value. One is more exposed to water and its lifetime becomes shorter from 231 ns with increasing charge density and the other is protected from water with the constant lifetime (290 ns) longer than 231 ns for the samples with lower charge density. Because of low pyrene content in polymers (0.1 mol%) and high solubility of homopolymers of AMPS and DMAA, these two different microenvironments appear to be the contribution of electrostatic interaction in the salt-free aqueous solution. However, the averaged lifetime $<\tau>$ decreases with increasing charge density because the whole polyelectrolyte chain becomes more extended, resulting in an obvious reduction of pyrene label's lifetime.

Fluorescence Quenching

The existence of different microenvironments for pyrene labels on polymers can be detected by the variation in their accessibility to fluorescence quencher. We quenched the fluorescence emission by various quenchers of cation Cu^{2+}, neutral nitromethane, and more hydrophobic neutral dinitrobenzene. The data were dealt with in the framework of Stern-Volmer plot as[25]

$$I_0 / I = 1 + k_q \tau_0 [Q] \qquad (2)$$

where I_0 and I are the emission intensity at about 397 nm without quencher and with quencher, respectively; k_q is the quenching rate constant, τ_0 is the fluorescence lifetime without quencher ($<\tau>$ in Table 2 was used in evaluating k_q), and [Q] is the quencher concentration. The Stern-Volmer plots of present samples are shown in Figures 2 through 4. The pyrene label concentration in these solutions ranges from 2.6 x 10^{-7} mol/L for ADPy-75 to 4.3 x 10^{-7} mol/L for ADPy-5 with the corresponding ionic monomer concentration of [AMPS] from 6.36 to 22.83 x 10^{-5} mol/L.

Figure 2 illustrates that quenching efficiency of Cu^{2+} increases with increasing charge density in the polymer and all data except that for ADPy-5 follow the curve bending downwards with increasing [Q]. There is only very weak quenching effect of Cu^{2+} for ADPy-5 with the lowest charge density.

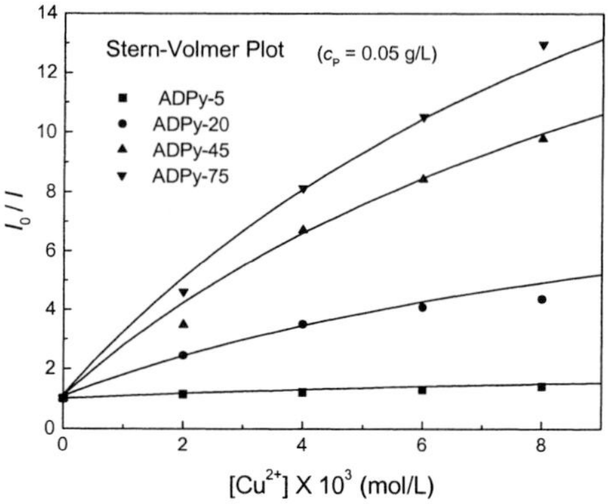

Figure 2. Fluorescence quenching by Cu^{2+} for pyrene labels on the samples with different AMPS contents in aqueous solutions. Curves: calculated from eq. 10.

Counterion condensation effect should be taken into account for interpreting the quenching of highly charged polyelectrolytes with cationic quenchers.[26]

The charge density parameter ξ is proportional to the AMPS mole fraction as[4]

$$\xi = F_{\text{AMPS}}\, q^2 / DkTb \qquad (3)$$

where q is the electronic charge in esu unit, D the dielectric constant of the solvent, k the Boltzmann constant, T the Kelvin temperature, and b the monomer length along the chain contour (usually b = 2.5 Å). Estimated ξ values of these samples are listed in Table 2. According to the Manning's modeling,[4] the criterion for divalent counterion Cu^{2+} to condense on polyanion is $\xi > 0.5$. Therefore, Cu^{2+} ions will condense on all samples of ADPy-20, ADPy-45, and ADPy-75, except ADPy-5. The condensed Cu^{2+} will induce the static quenching in addition to the dynamic quenching and the Stern-Volmer equation will be modified as the eq (2) of Ref. [26]

$$I_0/I = (1 + K_{SV}[Cu^{2+}]_{\text{eff}})\exp(n) \qquad (4)$$

Here, K_{SV} is the dynamic quenching constant, $[Cu^{2+}]_{\text{eff}}$ is the effective concentration of free Cu^{2+} ions which can actually quench the excited fluorophore by dynamic collisions since a large part of Cu^{2+} ions cannot be involved in quenching ultimately due to the $[Cu^{2+}]$ even two orders of magnitude higher than the [AMPS] in the present work. And n is the average number of condensed Cu^{2+} ions in one quenching effective sphere, given by

$$n = \alpha\,[Cu^{2+}]_{\text{con}} / [\text{pyrene}] \qquad (5)$$

where $[Cu^{2+}]_{\text{con}}$ is the concentration of condensed Cu^{2+} ions, [pyrene] is the concentration of pyrene residues, and α is the fraction of condensed Cu^{2+} ions existing in the quenching effective sphere which quench the fluorophore with unity efficiency. For the our ADPy solutions, the total concentration $[AMPS]_0$ of AMPS unit is much lower than $[Cu^{2+}]$, so that the $[Cu^{2+}]_{con}$ for satisfying the condition of $\xi = 0.5$ will be evaluated from $[AMPS]_0$ as

$$[Cu^{2+}]_{con} = [AMPS]_0(1 - 1/2\xi) / 2 \qquad (6)$$

Condensation of counterion H^+ has been ignored because the condensation preferentially occurs for divalent counterion Cu^{2+}. Only very small part of free Cu^{2+} ions can collide with the excited fluorophore effectively due to the extremely low fluorophore content in the samples. By analogy to the Langmuir isothermal absorption,[27] the ζ fraction of fluorophores already quenched by free Cu^{2+} ions is

$$\zeta = K[Cu^{2+}]_{free} / (1 + K[Cu^{2+}]_{free}) \qquad (7)$$

So, $[Cu^{2+}]_{eff}$ is assumed to be proportional to the fluorophore unquenched

$$(1 - \zeta) / \zeta = [Cu^{2+}]_{eff} / \{[Cu^{2+}]_{free} - [Cu^{2+}]_{eff}\} \qquad (8)$$

$$[Cu^{2+}] = [Cu^{2+}]_{free} + [Cu^{2+}]_{con} \approx [Cu^{2+}]_{free} \qquad (9)$$

because $[Cu^{2+}]_{free} >> [Cu^{2+}]_{con}$. Here, K is the equilibrium constant for the free Cu^{2+} ion entering the counterion atmosphere surrounding the excited fluorophore by dynamic collision and $[Cu^{2+}]_{free}$ the free Cu^{2+} ion concentration. Substitute eqs (5)-(9) into eq (4),

$$I_0/I = \{1 + K_{SV}[Cu^{2+}] / (1 + K[Cu^{2+}])\} \exp\{\alpha[AMPS]_0(1 - 1/2\xi) / 2[pyrene]\} \qquad (10)$$

The curves in Figure 2 are calculated results with the K_{SV}, K, and α values in Table 2 estimated by the best fitting. With increasing charge density of the samples, K_{SV} increases due to the increase in electrostatic attraction of the polyanion to the free cationic quencher Cu^{2+}. K reflects the equilibrium of Cu^{2+} ion entering and leaving the quenching effective atmosphere surrounding the fluorophore, consequently independent of the charge density. The α value is a probability product of a condensed Cu^{2+} ion encountering the nearest neighbor to the excited fluorophore and static quenching efficiency. The former is of millesimal owing to the mole ratio about 1.1~2.2 x 10^{-3} for pyrene residue to AMPS unit in these samples and the 80% of pyrene more exposed to water easy to be accessed by the quencher as shown above. Therefore, the small α value estimated is reasonable, indicating that almost no change in static quenching occurs by increasing $[Cu^{2+}]$ in the observed $[Cu^{2+}]$ range because the counterion condensation is completed on the samples at the first Cu^{2+} concentration.

The upward bending of the curves in Stern-Volmer plot has been reported in some literature for cationic quenchers with the existence of anionic polyelectrolyte due to the increase in the static quenching,[23, 26, 28] since the quencher concentration is much lower than that required for condensation on the polyelectrolytes. However, the downward bending of the curves in

Figure 2 is based on the fact that addition of Cu^{2+} only increases $[Cu^{2+}]_{free}$, which in turn reduces the proportion of free Cu^{2+} ion effectively entering the counterion atmosphere surrounding the excited fluorophore.

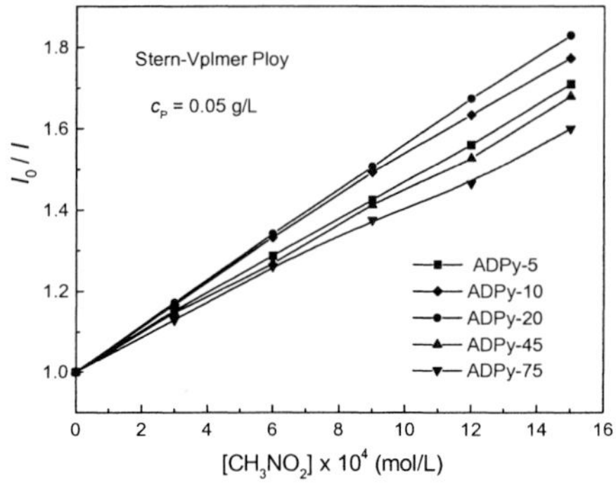

Figure 3. Fluorescence quenching by nitro-methane for pyrene labels on the samples with different AMPS contents in aqueous solutions.

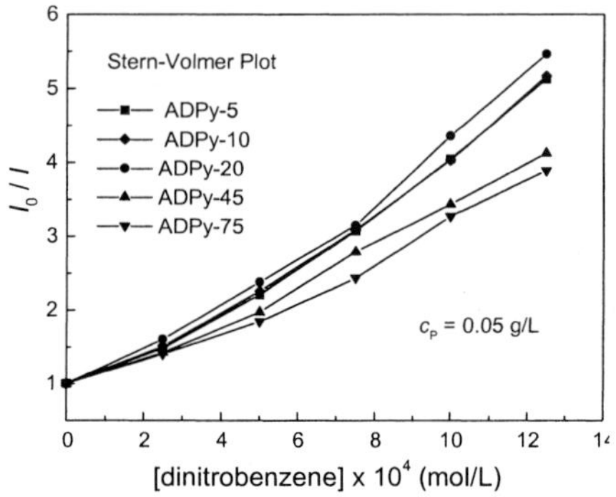

Figure 4. Fluorescence quenching by dinitro- benzene for pyrene labels on the samples with different AMPS contents in aqueous solutions.

As seen from Figures 3 and 4, the quenching efficiency of neutral quenchers nitromethane and dinitrobenzene does not change so much with the polymer charge density as the ionic quenchers. For easy comparison, we plotted the quenching rate constant k_q against F_{AMPS} in Figure 5. The curves for both quenchers are quite similar in shape, but the k_q value for dinitrobenzene is higher than that for nitromethane due to its higher hydro-phobicity and phenyl planar structure.

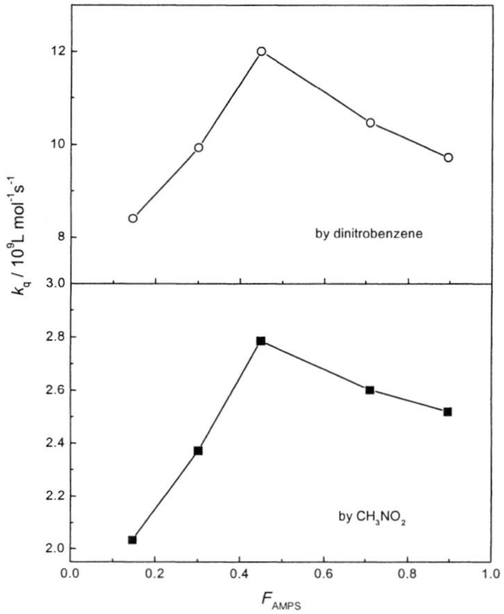

Figure 5. Plots of the quenching rate constant k_q for indicated quencher against the charge density F_{AMPS} of the sample.

The F_{AMPS} dependence of the k_q in Figure 5 manifests the change in accessibility of the neutral quenchers to the pyrene label with the polymer charge density. When F_{AMPS} increases from 0.147 to 0.35, where no counterion condensation occurs for univalent H^+, the actual charge density is increased monotonically. Consequently, the enhanced electrostatic repulsion between charged groups on the chain causes a more extended conformation. This effect, in turn, makes the fluorophore label be more easily approached by the quencher molecules. However, when the charge density exceeds 0.35, the counterion condensation occurs for univalent H^+ and the amount of condensed counterion increases with increasing charge density of the polymer. This appears to suggest that the decrease in k_q observed when F_{AMPS} exceeds 0.449 (for no sample of $F_{AMPS} = 0.365$ exactly) could be related to the counterion condensation.

Olea and Thomas[21] also found the decrease in the quenching rate constant k_q of molecular pyrene by nitromethane with increasing pH in aqueous solution of PMA when pH was higher than 9.0; they attributed this behavior to the deprotonation of nitromethane since their chromophore could be completely expelled into water in this pH range.

Relative Emission Intensity I_1/I_3

The relative emission intensity of the first peak to the third peak, I_1/I_3, of pyrene label on ADPy samples in water as well as in DMSO were plotted against F_{AMPS} in Figures 6 and 7, respectively. The data at the ordinate were obtained from the sample without ionic monomer AMPS. It is obvious that I_1/I_3 in water does decrease with increasing F_{AMPS}, especially it drops down abruptly at the F_{AMPS} interval from 0.449 (ADPy-20) to 0.574 (ADPy-30). Similar I_1/I_3 behavior in organic solvent DMSO means that this decrease is not due to the solvation difference for the copolymer backbone. At the same time, increasing polyelectrolyte

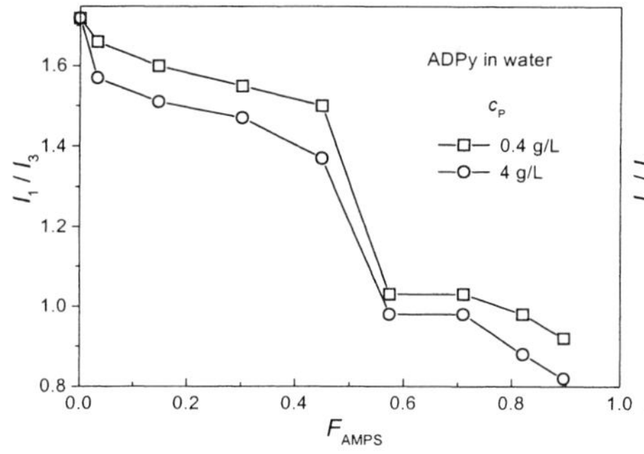

Figure 6. The relative emission intensity I_1/I_3 varies with mole fraction F_{AMPS} of AMPS in ADPy samples in salt-free aqueous solutions with indicated concentrations.

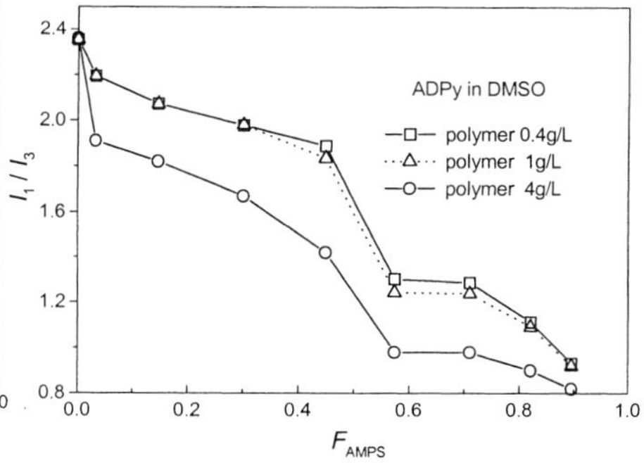

Figure 7. The relative emission intensity I_1/I_3 varies with mole fraction F_{AMPS} of AMPS in ADPy samples in DMSO solutions with indicated concentrations.

concentration also reduces the I_1/I_3 value, except for the sample without AMPS units. According to the polarity scale function of I_1/I_3, the decrease in I_1/I_3 with increasing F_{AMPS} suggests that the pyrene label experiences a decreased polarity in its microenvironment with increasing charge density on its host chains. This less-polar environment may be originated from the local "temporal aggregated" domain of polyelectrolyte chains,[29] and the aggregation degree is enhanced by increasing charge density of polyelectrolytes. The pyrene groups labeled to the polyelectrolyte have no contribution to this aggregation because no excimer has been observed. This aggregation is not due to the hydrophobic association of the copolymer components either because both homopolymers of AMPS and DMAA are completely dissolvable in water as well as in DMSO. This finding contradicts the common understanding of polyelectrolytes.

Based on the classical theory for polyelectrolytes, the repulsive force is enhanced with increasing charge density, resulting in a more extended conformation for single chain or more segregated distribution for multi-chains in solution.[1, 2] Consequently, the labeled fluorophore is exposed further to the polar solvent with increasing charge density of the chain, this will cause an increase in I_1/I_3. As reported by Thomas group,[21, 25] poly(methacrylic acid) (PMA) chains take more extended conformation at pH > 6 and this makes I_1/I_3 of the pyrene label and probe abruptly increase.

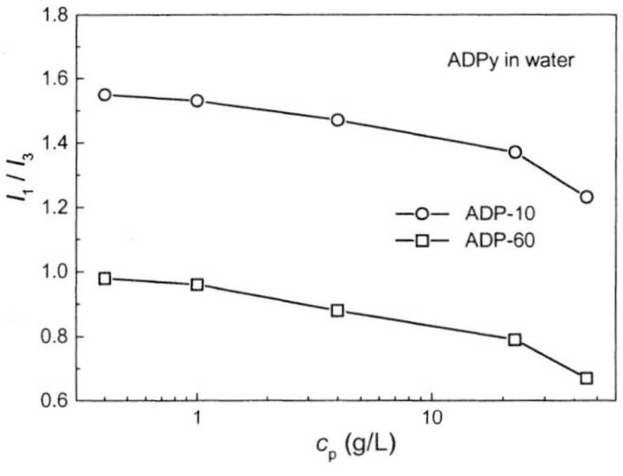

Figure 8. Polymer concentration c_P dependence of relative emission intensity I_1/I_3 of indicated samples in aqueous

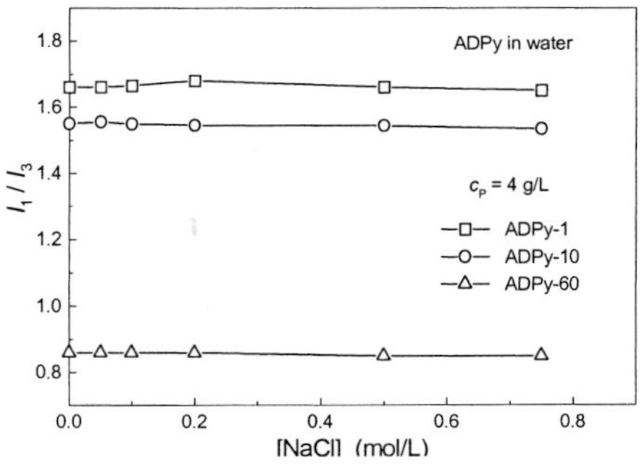

Figure 9. NaCl concentration dependence of relative emission intensity I_1/I_3 for indicated samples in aqueous solutions.

In order to revel the origin of this less-polar domain in dilute polyelectrolyte solutions, we determined polymer concentration c_P dependence of I_1/I_3 for two samples of ADPy-10 and ADPy-60 as shown in Figure 8. The gradual decrease in I_1/I_3 with increasing c_P indicates that this less-polar domain is relevant to the amount of polyelectrolyte chains. In other words, the aggregation driving force would be resulted from the polyelectrolyte chain itself. This less-polar domain still exits in aqueous NaCl solution as illustrated in Figure 9, where I_1/I_3 for three ADPy samples is independent of salt concentration up to 0.75 mol/L and just takes the same value as in pure water, i.e., the less-polar domain cannot be destroyed with adding salt to screen the charge interaction. This fact means that once the less-polar domain is formed in polyelectrolyte solutions, the excessive energy is required to separate these chains again. Similar phenomenon has been observed in the volume phase transition of the polyelectrolyte gel with the same components occurring in DMSO/THF mixtures.[30,31] After the gel is shrunken by adding THF just beyond the threshold composition, excessive amount of DMSO, much more than the threshold, is needed to make the gel swell again to its original volume, manifesting the swelling hysteresis during the process.

Discussion

The above results, to authors' knowledge, firstly demonstrate the existence of the less-polar domain in aqueous as well as in polar organic solutions of polyelectrolytes. This domain is relevant to the polymer charge density and polyelectrolyte concentration but independent of the concentration of added salt. With the concept of "temporal aggregation" proposed by Bruno and Mattice,[29] these fluorescence phenomena can be interpreted consistently.

The temporal aggregation in polyelectrolyte solutions increases the local polymer concentration and decreases the accessibility of the fluorophore label surrounded by the chain backbone. The k_q value gives the averaged information for the pyrene labels in different environments while fluorescence decay data in Table 2 obviously show that after counterion condensation there appear two species of pyrene labels in two different environments. The existence of pyrene labels in the temporal aggregation domain with a slower decay rate induces the decrease in k_q with increasing charge density.

We try to predict the appearance of this less-polar aggregation domain in polyelectrolyte solutions in terms of counterion condensation. When charge density of polyelectrolyte chain exceeds the Manning's criterion of $\xi \geq 1$ for univalent counterion, the counterion will condense on the polyelectrolyte chain. The amount of counterions condensed on one charge site of polyelectrolyte is θ, and $\theta = 1 - 1/\xi$. Then, the total concentration c_b^{o} of condensed counterion is

$$c_b^{o} = \theta F_{AMPS} \, c_P \qquad (11)$$

These condensed counterions will induce the aggregated dimmer of inter- or intra-polyelectrolyte chains as suggested by Bruno and Mattice.[29] Suppose that the condensed counterion outside the aggregation of concentration c_b is equilibrated with that in the aggregated dimmer cluster of concentration c_D, then

$$K_D = c_D / c_b^2 \qquad (12)$$

$$c_b = c_b^o - 2c_D \qquad (13)$$

Here, K_D is equilibrium constant for this "temporal aggregating" reaction and can be evaluated from the free energy ΔG. Ray and Manning have estimated the free energy for two infinitely long parallel linear polyelectrolytes and found attractive interaction when the distance between two polyelectrolytes is inside a Debye radius intermediately.[5-8] Because the counterion concentration in present aqueous solutions of about 2 x 10^{-3} mol/L (so that Debye radius κ^{-1} = 96 Å) is close to that used in Ray and Manning's simulation,[5] we adopt their free energy value of -1.25 kcal/mol at the minimum to calculate K_D value as 8.62 L/mol since the separation distance about 50 Å at the minimum is in consistent with the Förster transfer radius at which "temporal aggregates" were observed by Bruno and Mattice.[29]

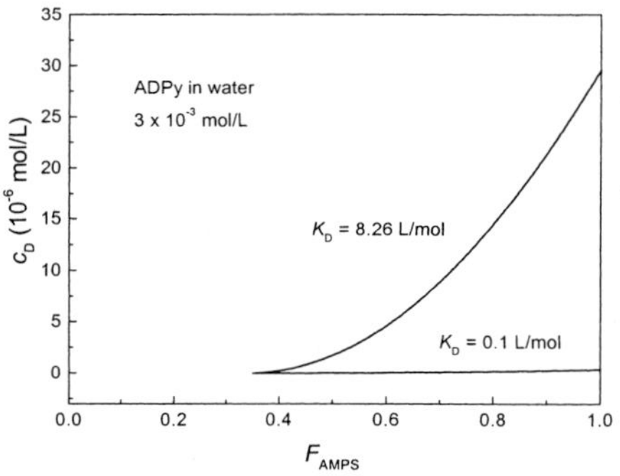

Figure 10. Calculated "temporal aggregated" dimmer concentration c_D plotted against mole fraction F_{AMPS} of AMPS in ADPy samples with indicated equilibrium constant K_D.

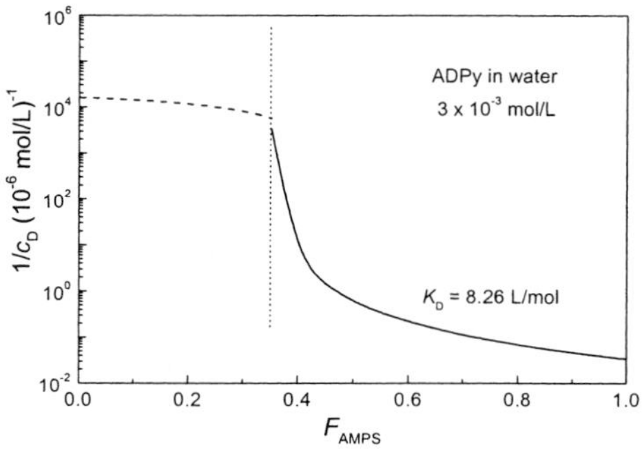

Figure 11. Reciprocal of calculated c_D (solid line) with dashed line guiding eyes for samples without counterion condensation, analogy to those in Figures 6 and 7.

Figure 10 depicts calculated c_D of the aggregated dimmer varying with F_{AMPS} for the ADPy samples of 3 x 10^{-3} mol/L. At K_D = 8.62 L/mol, c_D is greatly increased with F_{AMPS} after the counterion condensation occurs at $F_{AMPS} > 0.35$ but its highest value is less than 1% of c_b^o. However, if the K_D is small (saying 0.1 L/mol) there are almost no aggregated clusters.

It is difficult to formulate I_1/I_3 with "temporal aggregated" dimmer concentration c_D directly.

However, following the idea suggested by Khokhlov and Kramarenko in their simulation for the volume phase transition in polyelectrolyte gels,[32] the effective dielectric constant is composed of those of the solvent and polymer and the polarity of a polyelectrolyte backbone is much lower. Therefore, the formation of "temporal aggregated" dimmers increases the local polymer concentration and reduces the polarity of the microenvironment. We plotted the reciprocal of c_D against F_{AMPS} in Figure 11 by solid line with a dash line guiding eyes for the samples without counterion condensation. The change of I_1/I_3 with F_{AMPS} is reasonably reproduced by the simulation, indicating that the decrease in I_1/I_3 is caused by the "temporal aggregates" in polyelectrolyte solutions due to the attractive potential to the condensed counterions.

Acknowledgment

We are grateful to the NNSF of China (No. 29725411) and the NSF of Guangdong Province (No. 015036) for the financial support to this work.

[1] C. Tanford, "*Physical Chemistry of Macromolecules*", John Wiley & Sons, New York 1961, p. 457.
[2] H. Dautzenberg, W. Jaeger, J. Kötz, B. Philipp, C. Seidel, D. Stscherbina, "*Polyelectrolytes*", Hanser, Munich 1994, p.87.
[3] F. Oosawa, "*Polyelectrolytes*", Marcel Dekker, New York 1971, p. 23.
[4] G. S. Manning, in: "*Polyelectrolytes*", E. Selegny, Eds., Reidal Publisher, Dordrecht, Holland 1974, p. 9.
[5] J. Ray, G. S. Manning, *Langmuir* **1994**, *10*, 2450.
[6] J. Ray, G. S. Manning, *Macromolecules* **1997**, *30*, 5739.
[7] J. Ray, G. S. Manning, *Macromolecules* **1999**, *32*, 4588.
[8] J. Ray, G. S. Manning, *Macromolecules* **2000**, *33*, 2901.
[9] J. Bodycomb, M. Hara, *Macromolecules* **1995**, *28*, 8190.
[10] B. D. Ermi, E. J. Amis, *Macromolecules* **1996**, *29*, 2701.
[11] M. Sedlák, *J. Chem. Phys.* **1996**, *105*, 10123.
[12] R. Borsali, H. Nguyen, R. Pecora, *Macromolecules* **1998**, *31*, 1548.
[13] H. Matsuoka, Y. Ogura, H. Yamaoka, *J. Chem. Phys.* **1998**, *109*, 6125.
[14] H. Liu, L. Skibinska, J. Gapinski, A. Patkowski, E. W. Fischer, R. Pecora, *J. Chem. Phys.* **1998**, *109*, 7556.
[15] L. Skibinska, J. Gapinski, H. Liu, A. Patkowski, E. W. Fischer, R. Pecora, *J. Chem. Phys.* **1999**, *110*, 1794.
[16] B. D. Ermi, E. J. Amis, *Macromolecules* **1997**, *30*, 6937.
[17] Y. Morishima, *Prog. Polym. Sci.* **1990**, *15*, 949.
[18] M. A. Winnik, "*Photophysical and Photochemical Tools in Polymer Science*", Reidel, Dordrecht, Holland 1986, p. 15.
[19] F. Gao, Y. Yan, B. Ren, Z. Tong, *Chem. J. Chinese U.* (in Chinese) **2000**, *21*, 976.
[20] Y. Morishima, Y. Tominaga, M. Kamachi, T. Okada, Y. Hirata, N. Mataga, *J. Phys. Chem.* **1991**, *95*, 6027.
[21] A. F. Olea, J. K. Thomas, *Macromolecules* **1989**, *22*, 1165.
[22] C. E. Hoyle, P. W. Neilson, P. M. Chatterton, M. A. Trapp, *Macromolecules* **1991**, *24*, 2194.
[23] H. Yamamoto, M. Mizusaki, K. Yoda, Y. Morishima, *Macromolecules* **1998**, *31*, 3588.
[24] M. Suwa, A. Hashidzume, Y. Morishima, T. Nakato, M. Tomida, *Macromolecules* **2000**, *33*, 7884.
[25] D.-Y. Chu, J. K. Thomas, *Macromolecules* **1984**, *17*, 2142.
[26] Y. Morishima, H. Ohgi, M. Kamachi, *Macromolecules* **1993**, *26*, 4293.
[27] G. M. Barrow, "*Physical Chemistry*", 4th ed., McGraw-Hill, London 1979, p.
[28] J. A. Delaire, M. A. J. Rodgers, S. E. Webber, *J. Phys. Chem.* **1984**, *88*, 6219.

[29] K. R. Bruno, W. L. Mattice, *Macromolecules* **1992**, *25*, 327.
[30] X. Liu, Z. Tong, X. Cao, O. Hu, *Polymer* **1996**, *37*, 5947.
[31] X. Liu, Z. Tong, F. Gao, *Polym. Intern.* **1998**, *47*, 215.
[32] R. Khokhlov, E. Yu. Kramarenko, *Macromol. Theory. Simul.* **1994**, *3*, 4537.

Macromol. Symp. ***192***, *265–270 (2003)*

Surface Thermal Molecular Motion of Chain End-Modified Polystyrenes

Tisato Kajiyama, Keiji Tanaka, Atsushi Takahara*

Department of Applied Chemistry, Faculty of Engineering,
Kyushu University, Fukuoka 812-8581, Japan

Summary: Surface glass transition temperature, T_g^s, of three classes of monodisperse polystyrene, proton-terminated (PS-H), α,ω-diamino-terminated (α,ω-PS(NH_2)$_2$) and α,ω-dicarboxy-terminated (α,ω-PS(COOH)$_2$), were studied by scanning force microscopy. T_g^s of all samples was discernibly lower than the corresponding bulk glass transition temperature, T_g^b. Also, T_g^s value was strongly dependent on what the chemical structure of chain end groups was. Based on the time-temperature superposition principle, apparent activation energy of the α_a-relaxation process corresponding to surface segmental motion was evaluated to be approximately 230 kJ•mol^{-1}. This value was much smaller than the reported bulk ones, and was independent of the chemical structure of chain ends. This result implies that the cooperativity for the α_a-relaxation process at the PS surface is reduced in comparison with the bulk due probably to an existence of the free space presented to polymer segments at the surface. Hence, it was concluded that the surface α_a-relaxation process was activated by not only the chain end effect but also the reduced cooperativity at the surface.

Introduction

Molecular mobility of polymer chains at the surface has received a great deal of attention recently because of its importance in various functional applications such as permselective membranes, biomaterials, lubricant, adhesives, corrosion-resistant coatings, etc. Systematical understanding of surface properties, which are totally different from bulk ones, is crucial so that the performance of functional polymeric materials can be promisingly improved. This issue is also of importance and interest in the field as fundamentals of polymer science.

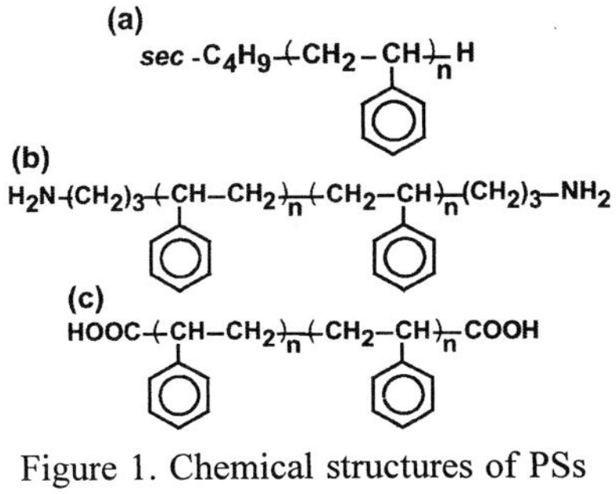

Figure 1. Chemical structures of PSs used; (a) PS-H, (b) α,ω-PS(NH_2)$_2$, (c) α,ω-PS(COOH)$_2$.

Since 1994, we have studied surface molecular

 CCC 1022-1360/00/$ 17,50+.50/0

motion of monodisperse proton-terminated polystyrene (PS-H) films based on scanning viscoelasticity microscopy (SVM) and lateral force microscopy (LFM).[1-3] SVM enables one to gain direct access to viscoelastic properties on solid surfaces by measuring an amplitude change in response force, and detecting a phase lag between a stimulus displacement and a response force. LFM relies on a measure of lateral force between a sample surface and a probe tip, and has been demonstrated to be a powerful tool to investigate viscoelasticity of various polymer surfaces as well. Based on both techniques, we have claimed that surface glass transition temperature, T_g^s, was much lower than the corresponding bulk glass transition temperature, T_g^b. [1-5] Although our conclusion has already emerged, it still opens why such active molecular motion takes place at the surface. For the moment, we have proposed the following two reasons to rationalize it; end group effect and reduced cooperativity of surface segmental motion. If the chain end effect is one of responsible factors on surface molecular motion, as we claim, the T_g^s value should be strongly dependent on what the chemical structure of chain end groups is. To strengthen our notion, surface glass transition behavior of PSs with hydrophilic chain ends, which might be depleted at the film surface so as to minimize the air/polymer interfacial free energy, is examined in this study.

Experimental

Proton-terminated polystyrene (PS-H), α,ω-diamino-terminated PS (α,ω-PS(NH_2)$_2$), and α,ω-dicarboxy-terminated PS (α,ω-PS(COOH)$_2$) were used as materials. Figure 1 shows the chemical structures of PS-H, α,ω-PS(NH_2)$_2$ and α,ω-PS(COOH)$_2$. PS films for SVM and LFM measurements were coated from toluene solutions onto cleaned silicon wafers with native oxide layer by a spin-coating method. The PS films were dried at room temperature for more than 24 h in air, and then annealed at 423 K for 24 h in vacuum. The film thickness evaluated by ellipsometric measurement was approximately 200 nm. T_g^s in PS films with various chain ends was evaluated by using SVM and LFM. The detail for both measurements has been reported elsewhere.[4-5]

Results & Discussion

To evaluate T_g^s, surface phase lag, δ^s, between imposed stimulus displacement signal and detected response force one was measured by SVM. Figure 2 shows a typical temperature dependence of δ^s for the PS-H film. The data was for PS-H film with number-average molecular weight, M_n of 140k. A peak observed in Figure 2 can be assigned to the α_a-relaxation corresponding to segmental motion at the surface. Following an empirical way, T_g^s was defined as an onset temperature, at which δ^s started to increase.

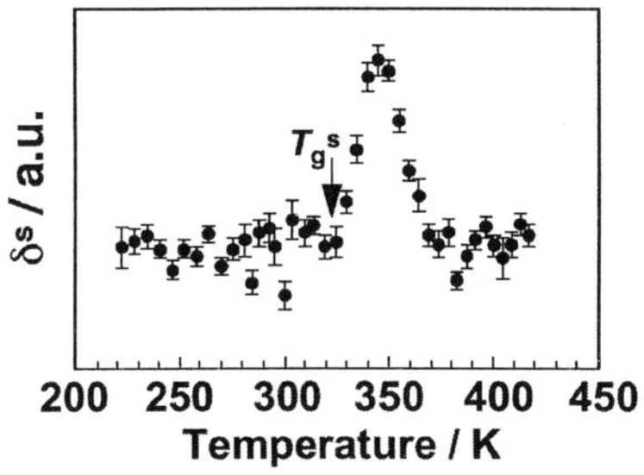

Figure 2. Temperature dependence of δ^s for PS-H with M_n of 140k.

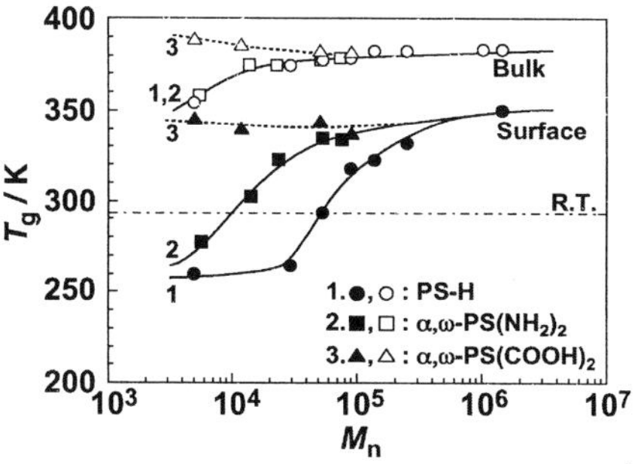

Figure 3. M_n dependence of T_g^s and T_g^b for the PS-H, α,ω-PS(NH_2)$_2$ and α,ω-PS(COOH)$_2$ films.

Figure 3 shows the M_n dependence of T_g^s and T_g^b for the PS-H, α,ω-PS(NH_2)$_2$ and α,ω-PS(COOH)$_2$ films. T_g^b was measured by differential scanning calorimetry (DSC). As a general trend, T_g^s was lower than the corresponding T_g^b. This result implies that thermal molecular motion at the surface is more active than that in the bulk, independent of chain end chemistry. However, the T_g^s value was strongly dependent on the chain end species at a given M_n. Hence, it seems reasonable to consider that the chain end effect is one of the responsible determining factors on the magnitude of T_g^s. In the case of the PS-H, the M_n dependence of T_g^s was more remarkable than that of T_g^b. Thus far, characteristic thermal molecular motion at the PS-H surface has been explained in terms of the chain end effect and the reduced cooperativity.[1-5] The chain end groups of the PS-H are preferentially segregated at the surface due to its lower surface free energy than the main chain part. Since the chain end groups have the larger freedom than the main chain part, an excess free volume is induced at the surface. Also, a segment existing at the surface is supposed to have a fewer neighbor segments to move cooperatively owing to the free space on the polymer phase. Thus, active molecular motion takes place at the PS-H surface. In the case of the α,ω-PS(NH_2)$_2$, the surface free energy of the chain end is relatively larger than that of the main chain part. Thus, it is inferred that the chain ends are depleted at the surface. Actually, angular-dependent X-ray photoelectron spectroscopic measurement revealed that this prediction would be true. This means that an excess free volume at the surface induced by chain end localization may not be counted for the α,ω-PS(NH_2)$_2$ films. Nevertheless, T_g^s was much lower than T_g^b, and the M_n dependence of T_g^s was clear, as shown in Figure 3. Therefore, it seems most likely that thermal molecular motion at the surface is activated by other factors in addition to the chain end effect, such as the reduced cooperativity. In contrast to the PS-H and α,ω-PS(NH_2)$_2$, neither distinct M_n dependence of T_g^s nor T_g^b was observed for the α,ω-PS(COOH)$_2$ film. Since the α,ω-PS(COOH)$_2$ are intermolecularly associated between chain ends via hydrogen bonding, the apparent M_n of the α,ω-PS(COOH)$_2$ increases in comparison with its own M_n.[6] Hence, it is difficult to discuss about the M_n dependence of T_g^s and T_g^b for the α,ω-PS(COOH)$_2$. In any event, it can be envisaged that the chain end chemistry is one of the responsible factors determining on the T_g^s value.

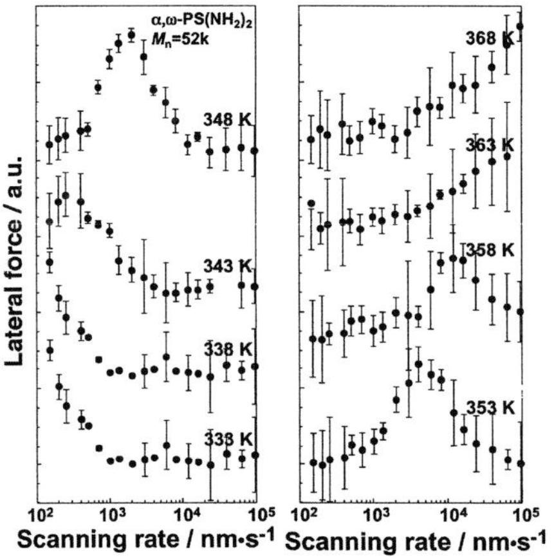

Figure 4. Scanning rate dependence of lateral force for the α,ω-PS(NH_2)$_2$ film with M_n of 52k at the temperature range of 333 to 368 K.

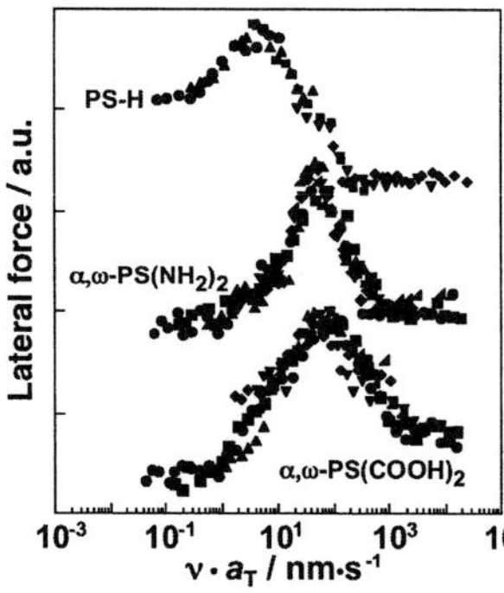

Figure 5. Master curves related to the scanning rate-lateral force relationship for the (1) PS-H with M_n of 140k, (2) α,ω-PS(NH_2)$_2$ with M_n of 52k and (3) α,ω-PS(COOH)$_2$ with M_n of 52k.

Next, apparent activation energy of the surface $α_a$-relaxation was examined on the basis of the time-temperature superposition principle. Since the measuring frequency of our SVM is limited in one decade at present, the complement technique, LFM, was applied. Figure 4 shows the scanning rate dependence of lateral force at various temperatures for the α,ω-PS(NH_2)$_2$ film with M_n of 52k. The lateral force was independent of the scanning rate at a temperature lower than 323 K. At higher temperatures, 343 to 358 K, a peak of lateral force was clearly observed. Then, the lateral force decreased with a decrease in the scanning rate at the higher temperature than 363 K. Eventually, the lateral force recovered the invariance with respect to the scanning rate. Thus, the overall profiles reflect a successive change of the surface molecular motion from the glassy state to the rubbery one via the transition with increasing temperature. Similar results were obtained for the PS-H and α,ω-PS(COOH)$_2$ films.

Figure 5 shows the master curves of the lateral force-scanning rate relation for the PS-H, α,ω-PS(NH$_2$)$_2$ and α,ω-PS(COOH)$_2$ films, and clearly indicates that the time-temperature superposition can be applied even to the surface α_a-relaxation process. Invoking that the relationship between shift factor, a_T, and measuring temperature is Arrhenius type, the apparent activation energy, $\Delta H^{\ddagger}$, of the surface α_a-relaxation process can be calculated from a linear slope in the plot of a_T against to reciprocal absolute temperature, as shown in Figure 6. The surface $\Delta H^{\ddagger}$ of the PS-H, α,ω-PS(NH$_2$)$_2$ and α,ω-PS(COOH)$_2$ films so obtained were 230±10, 230±10 and 220±10 kJ mol^{-1}, respectively. It is noteworthy here that the surface $\Delta H^{\ddagger}$ is much smaller than the reported values of the $\Delta H^{\ddagger}$ for bulk samples, ranging from 360 to 880 kJ mol^{-1}.[7] The discrepancy of $\Delta H^{\ddagger}$ between surface and bulk indicates that the size and/or energy barrier of the cooperative movement for the α_a-relaxation process are/is reduced at the surface in comparison with the bulk. The reduced cooperativity at the surface can be easily understood by taking into account a fact that the free space is presented to polymer segments at the surface. That is, the existence of the free space induces active micro-Brownian motion at the air/polymer interface. Thus, T_g^s was lower than T_g^b even in the case that the chain end effect can be negligible such as the α,ω-PS(NH$_2$)$_2$, α,ω-PS(COOH)$_2$ and PS-H films with infinite M_n, as shown in Figure 3. Finally, we conclude that the active thermal molecular motion occurs at the surface by the reduced cooperativity in addition to the chain end effect.

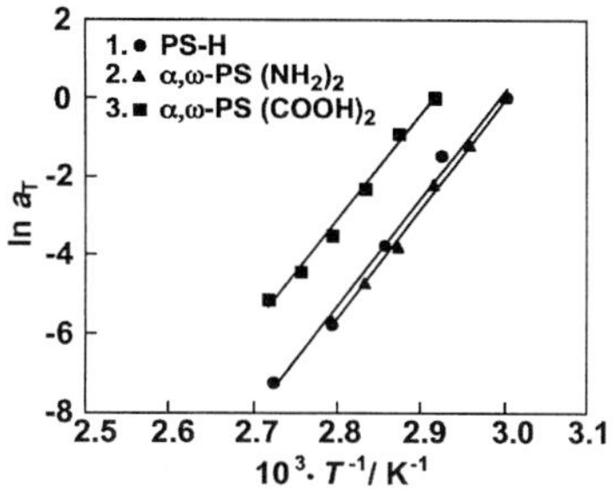

Figure 6. Semi-logarithmic plots of shift factor, a_T, versus reciprocal absolute temperature for the (1) PS-H, (2) α,ω-PS(NH$_2$)$_2$ and (3) α,ω–PS(COOH)$_2$ films.

Conclusions

Glass transition behaviors of the PS-H, α,ω-PS(NH$_2$)$_2$ and α,ω-PS(COOH)$_2$ films were studied by SVM in conjunction with LFM. It was found that T_g^s was lower than the corresponding T_g^b even in the case of the α,ω-PS(NH$_2$)$_2$ and α,ω-PS(COOH)$_2$ films. However, the T_g^s value was strongly dependent on chain end chemistry. The surface $\Delta H^{\ddagger}$ of PS-H, α,ω-$PS(NH_2)_2$ and α,ω-$PS(COOH)_2$ were 230±10, 230±10 and 220±10 kJ mol^{-1}, respectively. These values were well below the bulk $\Delta H^{\ddagger}$. It was concluded that the active thermal molecular motion at the surface was attained by the reduced cooperativity in addition to the chain end effect.

Acknowledgements

We greatly thank Profs. Seiichi Nakahama and Takashi Ishizone, Tokyo Institute of Technology, for the synthesis of α,ω-$PS(NH_2)_2$ samples. This work was in part supported by a Grant-in-Aid for Scientific Research (A) (#13355034) from the Ministry of Education, Science, Sports, and Culture, Japan.

[1] T. Kajiyama, K. Tanaka, I. Ohki, S.-R. Ge, J.-S. Yoon, A. Takahara, *Macromolecules* **1994**, *27*, 7932.
[2] T. Kajiyama, K. Tanaka, A. Takahara, *Macromolecules* **1997**, *30*, 280.
[3] T. Kajiyama, K. Tanaka, N. Satomi, A. Takahara, *Macromolecules* **1998**, *31*, 5150.
[4] K. Tanaka, A. Takahara, T. Kajiyama, *Macromolecules* **2000**, *33*, 7588.
[5] N. Satomi, K. Tanaka, A. Takahara, T. Kajiyama, T. Ishizone, S. Nakahama, *Macromolecules* **2001**, *34*, 8761.
[6] K. Tanaka, X. Jiang, K. Nakamura, A. Takahara, T. Kajiyama, T. Ishizone, A. Hirao, S. Nakahama, Macromolecules **1998**, *31*, 5148.
[7] N. G. McCrum, B. R. Read, "*Anelastic and Dielectric Effects in Polymeric Solids*", Dover, New York, 1967.

Real-Time AFM Study of Lamellar Growth of Semi-Crystalline Polymers

Yan-Hong Luo[1], *Yong Jiang*[1], *Xi-Gao Jin*[1], *Lin Li*[1]*, *Chi-Ming Chan*[2]*

[1]State Key Laboratory of Polymer Physics and Chemistry, Center for Molecular Science, Institute of Chemistry, Chinese Academy of Sciences, China
[2]Department of Chemical Engineering, Hong Kong University of Science and Technology, China

Summary: Atomic force microscopy (AFM) has been used to study the lamellar development during the crystallization and melting processes of poly(bisphenol A-co-alkyl ether) (BA-Cn) films. High-resolution and real-time AFM phase imaging enables us to observe the detailed growth process of the lamellae. At the early stage of the lamellar growth, embryos appeared firstly and some disappeared on the film surface after a period of time. The stable embryo developed into a single lamella. Then the lamella developed into a lamellar sheaf through branching and splaying. Our results revealed that the branches of the lamellae were formed by induced nucleation and it was also dependent on the crystallization temperature. Real-time AFM study of the melting, recrystallization and remelting processes of lamellae indicated that the thermal stability of different segments of a single lamella is different and that the thermal stability of the different lamellae is also different even if they develop at the same annealing temperature. The orientation and the development of the lamellae at the characteristic eyes and boundaries of the spherulites are observed in details.

Introduction

Spherulites are common crystalline structures observed not only in polymers but also in a large variety of inorganic substances. In the case of polymers, the detailed structure and formation mechanisms of spherulites have been studied using polarized optical microscopy (POM) and transmission electron microscopy (TEM)[1,2]. It is generally accepted that the three-dimensional geometry of spherulites develops from a stack of lamellae. During the growth process, the stacked lamellae splay apart continually and branch occasionally. Not until were AFM used to observe the growth process of spherulites, this process is hypothetical because the description is not based on direct observations of the formation of spherulites[3,4].

The invention of atomic force microscopy (AFM) made it possible to observe the lamellar growth of a semi-crystalline polymer in real time. The development of tapping-mode AFM (TM-AFM) has made it easier to map soft, adhesive and fragile samples. Many studies have been performed to interpret the height and phase images recorded by TM-AFM. Results

 CCC 1022-1360/00/$ 17,50+.50/0

clearly indicate that phase images can provide enhanced contrast on heterogeneous surfaces. TM-AFM has been shown to be a powerful tool to study the surfaces of polymer blends, copolymers, and semicrystalline polymers. Most recently, hot-stage atomic force microscopy has been applied to image crystallization and melting of polymers in real time.

By utilizing the advantages of AFM, spherulites and lamellae of various semi-crystalline polymers have been investigated[5-9]. It was found by Vancso et al. that polypropylene lamellar thickness and hedritic morphology were identical to the values in the literature as revealed by other techniques. The growth rate of a spherulite and the internal lamellar arrangement of a poly(hydroxy-butyrate-*co*-valerate) copolymer have been determined by Barham and colleagues using TM-AFM[7)]. Their results indicate that the overall gross growing front of a spherulite propagated at a constant rate, but the growth rate of internal lamellae were variable.

Poly(bisphenol A-co-alkyl ether)s (BA-Cn) are the ideal candidates to investigate the fundamental issues associated with the development of lamellae and spherulites by real-time AFM due to their slow crystallization rate at room temperature and relatively low melting temperatures (about 90 °C). Here, we summarized our results of real-time AFM imaging of the lamellar growth, branches and melting of the BA-Cn spherulites. We believe that the studies on the detailed lamellar structure and development in this paper will provide a better understanding of crystallization mechanisms of polymers.

Experiment

The BA-Cn polymers were prepared as described previously[3, 4]. The thin films for AFM observations were prepared by spin coating a 20 ~ 40 mgml^{-1} polymer-chloroform solution at 3000 rpm onto silicon wafer surface or freshly cleaved surface of mica. The samples were dried in a vacuum at room temperature for 30 minutes. The thickness of the amorphous BA-Cn films was estimated to be 300 to 400 nm by a profilometer.

Real-time AFM measurements were carried out using the tapping-mode AFM (Digital Instrument Nanoscope IIIA) equipped with a hot stage. Both height and phase images were recorded simultaneously using the retrace signal. Si tips with a resonance frequency of approximately 300 kHz and a spring constant of about 40 Nm^{-1} were used, and the scan rate and the scanning density were 0.5-1 Hz and 512 lines/frame, respectively. A set point amplitude ratio was around 0.8.

Results and Discussion

Embryos and Primary Nuclei

The development of a stable embryo into a single lamella can be clearly observed, as shown in a series of AFM phase images (c.f., Figure 1), where polymer of BA-C8 crystallized at 22 °C. Lamellar embryos, as the fine dots with a dimension smaller than 10 nm, appeared and disappeared on the surface of an initially amorphous polymer film during scanning (Figures 1a and 1b). Figure 1a shows the presence of two embryos. In Figure 1b, which was obtained approximately 8.7 min after the image shown in Figure 1a, one of the embryos disappeared. As predicted by thermodynamics, embryos smaller than a critical dimension are unstable and may not ultimately grow. The stable embryo, a primary nucleus, can grow along the length of the lamella at both ends (Figure 1c, d). We name this lamella, which originates from a primary nucleus, as the founding lamella because all other lamellae presented in the spherulite are its descendants.

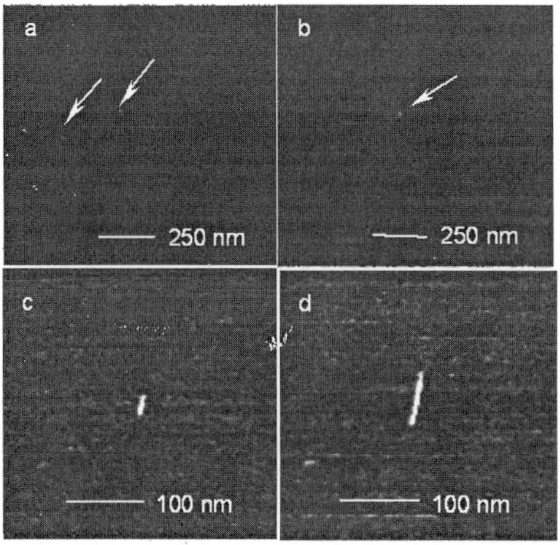

Figure 1 The formation of embryos and the growth of a stable embryo to single lamellae

In our experiments, the dimension of primary nuclei was measured to be smaller than 10 nm, however, the critical size of a stable embryo could not be accurately measured with AFM.

Induced Nuclei and Lamellar Branches

After the founding lamella grew to a certain length, it would develop into a lamellar sheaf through extensive branching and splaying. Usually, the branching of the parent lamellae appeared at some distance behind its growing tip. The effect of crystallization temperature on the formation of lamellar branches was studied using AFM with a hot stage. The branching of parent lamellae at different temperatures (16, 22 and 28 °C, respectively) is presented in Figure 2. It can be clearly observed that crystallization temperature strongly influenced the formation of lamellar branches. At the lower temperature, such as 16 and 22 °C as shown in Figures 2a-d, there was no branching as the lamellae grew to about 0.5 μm in length. However, branching appeared before the parent lamellae grew to 0.5 μm at 28 °C. This result suggests that the branching of parent lamellae can occur easier at the higher crystallization temperature. However, the rate of branching was found to decrease at temperatures near the melting point of the polymer,

In our earlier papers[3, 4], we proposed that branching of a parent lamella is initiated by induced nucleation, which is caused by trapping one end of a polymer chain in the parent lamella. The reduced mobility and the accumulation of the stress of the trapped chain segments induce the formation of a nucleus near the parent lamella. The induced lamellar embryos could also appear and disappear. This result suggests that the screw dislocation at the parent lamella does not cause the formation of an induced nucleus. Induced nuclei did not grow only in one direction; they also could grow in the opposite direction (Figure 2f). In this way, the parent lamellae gradually branch and splay apart to develop into a lamellar sheaf[3].

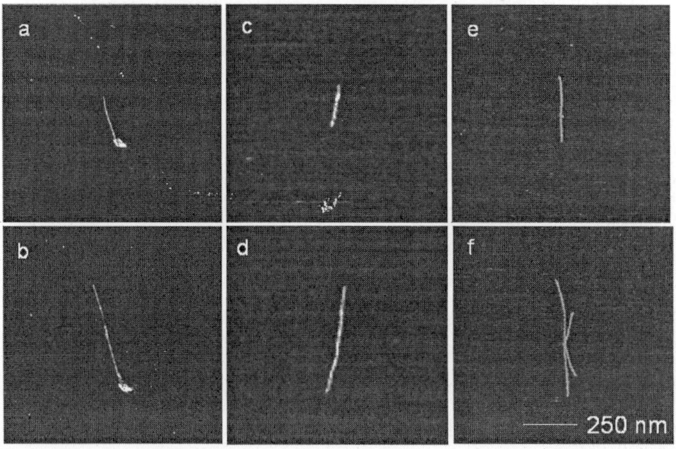

Figure 2 AFM phase images of the lamellae grown at a-b: 16 °C c-d: 22 °C and e-f: 28 °C

Lamellae at the Eyes of Spherulites

Spherulites with two characteristic eyes at the center can be clearly seen on the BA-C8 films, as shown in Figure 3. The top column displays the AFM topographic images and the bottom column shows the AFM phase images. The data scales of the AFM height images are 20 nm. The time interval between each consecutive image of Figures 3a-c is about 15 minutes. From the AFM height images, it is clear that the height of the eye area is different from the other areas of the film surface. This result suggests that the lamellar structure of the eyes may also be different from that in other parts of the spherulite. However, the lamellae at the eyes, as labeled by "e" in the AFM phase images, were formed at a relatively late stage of the spherulitic growth. Comparing the AFM height and phase images of Figure 3a, the height changes of the eyes can be seen clearly, but the lamellae of the eyes have not been formed yet. During the growth of the spherulite, the "flat-on" lamellae were formed at the area of the eyes. The growth of "flat-on" lamellae at the eyes was limited and stopped by surrounding "edge-on" lamellae. The boundaries between the "flat-on" and the "edge-on" lamellae at the eyes can also be identified. This observation suggests that the chain segments at the boundaries of the eyes are difficult to fit perfectly into the different orientations of the "flat-on" and "edge-on" lamellae.

The detailed lamellar structures of the eyes show that the central area of the eyes is lower than the boundary areas (Figure 3d). The layer structure of the "flat-on" lamellae surrounded by the "edge-on" lamellae can be seen in the AFM phase images. However, the formation mechanism of the eyes at the center of a spherulite is not fully understood. For example, the height changes within the eye areas occurred at the initial stages of the spherulitic growth, but the layer structures of the "flat-on" lamellae within the eye areas were formed at a later stage.

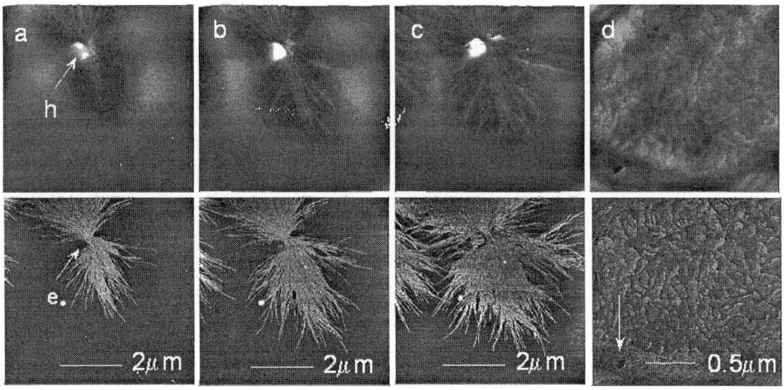

Figure 3 A series of AFM images showing the formation of the eyes at the center of a spherulite

Lamellae at the Boundaries of Spherulites

It is well known that lamellae of spherulites can penetrate into and impinge each other during the later stages of crystallization. However, when lamellae impinge each other at the boundaries of spherulites more defects are created, resulting in less constriction of the film. Therefore, the boundaries are higher than the other parts of spherulites. Figure 4 shows a series of AFM phase images of the lamellar propagation at the boundaries of spherulites. When the lamellae from different spherulites grew closer to each other, the lamellae that growing in opposite directions could penetrate each other, as indicated by the arrows in Figures 4a and 4b. The penetrated lamellar sheaf, which consisted of a stack of parallel lamellae from different spherulites, formed a "lamellar bridge" connecting the spherulites. Many "lamellar bridges" could be formed at the boundaries of the spherulites, as shown in Figures 4c and 4d. However, when the lamellae from different spherulites grew in different directions, they met and stopped one other at the boundaries of the spherulites producing many defects, as indicated by arrows in Figures 4e and 4f.

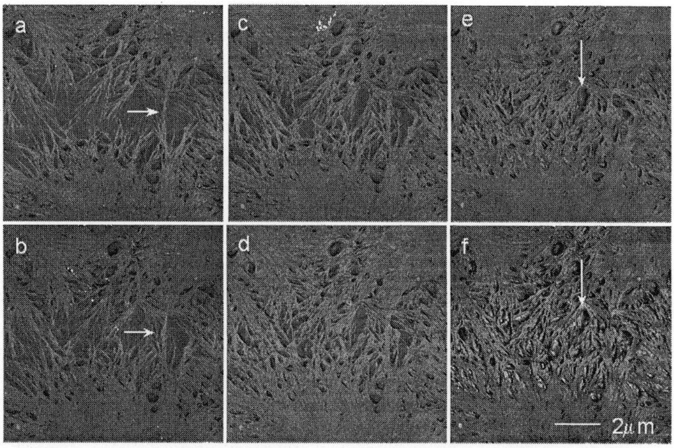

Figure 4 A series of AFM phase images showing lamellae propagation at the boundaries of spherulites

Due to the penetration and impinging of the growing "edge-on" lamellae at the boundaries, some areas in these regions were divided into many small "island-like" areas with various shapes, which contained amorphous materials. The size of these small islands of amorphous phase decreased with the propagation of the lamellae at the boundaries, as indicated by the arrows shown in Figure 4. It is also possible that the polymer crystals inside the small areas were formed by the induced nucleation of the "edge-on" lamellae around the small areas. Hence, the boundary areas among the spherulites of the BA-C8 polymer thin films consisted of amorphous polymer and various orientations of lamellae, such as "edge-on", "flat-on" and "tilt" orientations.

Melting and Recrystallization of Lamellae

The melting process, which usually is accompanied by lamellar thickening, perfection and recrystallization, has been extensively studied by differential scanning calorimetry (DSC), wide angle x-ray scattering (WAXS) and small angle x-ray scattering (SAXS). Recently, AFM with a hot stage has been applied to image the crystallization and melting processes of polymers in real time.

Figure 5 presented a series of images of the melting and recrystallization and remelting processes of lamellae. All temperature measurements were accurate to ± 1°C. Figure 5a is the image of the lamellae crystallized at 35 °C. The lamellae with edge-on orientation can be clearly observed. All lamellae are uniform in appearance. As seen in Figure 5b, before the temperature was increased to 80 °C there was no obvious sign of melting. When the annealing temperature was increased to 80°C, (c.f., Figure 5c), some parts of the lamellae melted. With an increase of the annealing time, some of the broken lamellar segments joined together with the formation of new lamellar segments (c.f., Figures 5d). Different lamellae showed different

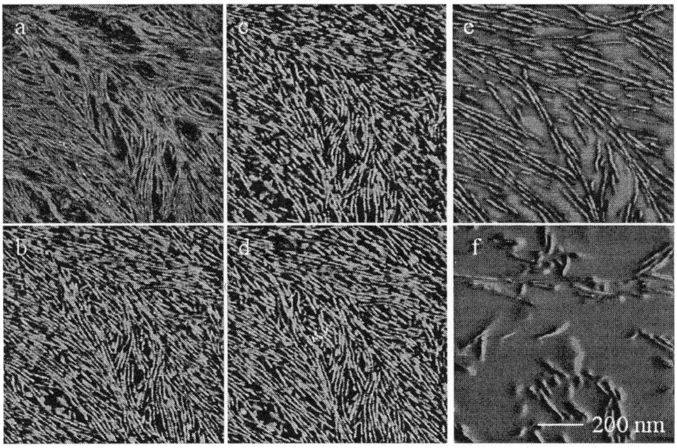

Figure 5 AFM phase images of the lamellar structure of the BA-C10 polymer thin film annealing at a: 35 °C; b: 78 °C; c: 80°C; d: 80 °C, about 9 min after c; e: about 9 min at 90°C; and f: about 9 min at 92 °C

melting and recrystallization behaviors. These results suggest that the stability and the degree of perfection of the lamellar crystals are very different in different lamellae and even in the different locations of one lamella. When the annealing temperature was continually increased, more and more lamellae melted and the broken segments did not join together again (Figure 5e,f). The lamellar thicken also can be seen clearly (Figure 5e).

Conclusion

In summary, our *in-situ* AFM investigations revealed some important features of the crystallization process of polymers, such as, embryos appeared and disappeared when they were smaller than a certain critical size. A stable embryo grew into a founding lamella. The branching of parent lamellae was a result of induced nucleation and then the induced nuclei developed into daughter lamellae or subsidiary lamellae. Crystallization temperature has strong influence on the rate of lamellar branching. When lamellae were heated to near their melting temperature, melting, re-crystallization and remelting occurred simultaneously. The areas of the two characteristic eyes at the center of a spherulite and boundaries among the spherulites were higher than the other areas of the semi-crystalline film surfaces. The "flat-on" lamellae inside the eyes could be observed clearly using AFM phase imaging. There were many "lamellar bridges" formed by "edge-on" lamellar sheaves at the spherulitic boundaries. This "lamellar bridges" should have a major contribution to the mechanical properties of the semi-crystalline polymers.

Acknowledgements

We are grateful for the support of National Science Foundation of China (Grant Nos. 20044005, 20174049 and 20131160730) and National Science Foundation of China and the Hong Kong Government Research Grants Council Joint Research Scheme under Grant No. N_HKUST 618/01.

[1] F.J. Padden, H.D. Keith, *J. Appl. Phys.* **30**, 1479 (1959)
[2] D.C. Bassett, *Principles of Polymer Morphology*; Cambridge University Press: Cambridge (1981)
[3] L. Li, C.M. Chan, J.X. Li, K.M. Ng, K.L. Yeung, L.T.Weng, *Macromolecules*, **32**, 8240 (1999)
[4] L. Li, C.M. Chan, K.L. Yeung, J.X. Li, K.M. Ng, Y.G. Lei, *Macromolecules*, **34**, 316 (2001)
[5] D.T.V. Haeringen, J. Varga, G.W. Ehrenstein, G.J. Vancso, *J. Polym. Sci., Polym. Phys.*, **38**, 672 (2000)
[6] D. Trifonova, J. Varga, G.J.Vancso, *Polym. Bull.*, ***41***, 341 (1998)
[7] J.K. Hobbs, T.J. McMaster, M.J. Miles, P. J. Barham, *Polymer*, **39**, 2437 (1998)
[8] J. M. Schultz, M.J. Miles, *J. Polym. Sci., Polym. Phys.*, **36**, 2311, (1998)
[9] Y.K. Godovsky, S. N. Magonov, *Langmuir*, **16**, 3549 (2000)
[10] L.Li, K.M. Ng, C.M. Chan,; J.Y. Feng, X.M. Zeng, L.T. Weng, *Macromolecules*, **33**, 5588 (2000)

Production of Nanocomposites of PET-Montmorillonite Clay by an Extrusion Process

A. Sanchez-Solis, A. Garcia-Rejon†, O. Manero*

Instituto de Investigaciones en Materiales. Universidad Nacional Autónoma de México. Apartado Postal 70-360. Coyoacán, 04510. México, D.F Email:sancheza@servidor.unam.mx
†Industrial Materials Institute. National Research Council of Canada. 75 de Mortagne J4B 6Y4. Boucherville, Quebec. Canada

Summary: Polyethylene terephthalate (PET) was mixed by extrusion with clay (montmorillonite) to produce nanocomposites. Additives (pentaerythrytol and maleic anhydride) were added to induce compatibility between the particles and the matrix. Thermal, mechanical and rheological properties of the nanocomposites were evaluated. The nanocomposites with additives presented improved mechanical strength. In shear, the viscoelastic properties decrease with rising particle concentration and additives content. The resulting new materiales present processing advantages and potential improvements in performance.

Introduction

Polymeric nanocomposites are a class of relatively new materials with ample potential applications. Products with commercial applications appeared during the last decade[1], and much industrial and academic interest have produced. Reports on the manufacture of nanocomposites include those made with polyamides[2-5], polyolefines[6-9], polystyrene and polystyrene copolymers[10-11], ethylene vinyl alcohol[12-15], acrylics[16-18], polyesters[19-20], polycarbonate[21-22], liquid crystalline polymers[23-26], fluoropolymers[27-29], thermoset resins[30-32], polyurethanes[33-38], ethylene-propylene oxide[39], vinyl carbazole[40-41], polydiacetylene[42], and polyimides[43], among others.

Generally, polymer nanocomposites can be obtained through two routes: the first one is the polymerization of monomers in contact with the exfoliated clay, and the second one uses existing transformation processes to produce nanocomposites, as for example, by a reactive extrusion. There are, however, problems present due to the lack of affinity of the clay-polymer system because of the hydrophilic character of the particles. It is then necessary to treat the clay chemically to increase its affinity with the polymer matrix. This constitutes another whole area of research in the nanocomposites production.

 CCC 1022-1360/00/$ 17,50+.50/0

The mixing of the nano-particle with the polymer requires an intercalation process of the macromolecule into the clay galleries gap. This is a diffusion-controlled process which requires long times of contact between the polymer and the clay under the pressure produced inside the extruder. The intercalation process leads to the exfoliation of the clay. However, low screw speeds and long residence times in the extruder may cause polymer degradation. To avoid such problem that would enable the complete clay exfoliation, it would be necessary to change the screw configuration or to consider the chemical modification of the clay. Moreover, the clay exfoliation is not a sufficient condition to obtain optimum systems. The problem of compatibility of the clay-polymer matrix is an outstanding one, and the thermodynamic processes involved in the synthesis of these materials[44-45], constitute areas which are currently under investigation.

The macroscopic effects of the integration of nanoparticles to the polymer matrix are quite remarkable. For example, the barrier properties of the new systems are enhanced, because the diffusion of gas molecules through the material is largely retarded. A lower permeability of O_2 is obtained without substantially modifying the production method of films, containers, bottles, etc. Oxygen permeability decreases drastically with nanoparticle concentration.

In this work, the production process of a polymeric nanocomposite made of polyethylene terephthalate (PET) and a commercial clay (montmorillonite) is described. The clay is chemically modified with a quaternary ammonium salt, using maleic anhydride (MAH) and pentaerythrytol (PENTA) as compatibilizing agents. Thermal, mechanical and rheological properties of PET are substantially modified. In addition, the viscoelastic behavior of the new systems is discussed.

Experimental Procedure

Materials

PET Eastapack 9921 from Eastman Chemical had a density of 1.40 g/cm^3 and melting point of 269°C. A glass transition temperature (Tg) of 61.2°C was determined. MAH and PENTA are analytic reagents from Aldrich with melting points of 55°C and 192.8°C, respectively. These reagents were used as received. Thermal properties were determined by differential scanning calorimetry (DSC) with a heating speed of 20°C/min under nitrogen atmosphere. No additives to prevent oxidation were used in the formulations. Clay montmorillonite (Cloisite 15A) from Southern Clay Co., is a natural clay modified with a quaternary ammonium salt, had a density of 1.66g/cm^3 and a ionic interchange capacity of 125 meq/100g.

Equipment

Blends were made in a co-rotating twin-screw extruder (Werner & Pfleiderer ZSK30) equipped with a volumetric band feeder under nitrogen atmosphere to prevent moisture absorption. Moisture determinations were carried out in a Moisture Analyzer Omnimark Mark2. Interlayer distance among clay platelets was determined in a Bruker-AXSD8-Advance Diffractometer with a wavelength of CuKα = 1.5418 Å at 35kv and 30 mA with scanning speed of 1° 2θ/min. Bottles were made in an injection-stretch blow-molding machine AOKI on one step with a temperature profile of 275°C at the feeder and 320°C at the injection nozzle, with perform temperature of 112°C and blow mold temperature of 16°C. Determination of the rheological properties was carried out in a TA-Instruments AR 1000-N controlled stress rheometer with parallel plates of 25 mm diameter.

Results

Thermal Properties

Clay concentrations larger than 2 wt.% cause a decrease in the glass transition temperature (Tg), and therefore only clay concentrations of 1% and 2% were considered. Raw materials were previously mixed before placing them in the extruder feeder. Extruded threads were cooled and cut into pellets of 5 mm length. Thermal properties of blends made with PET, clay, MAH and PENTA are shown in figures (1) and (2). Crystallinity of the samples was calculated using the relation $\%crystallinity = \left(\Delta H_m / \Delta H_m^0\right) * 100$ where ΔH_m^0 is the theoretical melt enthalpy of the PET crystal (117.6 J/g) and ΔH_m is the sample enthalpy[47].

Concentrations of MAH and PENTA used were 0.12% and 0.03 %, respectively, for a clay content of 1 wt%. They were increased to 0.24% and 0.06 %, respectively, for clay a content of 2%. The large additive concentration systems include concentrations of 0.24% and 0.06 % when the clay content is 1%, and 0.48% and 0.12 %, respectively, for a clay content of 2%.

Figure (1) shows that the crystallization temperature diminishes with clay concentration for the system with no additives. As the clay-quaternary ammonium salt, MAH and PENTA are added; the crystallization temperature (Tc) diminishes even more with clay concentration (the melting temperature, Tm, remains practically constant). For the system with large MAH and PENTA concentrations, the decrease in the crystallization temperature is much smaller than that observed in the system with no additives and in the system with small concentrations of additives.

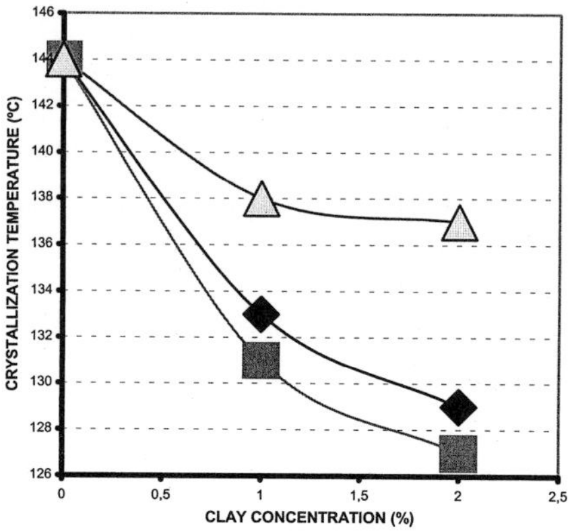

Figure 1. Crystallization temperature as a function of clay concentration for the system without additives (◆), with MAH and PENTA (small concentrations) (■) and with MAH and PENTA (large concentrations) (▲)

As observed in figure (2), for increasing clay concentration larger crystallinity is attained, since clay particles may act as nucleation agents. The system with large additive concentrations shows a remarkable increase of the crystallinity with clay concentration. In addition, ΔH_m increases with clay content, suggesting that a large amount of energy is necessary to melt the crystals.

When nanocomposites are produced by polymerization *in situ,* it was found[19] that Tc and Tm decrease slightly as a function of clay concentration (5%), but the melting enthalpy decreases with clay concentration. These discrepancies arise because the different production procedures.

Clay Exfoliation

The clay interlayer distance (d_{001}, referred to the plane 001) was determined by X-rays diffraction. The diffraction area is located at low angles, such as 2θ (the diffraction position) is smaller than 10°, which corresponds to an interlayer distance > 8.84 Å (according to the Bragg equation: $2d_{001} \sin\theta = \lambda$). λ is the wavelength of radiation (1.5418 Å). The clay has an interlayer distance equal to 31.5 Å. All nanocomposites analyzed present similar values, which means that none of the samples reached a complete clay exfoliation. It was concluded that it is necessary to use different screw configurations to produce exfoliation.

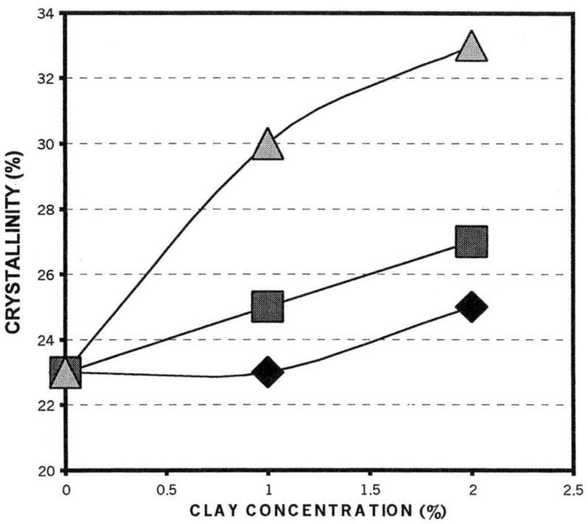

Figure 2. Crystallinity versus clay concentration for the three systems shown in figure (1).

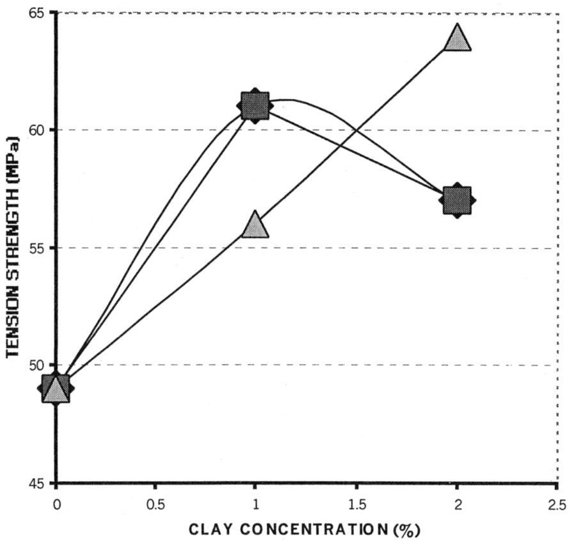

Figure 3. Tension strength as a function of clay concentration, for same systems shown in figure (1). Temperature of the test is 270^{0}C

Tensile Mechanical Properties

Tensile mechanical properties were performed on specimens cut from bottles made by the injection-stretch blow-molding process. PET was crystallized during 48 hours at 130°C previously to processing, to avoid possible jamming in the machine due to the amorphous content of the material. The bottles produced from PET-clay are transparent with slight coloring (white to brown). It is necessary to mention that the mechanical properties were measured on systems with the same thermomechanical history, and therefore they have similar degrees of degradation due to processing.

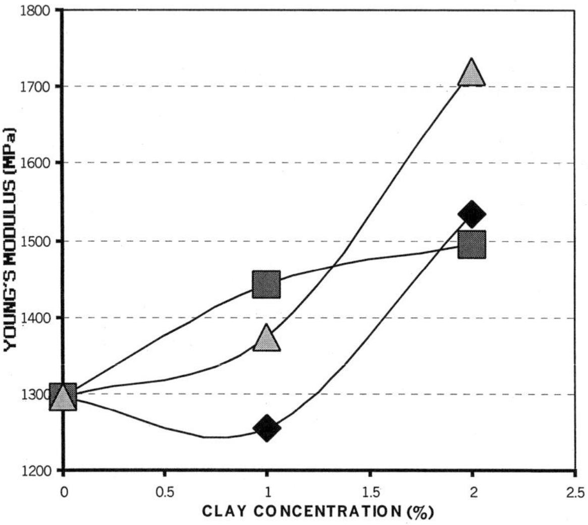

Figure 4. Variation of the Young modulus with clay concentration for same systems of previous figures. T = 270^{0}C

In general, it is observed that mechanical properties of nanocomposites are largely improved when the clay particles are totally exfoliated and good phase affinity is attained. The mechanism by which these improvements are obtained considers the absorption of macromolecular layers on the clay platelet surface. As a consequence, the diffusion of molecules is impeded and retarded, and hence the material acquires additional crystallinity. This effect produces increases in the material strength and modulus.

However, it is not necessary to obtain a complete exfoliation of the particles, since non-exfoliated nanocomposites may present improvements in mechanical properties as well. For instance, in this study, complete clay exfoliation was not attained, but the compatibility and particle dispersion reached through the action of the additives lead to an increase in the mechanical properties, up to 30% and 32% in strength and modulus, respectively (see figures

3 and 4). In figure (3), the tension strength presents a maximum in PET with 1% clay concentration, in systems with no additives or with small additive concentration. For larger additives concentration, there is a continuous rise in the strength up to 2 % clay content. A similar increase in the Young's modulus is also obtained in the system with the larger additive concentration, as shown in figure (4). The mechanism by which this improvement has been reached is not completely clear, because the lack of understanding on the compatibilizers (additives) action, and so additional work is necessary.

Rheological Properties

As in the mechanical properties, the rheological properties were measured on systems with the same thermorheological history, so the measurements shown correspond to systems with same degradation due to processing. The viscosity variation with shear rate for processed PET, PET with clay and PET with clay plus additives is shown in figure (5). In addition, in figure (6) the zero shear-rate viscosity for the seven systems of figure (5) is plotted with clay concentration.

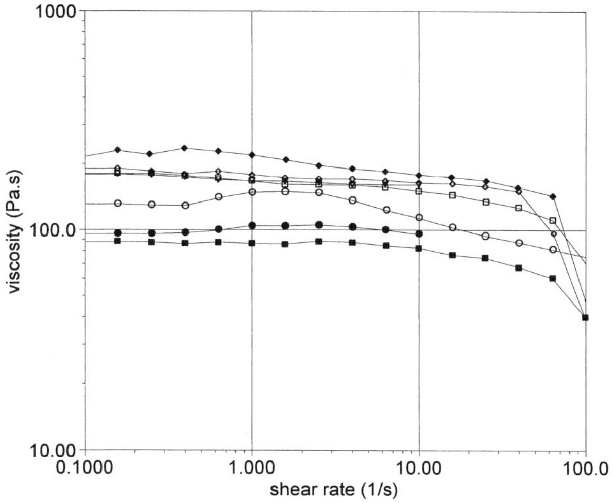

Figure 5. The shear viscosity is plotted versus shear rate for various systems: ♦ processed PET, ∇ 1% clay without additives, □ 1% clay with MAH and PENTA, ◊ 1% clay with high MAH and PENTA contents,O 2% clay without additives, ● 2% clay with MAH and PENTA, + 2% clay with high MAH and PENTA contents. T = 270^{0}C

The basic behavior shows a near-linear decrease of the zero shear-rate viscosity with clay concentration. The viscosity lowers to values less than a half of the PET viscosity for 2% clay content. Additives promote an even lower viscosity at high clay contents, but they do not have any substantial effect at low clay contents.

As is well known in injection processes, smaller viscosities of the polymer in the melt state results in better filling of the mold cavities. On the other hand, in extrusion processes, the loss of melt strength caused by lower viscosity, requires that the material should be processed at lower temperatures to preserve consistency and to avoid sagging, which means smaller thermal degradation.

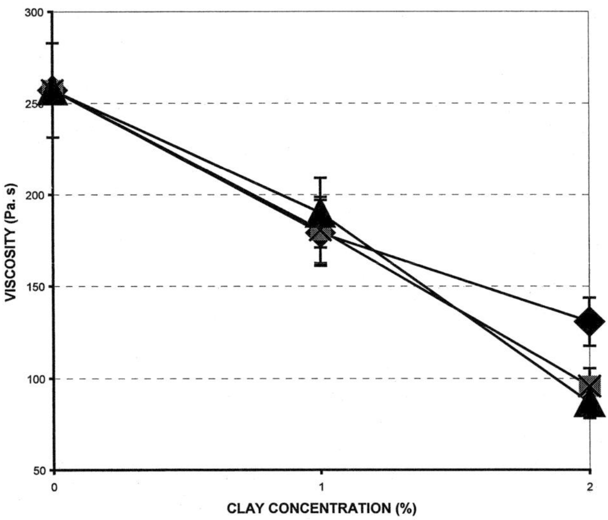

Figure 6. Zero shear-rate viscosity versus clay concentration. Symbols are the same as in figure 1

The reduction of the complex viscosity with clay concentration and the effect of the additives on the viscosity follow the same tendency to that observed in the shear viscosity, as shown in figure (7). However, there are substantial differences for shear rates higher than 10 s^{-1}, specially when the clay content is 2%. Comparison of figures (5) and (7) illustrates that for high clay contents, particle alignment with the flow induce the shear thinning observed in the high shear rate region. On the other hand, at low shear rates particles are not aligned and a region of shear thickening is observed in the system with 2% clay content without additives. The shear thickening effect decreases with addition of additives, which clearly reflects the consequences of increased compatibility of the particles with the matrix. The onset point for shear-thinning, which is closely related to the relaxation time in shear, is shifted to lower deformation rates with respect to those of the complex viscosity. This indicates that as the clay content increases, the shear viscosity is governed by a relaxation time which is larger than that which governs the complex viscosity. This increase in the characteristic time of the system is ascribed to the effect of the particles on the flow, and this effect is larger with increasing compatibility caused by rising additive concentration.

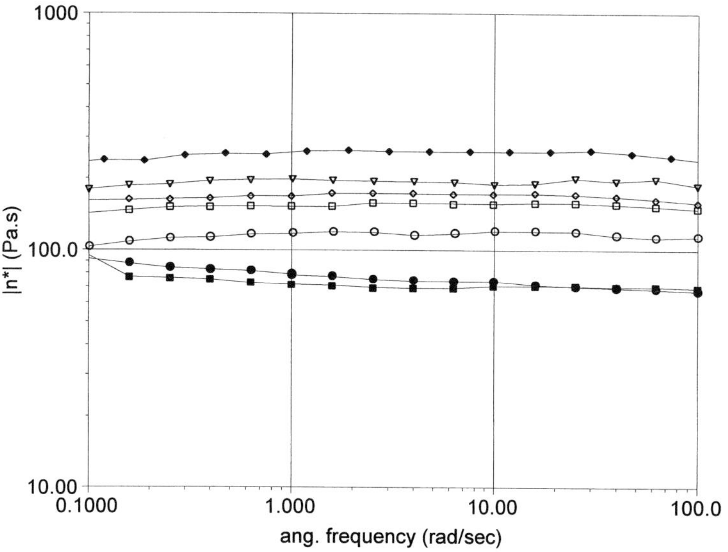

Figure 7. The complex viscosity is plotted versus angular frequency for the seven systems shown in figure 5. T = 270^{0}C

To assess the effect of increasing clay and compatibilizing agent concentration on the linear viscoelastic properties of the systems, the elastic modulus (G') is shown as a function of frequency (figure 8) and also as a function of clay concentration (figure 9) at a fixed frequency (100 s^{-1}) . Temperature of the tests was set at 270°C.

The variation of the elastic modulus with frequency indicates that for large frequencies, the overall effect of the addition of clay is similar to that observed in the shear viscosity, i.e., a continuous decrease of the viscoelastic properties with clay concentration. The decrease is larger when additives are added. On the other hand, at low frequencies the modulus does not approach a slope of two, and hence additional relaxation modes exist besides the main relaxation mechanism. The effect of high additive contents is more evident at low frequencies, where the systems depict higher modulus. In this case compatibility is reflected in the widening of the relaxation spectrum of the polymer due to the presence of the particles.

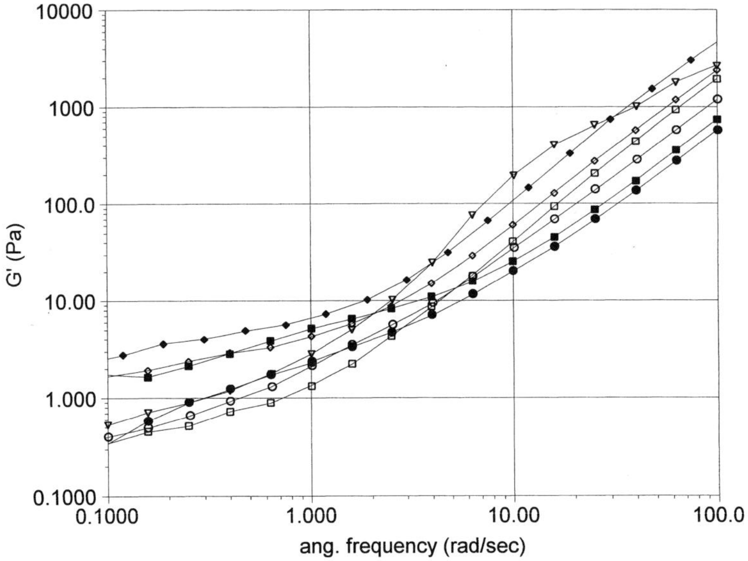

Figure 8. Storage modulus versus angular frequency for the seven systems indicated in figure (5). T = 270^{0}C

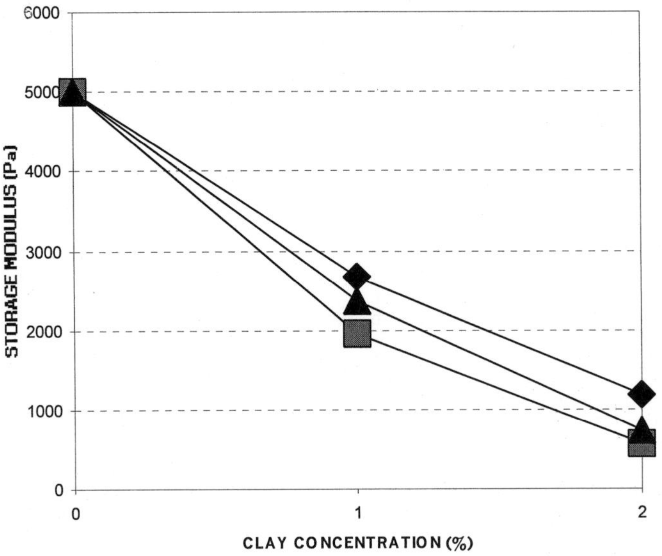

Figure 9. Storage modulus versus clay concentration. Symbols are the same as in figure 1

Discussion and Conclusions

The addition of clay nanoparticles to polyethylene terephthalate (PET) to produce a nanocomposite, affects the thermal properties of PET, such as the crystallization temperature. The additives used here (MAH and PENTA) produce an increased crystallinity with higher modulus and material strength. These nanocomposites present smaller shear viscosity and lower magnitudes of the linear viscoelastic properties than those of PET. The viscoelastic properties are affected strongly by the interaction between polymer and clay particles, due to increasing polarity.

Observation of the results of the mechanical and rheological properties of the systems analyzed in this study, it is remarkable that while in simple shear and oscillatory flow the rheological properties show a substantial decrease with increasing particle concentration (in the systems with large additives concentration), in extensional flow the mechanical properties and materials strength increase. This effect reflects, on the one hand, a diminishing particle-matrix interaction as the clay concentrations rises. On the other hand, the compatibility of the particles with the matrix induced by the additives and the increase in crystallinity, leads to materials with increasing strength and modulus as the clay concentration rises.

These results show the advantages of nanocomposites in the processing operations and final properties of these new materials. When polymeric nanocomposites are undergoing a deformation by shear in the processing operation, the modified rheological behavior due to the presence of the particles induce better surface properties with smaller shrinkages when the polymer solidifies inside the mold (in injection-molding) or outside the die (in extrusion). These results show potential advantages in processing and performance of these new materials.

Acknowledgements

Authors wish to thank C. Granpre and Y. Zimmard for their collaboration in the injection blow-molding and extrusion work, respectively. Also, we thank Dr. M. Estrada for the X-Ray measurements, Dr. A. Maciel for the mechanical measurements and Mr. F. Calderas for the rheometric work. Support from project G-27837U (CONACYT) is also acknowledged.

[1] Y. Kojima, A. Usuki, M. Kawasumi, A. Okada, Y. Fukushima, T. Karauchi, O. Kamigaito. *J. Mater. Res.* **1993**, 8, 1185.

[2] A. Okada, A. Usuki. *Mater. Sci. Eng.* **1995**, C3, 109.

[3] M. K. Akkapeddi. *Polym. Comp.* **2000**, 21(4), 576.

[4] L. Liu, Z. Qi, X. Zhu. *J. Appl. Polym. Sci.* **1999**, 71(7), 1133.

[5] L. P. Cheng, D.J. Lin, K.C. Yang. *J. Membranes. Sci.* **2000**, 172, 157.

[6] M. Kawasumi, N. Hasegawa, M. Kato, A. Usuki, A. Okada. *Macromolecules.* **1997**, 30, 6333.

[7] M. Kato, A. Usuki, A. Okada. *J. Appl. Polym. Sci.* **1997**, 66(9), 1781.

[8] J. Heinemann, P. Reichert, R. Thomann, R. Mulhaupt. *Macromol. Rapid. Commun.* **1999**, 20(8), 423.

[9] H. G. Jeong, H.T. Jung, S.W. Lee, S.D. Hudson. *Polym. Bull.* **1998**, 41(1), 107.

[10] M. H. Noh, D. C. Lee. *J. Appl. Polym. Sci.* **1999**, 74(12), 2811.

[11] X. Fu, S. Qutubuddin. *Polymer.* **2001**, 42(2), 807.

[12] P. B. Messersmith, E. P. Giannelis. *Chem. Mater.* **1993**, 5(8), 1064.

[13] R. A. Vaia, S. Vasudevan, W. Krawiec, L. G. Scanlon, E. P. Giannelis. *Adv. Mater.-Weinheim.* **1995**,7(2), 154.

[14] R. Levy, C.W. Francis. *J. Coll. Inter. Sci.* **1975**, 50(3), 442.

[15] D.J. Greenland. *J. Collo. Sci.* **1963**, 18, 647.

[16] C.D. Muzny, B.D. Butler, H.J.M. Hanley, F. Tsvetkov, D.G. Peiffer. *Mater. Letters.* **1996**, 28(4), 379.

[17] M. Okamoto, S. Morita, H. Taguchi, Y. H. Kim, T. Kotaka, H. Tateyama. *Polymer.* **2000**, 41(10), 3887.

[18] F. Dietsche, R. Mulhaupt. *Polym. Bull.* **1999**, 43(4-5), 395.

[19] Y. Ke, C. Long, Z. Qi. *J. Appl. Polym. Sci.* **1999**, 71(7), 1139.

[20] J. C. Matayabas, S. R. Turner, B. J. Sublett, G. W. Connell, J. W. Gilmer, R. B. Barbee. Patent # 6084019 USA, 2000.

[21] N. R. Rosenquist, K. F. Miller. Patent # 4786710 USA, 1988.

[22] X. Huang, S. Lewis, W. J. Brittain, R. A. Vaia. *Macromolecules.* **2000**, 33(6), 2000.

[23] J. Heinemann, P. Reichert, R. Thomann, R. Mulhaupt. *Macromol. Rapid. Commun.* **1999**, 20, 423.

[24] E. P. Giannelis. *Adv. Mater.* **1996**, 8(1), 29.

[25] P. Aranda, E. Ruiz-Hitzky. *Appl. Clay. Sci.* **1999**, 15, 119.

[26] A. Usuki, A. Koiwai, Y. Kojima, M. Kawasumi, A. Okada, T. Kurauchi, O. Kamigaito. *J. Appl. Polym. Sci.* **1995**, 15, 119.

[27] M. J. Michalczyk, K. G. Sharp, C. W. Stewart. Patent # 5726247 USA,1998.

[28] S. Badesha , A. Henry, J. Maliborski, E. Clifford . Patent # 5840796 USA, 1998.

[29] M. W. Ellsworth. Patent # 5962553 USA, 1999.

[30] M. Wang, T. Pinnavaia. *Chem. Mater.* **1994**, 6(4), 468.

[31] C. Zilg, R. Mulhaup, J. Finter. *Macromol. Chem. Phys.* **1999**, 200(3), 661.

[32] X. Kornmann, L. A. Berglund, J. Sterte, E. P. Giannelis. *Polym. Eng. Sci.* **1998**, 38(8), 1351.

[33] L. Chen, K. Liu, C.Z. Yang. *Polym. Bull.* **1996**, 37(3), 377.

[34] T. Hirai, M. Miyamoto, I. Komasawa. *J. Mate. Chem.* **1999**, 9(6) 1217.

[35] Z. Wang, T. J. Pinnavaia. *Chem. Mater.* **1998**, 10(12), 3769.

[36].- C. Zilg, R. Thomann, R. Mulhaupt, J. Finter. *Adv. Mater.* **1999**, 11(1), 49.

[37].- T.K. Chen, Y.I. Tien, K.H. Wei. *J. Appl. Polym. Sci.* **1999**, Part A: Polym. Chem. 37(13), 2225.

[38] T. K. Chen, Y.I. Tien, K.H. Wei. *Polymer.* **2000**, 41(4), 1345.

[39] G. Spenleuhauer, D. Bazile, M. Veillard, C. Prud'Homme, J. P. Michalon. Patent # 5766635 USA,1998.

[40] M. Biswas, S. Sinha Ray. *Polymer.* **1998**, 39(25), 6423.

[41] S. Sinha Ray, M. Biswas. *J. Appl. Polym. Sci.* **1999**, 73(14), 2971.

[42] T. Srikhirin, A. Moet, J. B. Lando. *Polym. Adv. Tech.* **1998**, 9(8), 491.

[43] H. Tyan, Y. Liu, K. Wei. *Chem. Mater.* **1999**, 11(7), 1942.

[44] R. A. Vaia, E. P. Giannelis. *Macromolecules.* **1997**, 30, 7990.

[45] R. A. Vaia, E. P. Giannelis. *Macromolecules.* **1997**, 30, 8000.

[46] L. Quintanilla, M. Alonso, J. Rodriguez, M. Pastor. *J. Appl. Polym. Sci.* **1996**, 59, 769.